Cognitive Enhancement

Trends in Augmentation of Human Performance

Series Ed.: Cutsuridis, Vassilis

Human performance enhancement refers to the augmentation of human skills, attributes and competencies through the use of technology, medicine and therapy designed to replace or increase human performance capability. In the turn of the second decade of the 21st century the convergence of the neurosciences including neuroimaging, genetics and cognitive science, neurotechnology and informatics is creating powerful tools that have the potential to significantly enhance human performance.

Three broad domains of human performance enhancement will be covered by the book series: (1) Enhancement of human performance to restore normal human capability from the disabled or dysfunctional, (2) enhancement of human performance to superhuman levels, and (3) Ethical consequences of the first two domains.

Examples of the first domain will include but not limited to restoration of sight, hearing, memory, mobility, reversing aging effects, prosthesis of limbs, brain-machine interface and genetic manipulation. Examples of the second domain will include, but not limited to enhanced memory, augmentation of vision, hearing, strength, mobility, brain-machine and machine-brain interfaces.

This series will publish works of the highest quality that advance the understanding and practical application of human performance enhancement. Research monographs, introductory and advanced level textbooks, volume editions and proceedings will be considered. Please contact the editors with your book ideas.

Finally, an advisory board of distinguished scientists from the fields on neurotechnology, neurosciences, cybernetics, cognitive sciences, psychophysics, psychiatry, genetics, neuropharmacology, neuroimaging and ethics will assist the book series editor with the selection of future book proposals. The advisory board members will also serve as reviewers of future book proposals.

For further volumes:
http://www.springer.com/series/10849

Elisabeth Hildt • Andreas G. Franke
Editors

Cognitive Enhancement

An Interdisciplinary Perspective

Springer

Editors
Elisabeth Hildt
Department of Philosophy
University of Mainz
Mainz
Germany

Andreas G. Franke
Psychiatry and Psychotherapy
University Medical Center Mainz
Mainz
Germany

ISSN 2213-1310 ISSN 2213-1329 (electronic)
ISBN 978-94-007-6252-7 ISBN 978-94-007-6253-4 (eBook)
DOI 10.1007/978-94-007-6253-4
Springer Dordrecht Heidelberg New York London

Library of Congress Control Number: 2013936404

Printed on acid-free paper

Springer is part of Springer Science+Business Media (www.springer.com)

Preface

This book is a collection of contributions to the conference "Cognitive Enhancement," which was held from 21 February until 1 March 2011 at the University of Mainz, Germany. During this interdisciplinary conference, young scholars from Europe, Canada and Australia met to present their work and to discuss issues related to cognitive enhancement with renowned international experts. The scholars stem from various disciplines such as medicine, psychiatry, neuroscience, neurotechnology, philosophy, medical ethics, neuroethics, sports science, social sciences and law.

Several persons rendered the realization of this book possible. First, we would like to thank the following individuals for their continuous support, for chairing the conference sessions and for stimulating discussions: Prof. Dr. Ruth Zimmerling, Political Science, University of Mainz; Prof. Dr. Klaus Lieb, Department of Psychiatry and Psychotherapy, Mainz University Medical Center; Prof. Dr. Thomas Metzinger, Department of Philosophy, University of Mainz; and Prof. Dr. Perikles Simon, Department of Sports Medicine, Mainz University Medical Center.

In addition, we would like to thank Prof. Dr. Hartmut Lüddens and Prof. Dr. Ulrich Schmitt, who, during the conference, held a very inspiring hands-on day at the Department of Psychiatry and Psychotherapy, Mainz University Medical Center. We would also like to thank Maike Reinerth, Department of Film Studies, University of Mainz, for her thought-provoking film presentation.

We are particularly grateful to Sheila Madary for her formidable organizational support before, during and after the conference. Her thoughtful English editing of the manuscripts contributed considerably to this book.

In addition, we thank the German Federal Ministry of Education and Research (Bundesministerium für Bildung und Forschung, BMBF) for funding the international conference and the publication of the conference proceedings.

Last but not least, we would like to thank all of the conference participants, who, quite literally, made this book possible by contributing their respective chapters.

Mainz, Germany
October 2012

Elisabeth Hildt
Andreas G. Franke

Contents

Contributors

Kirsten Brukamp Institute for History, Theory, and Ethics of Medicine, RWTH Aachen University, University Hospital Aachen, Aachen, Germany

Jan-Christoph Bublitz Faculty of Law, University of Hamburg, Hamburg, Germany

Emma Peng Chien Department of Philosophy, University of Alberta, Edmonton, AB, Canada

John Danaher School of Law, Keele University, Keele, UK

Pavel Dietz Department of Sports Medicine, Rehabilitation and Disease Prevention, Faculty of Social Science, Media and Sports, Johannes Gutenberg-University of Mainz, Mainz, Germany

Colleen A. Dockery Institute of Medical Psychology and Behavioral Neurobiology, University of Tuebingen, Tuebingen, Germany

Martin Dresler Max Planck Institute of Psychiatry, Munich, Germany

Cynthia Forlini Neuroethics Research Unit, Institut de Recherches Cliniques de Montréal (IRCM), McGill University of Montreal, Montreal, QC, Canada

Lisa Forsberg Centre for Medical Law and Ethics, King's College London, London, UK

Andreas G. Franke Department of Psychiatry and Psychotherapy, University Medical Centre, Johannes-Gutenberg University of Mainz, Mainz, Germany

Frédéric Gilbert School of Philosophy, Faculty of Arts, Australian Centre of Excellence for Electromaterials Science (ACES), University of Tasmania, Hobert, TAS, Australia

Armin Grunwald Institute for Technology Assessment and Systems Analysis (ITAS), Karlsruhe Institute of Technology (KIT), Karlsruhe, Germany

Michael Hauskeller Faculty of Sociology and Philosophy, University of Exeter, Exeter, UK

Elisabeth Hildt Department of Philosophy, Johannes Gutenberg University of Mainz, Mainz, Germany

Roland Kipke Internationales Zentrum für Ethik in den Wissenschaften (IZEW), University of Tübingen, Tübingen, Germany

Klaus Lieb Department of Psychiatry and Psychotherapy, University Medical Centre, Johannes-Gutenberg University of Mainz, Mainz, Germany

Bradley Partridge University of Queensland, Centre for Clinical Research (UQCCR), Herston, QLD, Australia

Eric Racine Neuroethics Research Unit, Institut de Recherches Cliniques de Montréal (IRCM), McGill University of Montreal, Montreal, QC, Canada

Peter B. Reiner National Core for Neuroethics, University of British Columbia Hospital, Vancouver, BC, Canada

Dimitris Repantis Department of Psychiatry, Charité-Universitätsmedizin Berlin, Berlin, Germany

Elizabeth Shaw School of Law, University of Edinburgh, Edinburgh, UK

Carlos Trenado Systems Neurosciences and Neurotechnology Unit, Saarland University Hospital, Saarbrücken, Germany

Department of Kinesiology, School of Public Health, University of Maryland College Park, College Park, MD, USA

Greta Wagner Department of Social Sciences, Institute of Sociology, Goethe-University Frankfurt am Main, Frankfurt, Germany

Chapter 1
Cognitive Enhancement – A Critical Look at the Recent Debate

Elisabeth Hildt

Abstract Cognitive enhancement, which can be characterized as the attempt to increase cognitive functions such as attention or memory in healthy individuals, has received considerable attention during the last decade, both in the general public and in academic discourse. In spite of a very active interdisciplinary debate which has provided helpful reflections, categorizations and clarifications, the numerous questions and problems related to cognitive enhancement are far from having been exhaustively discussed or even solved. Without any doubt, there are several aspects within this field that require more reflection and further clarifications. In this chapter, which serves as an introduction to the following book, I'll discuss three of them. The first aspect concerns conceptual issues. Specifically, it concerns the question of what we are talking about when we use the term cognitive enhancement. The second aspect regards the question of how issues in cognitive enhancement should adequately be discussed within society. And the third aspect regards the interplay between individual autonomy and society. The last section of this chapter includes an overview of the contributions to the present book.

Keywords Cognitive enhancement • Neuroenhancement • Cognition • Autonomy • Society • Ethics

Cognitive enhancement (CE) has received considerable attention during the last decade, both in the general public and in academic discourse. Cognitive enhancement can be characterized as the attempt to increase cognitive functions such as attention or memory in healthy individuals. There are various strategies that aim at enhancing cognitive functions. Pharmacological cognitive enhancement is in the

E. Hildt (✉)
Department of Philosophy, Johannes Gutenberg University of Mainz, Mainz, Germany
e-mail: hildt@uni-mainz.de

E. Hildt and A.G. Franke (eds.), *Cognitive Enhancement: An Interdisciplinary Perspective*, Trends in Augmentation of Human Performance 1,
DOI 10.1007/978-94-007-6253-4_1,

current focus of attention (de Jongh et al. 2008; Repantis et al. 2010b; Morein-Zamir and Sahakian 2011). It encompasses the use of caffeine, prescription stimulants (e.g. methylphenidate, modafinil) or illicit drugs (such as illicit amphetamines) (de Jongh et al. 2008; Repantis et al. 2010b; Morein-Zamir and Sahakian 2011). Non-pharmacological strategies for cognitive enhancement include non-technological forms, such as physical exercise, meditation, and mnemonics as well as technological ones, such as transcranial magnetic stimulation (TMS) or transcranial direct current stimulation (tDCS) (Hamilton et al. 2011; Dresler et al. 2013).

Over the past decade, there has been a very active interdisciplinary debate on the various strategies for cognitive enhancement and the individual, social and ethical implications of cognitive enhancement; this debate has provided helpful reflections, categorizations and clarifications. Issues analyzed and discussed include safety, efficacy, risk-benefit-ratio, autonomy, cognitive liberty, identity, authenticity, pressure to perform, fairness, justice, human nature and medicalization of human life (cf. President's Council on Bioethics 2003; Farah et al. 2004; Sahakian and Morein-Zamir 2007; Gordijn and Chadwick 2008; Greely et al. 2008; Bublitz and Merkel 2009; Savulescu and Bostrom 2009; Nagel 2010; Metzinger and Hildt 2011; Metzinger 2012; Outram 2012).

However, the numerous questions and problems related to cognitive enhancement are far from having been exhaustively discussed or even solved. Without any doubt, there are several areas that need more reflection and further clarifications. In this chapter, which serves as an introduction to this book, I will discuss three of these areas. The first regards conceptual issues. It concerns the question of what we talk about when we talk about cognitive enhancement. The second concerns the question of how issues in cognitive enhancement should adequately be discussed within society. And the third deals with the interplay between individual autonomy and society.

1.1 What Do We Mean by the Term Cognitive Enhancement?

According to a widespread definition, enhancement is a term used to "characterize interventions designed to improve human form or functioning beyond what is necessary to sustain or restore good health" (Juengst 1998, 29). There are various fields of enhancement, such as those that aim at improvements in bodily appearance (cosmetic surgery), athletic performance (doping in sports), genetic make-up (genetic enhancement) or brain functions (neuroenhancement) (Gordijn and Chadwick 2008; Savulescu and Bostrom 2009).

Usually, however, the term "enhancement" does not serve to specify a certain method or technology but to specify the context of its use, for whether a certain method or technology is used as a treatment or as an enhancement depends on the concrete situation. A typical example of such diverging contexts is the use of methylphenidate (Ritalin®) in the treatment of attention deficit hyperactivity disorder (ADHD), as opposed to the enhancement use of methylphenidate by healthy individuals who attempt to increase mental performance.

Unlike medical treatments, enhancements aim at some kind of betterment in healthy individuals. Eric T. Juengst (1998) described the concept of enhancement to be a "moral boundary concept". Whereas on the descriptive level, enhancement serves to characterize a certain measurement to lead to some form of improvement, on the normative level, enhancement could be described as dwelling outside the field of medicine and beyond medical obligation, a measurement not legitimized by medical needs. In this distinction between treatment and enhancement, the concepts of health, disease and normality and the aims of medicine are crucial.

This distinction between treatment and enhancement is broadly accepted for pragmatic reasons, in spite of having several drawbacks. When it comes to concrete applications, it is often problematic to draw a clear line between medical uses of a certain technology and non-medical ones, for, in many contexts, there is a grey zone that renders the distinction very problematic. In addition, the normative implications of the treatment-enhancement distinction are far from being clear. For what does it imply to consider a certain practice to be outside the field of medicine – beyond the claim that the health care system will not pay for it?

The above sketched view on enhancement clearly focuses on technological forms of enhancement and on the question of how far new technologies should be used in order to improve human capacities or traits in healthy individuals (cf. Juengst 1998; Parens 1998). In contrast, another view on enhancements starts from a totally different point in that it stresses that all attempts of humans to strive for betterment can be considered to be enhancements (Caplan 2002; Greely et al. 2008; Bostrom and Sandberg 2009). A typical example of this strategy can be found in a commentary by Greely et al. (2008, 702):

> Human ingenuity has given us means of enhancing our brains through inventions such as written language, printing and the Internet. Most authors of this Commentary are teachers and strive to enhance the minds of their students, both by adding substantive information and by showing them new and better ways to process that information. And we are all aware of the abilities to enhance our brains with adequate exercise, nutrition and sleep. The drugs just reviewed, along with newer technologies such as brain stimulation and prosthetic brain chips, should be viewed in the same general category as education, good health habits, and information technology – ways that our uniquely innovative species tries to improve itself.

In this second approach, the term "enhancement" is used in a much broader way. What at first sight seems to be nothing but a question of definition, to be simply a disagreement about what we talk about when we talk about enhancements, in fact has much wider implications. For depending on how the term is used, reasoning with regard to cognitive enhancement differs considerably. Whereas the first approach aims to contextualize technological enhancements in that it compares them to current technology use in medicine, the second approach tends to trivialize technological forms of enhancement.

Undoubtedly, there is nothing wrong with saying that people have always striven to improve themselves in various ways. There is nothing new with this idea of improvement, either. Anthropological thinking over the past centuries has been dominated by the idea that it is characteristic for humans to strive for the better. What definitely is new, however, is the use of current technologies to strive for this end.

Compared to traditional strategies, biotechnologies undoubtedly provide different means to improve human capacities. This does not mean that enhancements via biotechnologies differ categorically from enhancements via non-technological strategies such as sleep, training or nutrition. But means undoubtedly do matter: Different strategies may differ with regard to the mechanisms involved, with regard to efficacy, benefits, risks and numerous other aspects.

In the following, I will not go into a detailed argument that means matter (cf. Cole-Turner 1998; Parens 1998). My point merely is that means matter factually. What makes technological enhancements factually different is that they rely on highly specific additional factors – medical or technical support, drugs or technical devices. Some of them need the assistance of a physician or some other medical professional. Others do not need any medical assistance but depend on the provision of or the access to technical devices or psychoactive drugs. All of these highly specific components are to be seen in the context of an established medical or social practice in which access and practical procedure underlie general regulations.

In contrast, non-technological strategies, such as sleep or meditation, are primarily employed in the private domain in so far as they usually do not depend on any specific external factor that has to be provided by a third party. Furthermore, the risks going along with the non-technological strategies are rather low, in general, and, therefore, do not legitimize any external influence.

That's why the general conditions of the various technological and non-technological forms of enhancement will differ considerably. When talking about how to handle cognitive enhancement in society, the concrete context of the enhancement strategy in question should always be taken into account. For without any contextualization, there is nothing that can be said except the rather general statement that people have always attempted to improve themselves – a statement that sounds liberal but is rather trivial and tells us absolutely nothing about how to use a particular technology.

Whereas the above reflections pertain to all kinds of enhancements, there are conceptual questions that pertain specifically to the field of cognitive enhancement, the most central one being: What is *cognitive* enhancement? A plausible direct answer is: Cognitive enhancement aims at increasing cognitive functions. Cognitive functions are information-processing functions such as learning, planning, concept formation, perception, attention, memory, reasoning and problem solving.

If you take a closer look at the interdisciplinary debate, it is far from clear, however, what is meant when people talk about cognitive enhancement. Whereas some authors consider the aim of cognitive enhancement to "improve the performance of the healthy" (Greely et al. 2008, 702) or to "augment the minds of the healthy" (Cakic 2009, 611), according to a definition by Nick Bostrom and Anders Sandberg (Bostrom and Sandberg 2009, 311), cognitive enhancement is "the amplification or extension of core capacities of the mind through improvement or augmentation of internal or external information processing systems [. . .]." Another characterization of the term "cognitive enhancement" is that it "encompasses a number of theoretical and empirical observable phenomena broadly inclusive of enhancements or improvements (observed or theoretical) in memory, cognitive performance, and intelligence" (Outram and Racine 2011, 324).

The vagueness of these characterizations may at least partly be due to the fact that there is a considerable lack of knowledge concerning the question of in how far current technologies actually succeed to enhance *cognitive* functions.

In view of these difficulties, several authors prefer to use the term "neuroenhancement" instead of "cognitive enhancement". Neuroenhancement is a broader term to characterize all kinds of interventions intended to improve brain functions in healthy individuals. It encompasses different types of enhancement, such as cognitive enhancement, mood enhancement, moral enhancement or memory blunting. Mood enhancement has been characterized as aiming at feeling "better than well." It is the attempt to promote subjective well-being in otherwise healthy individuals who suffer from poor self-esteem or who feel down, alienated or socially isolated (Kramer 1993; Elliott 2000; DeGrazia 2000; Stein 2012; Synofzik et al. 2012). In the past, pharmacological forms of mood enhancement, in particular antidepressants such as selective serotonin reuptake inhibitors (for example Prozac®), have been the focus of attention. In contrast to this, reflections on moral enhancement, i.e. on interventions "that may reasonably be expected to result in [a person] having morally better future motives, taken in sum, than she would otherwise have had" (Douglas 2008, 229), are currently fictitious for the most part (Douglas 2008; Harris 2011). In addition, there are attempts to selectively blunt emotionally-laden extremely negative or traumatic memories (de Jongh et al. 2008; Parens 2010).

These distinctions between different types of neuroenhancement may seem to be a little bit artificial. Apart from the consideration of how far positive effects on these functions can actually be achieved in healthy individuals, there is the question of how far it is possible to *selectively* modify mood, cognition or motives since interventions that improve mood may imply an increase in motivation, which, in turn, may result in positive effects on cognitive functions. Also, cognitive enhancement may imply some sort of feeling better brought about by an increase in alertness. And – one could speculate – moral enhancements will probably imply modifications in cognitive and emotional functions.

Nevertheless, in spite of these difficulties to draw a clear line, in my opinion, it is helpful to distinguish between these different forms of enhancements. The reason is that the context in which they are sought, the aims they are heading at, and the consequences going along with them differ considerably. Extremely simplified, mood enhancement aims at making people happier and moral enhancement aims at making people better from a moral point of view; whereas, cognitive enhancement aims at making people capable of higher mental performance.

The term cognitive enhancement serves to characterize strategies that are expected to improve cognitive performance and thereby to confer an advantage in certain situations. Notwithstanding the fact that an increase in cognitive functions may also be considered beneficial in order to enjoy music or to have fun, the social reality in which cognitive enhancement is most often sought is to have an advantage in competitive situations or to perform better in situations where there is pressure to perform.

When people talk about cognitive enhancement, the focus of interest is not so much on the question of which particular kinds of brain functions are being improved. Rather, the focus is more on the question of why enhancement is sought, whether it is profitable and what the implications are. For the debate on cognitive enhancement is not so much about an increase in cognitive functions *per se* but about an increase in cognitive functions that aims at an increase in mental performance in competitive situations. The term cognitive enhancement is used to emphasize this social context; whereas, the broader term neuroenhancement is much less specific in that it refers to all kinds of interventions involving improvements of brain functions.

1.2 An Empirically-Based Broad Societal Debate Is Needed

After having discussed some conceptual issues relating to cognitive enhancement, let me now say a few words on the current debate on cognitive enhancement. In spite of considerable enthusiasm with regard to cognitive enhancement and in spite of various reports suggesting that prescription drugs are currently being widely used in order to improve cognitive functioning in healthy individuals (Forlini and Racine 2009; Partridge et al. 2011), there is a remarkable lack of knowledge concerning safety and efficacy of the purported cognitive enhancers and concerning the distribution among society.

A central question is: To what extent do the biomedical approaches considered to provide cognitive enhancement actually improve cognitive functions in healthy individuals? With regard to psychoactive drugs, current evidence suggests that cognition-enhancing effects of the putative cognitive enhancers in healthy individuals are at best very modest (Franke and Lieb 2010; Repantis et al. 2010a, b; Husain and Mehta 2011). In addition, further consideration is required regarding the questions of negative side effects and long-term effects, in particular, whether there is a risk of addiction that is linked to the use of stimulants for cognitive enhancement in healthy individuals.

Furthermore, prevalence rates are far from clear at the moment. In the literature, prevalence rates concerning non-medical use of prescription stimulants – which includes all forms of purposes, among them cognitive enhancement, but also partying, recreation etc. – vary widely. For example, in students, past-year prevalence rates between 5 and 35 % were reported in a meta-analysis published in 2008 (Wilens et al. 2008). Up to now, there are only a few studies that selectively assess the use of stimulants (prescription stimulants and illicit drugs) for cognitive enhancement (Franke et al. 2011).

Besides, with regard to cognitive enhancement use, societal and cultural differences may play a role, so that there may be considerable differences between countries. Data stemming from the USA cannot be directly transferred to other countries with different cultural traditions and social contexts and with different regulations pertaining to psychoactive substances.

In a situation like this, it is important to avoid favoring a rather unfounded cognitive enhancement euphoria by carelessly reporting about putative high benefits and putative high prevalence rates of cognitive enhancement. Researchers and journalists alike have to be very careful when they interpret and discuss study results in order to avoid an overoptimistic picture. Otherwise, people may be led to believe that part of the peer group might profit from cognitive enhancement – which may draw them towards using cognition-enhancing drugs.

However, as with assuming high putative benefits, it is equally problematic to understate the effects that may go along with putative cognitive enhancers. For even if the only desired effect of some drug were wake-promotion, this effect could lead to an increase in performance—an effect that could be considered beneficial in particular in situations dominated by time pressure or sleep deprivation. In addition, it might be expected that future substances might be more effective enhancers. So, to stress that current drugs do not seem to efficiently enhance cognition is important in that it shows that their usefulness for enhancement purposes is limited at the moment. In a substantial sense, however, it does not render reflections on cognitive enhancement obsolete.

In sum, more detailed empirical data is absolutely needed on the safety and efficacy of purported cognitive enhancers in healthy individuals, on the risk of addiction, on their distribution in society, on the life context in which the substances are used and on the social implications. All of this is necessary in order to enable a realistic, empirically-based discussion of the medical, social and ethical issues in cognitive enhancement, which, in turn, is an important precondition for any kind of attempt to regulate the use of cognition-enhancing technologies in society.

There is another aspect that is important for a fruitful societal debate on the medical, social and ethical issues in cognitive enhancement. According to an investigation by Cynthia Forlini and Eric Racine, in the discussion of the non-medical use of prescription stimulants, three different discourses can be distinguished, each of these discourses being dominated by a different paradigm (Forlini and Racine 2009; Racine and Forlini 2010): A bioethics discourse, a public health discourse, and a (print) media discourse. According to this analysis, the public health discourse is characterized by a "prescription drug abuse" framework which clearly distinguishes between the use of freely-available substances such as caffeine and the non-medical abuse of prescription stimulants, the latter being a practice that is highly criticized. In contrast, the (print) media discourse is dominated by a "lifestyle choice" framework. Here, the focus is on individual lifestyle and individual choice. Characteristic of the third discourse, the bioethics discourse, is a "cognitive enhancement" framework, which "focuses on the ethical issues arising from presumed benefits of non-medical use of neuropharmaceuticals by healthy individuals" (Forlini and Racine 2009).

This clearly is an interesting analysis which reflects the different perspectives in society with regard to the non-pharmacological use of prescription stimulants and its pros and cons. There is an important fourth point of view, however, which has not been taken adequately into account up to now: The perspective of those availing themselves of psychoactive substances for cognitive enhancement – a perspective that might be called "user discourse": In a recent interview study 18 university students experienced with the use of prescription or illicit stimulants for cognitive enhancement were asked about their views concerning the differences between using caffeine and illicit/prescription stimulants for cognitive enhancement (Franke et al. 2012). In short, their perspective can be characterized as dominated by a "function-oriented" framework. In this, the users stress the usefulness and the potential benefits and harms for themselves that result from substance use.

In my view, it is extremely important not to consider these various discourses within society to be separate discourses but to mix them up. For as long as they are separate discourses, they each mainly represent one particular perspective in the complex field of cognitive enhancement. Instead, an approach that integrates the various discourses would help to adequately consider various views held by society. A broad, empirically-based rational debate encompassing the various institutions, groups and opinions in society is a suitable basis for policy recommendations concerning cognitive enhancement technologies.

1.3 Individual Autonomy, Cognitive Liberty and Society

One of the central and most controversial issues in the debate on cognitive enhancement concerns the relationship between individuals and society. With regard to this question, the concept of cognitive liberty is crucial. Wrye Sententia characterizes cognitive liberty to be "every person's fundamental right to think independently, to use the full spectrum of his or her mind, and to have autonomy over his or her own brain chemistry" (Sententia 2004, 223). She then goes on to describe two fundamental principles of cognitive liberty (Sententia 2004, 227):

1. As long as their behavior doesn't endanger others, individuals should not be compelled against their will to use technologies that directly interact with the brain or be forced to take certain psychoactive drugs.
2. As long as they do not subsequently engage in behavior that harms others, individuals should not be prohibited from, or criminalized for, using new mind-enhancing drugs and technologies.

The concept of cognitive liberty stresses individual decision-making and the individual's right to decide for him- or herself, in particular to decide on whether or not to use technologies that modulate brain functions. This can be considered as analogous to the concept of informed consent, according to which, in medical contexts, it is up to the patient to decide on the medical treatment to be applied. A valid informed consent presupposes a competent person to decide voluntarily after having been supplied with the information necessary for autonomous decision-making. Central to the concept is the thorough understanding of the relevant circumstances and the absence of external or internal constraints limiting free decision-making (Faden and Beauchamp 1986).

Without any doubt, cognitive liberty is central to the use of any cognition-enhancing technology: It is up to the individual to decide whether or not to take substances that influence his or her brain. Nobody should be allowed to tamper with another person's brain. With regard to pharmacological cognitive enhancement, however, there are several aspects that may limit cognitive liberty – for it requires that a person have thorough information on the substances used and to freely decide, without any constraints. Both of these presuppositions can be questioned since currently there is not much knowledge available concerning the effects of so-called cognition-enhancing drugs, nor can it always be assumed that there are no constraints. Constraints may include subtle and indirect social pressure to perform, but also direct instructions at the workplace, for example in the armed forces. Particularly in minors, given the influence of third persons, the voluntariness of taking cognition-enhancing drugs is questionable.

Cognitive liberty does not imply that there is a societal obligation to provide technological options that allow individuals to modify their brain activity. In particular, it does not imply that society is bound to ensure free access to existing technologies since the general context in which technologies are used within a society has to be taken into consideration. It is not realistic, for example, to claim free access to certain psychoactive substances for cognitive enhancement when, in therapeutic contexts or with regard to drug consumption, access to these substances is strictly limited by formal regulations or medical law. In order to reflect on an adequate use of cognition-enhancing technologies within the existing societal framework, it will be important to characterize the relevant analogies and disanalogies with current existing practice. Concerning pharmacological cognitive enhancement, these are the analogies and disanalogies between drug use for enhancement purposes, for medical purposes and for illicit consumption.

In addition, when discussing cognitive liberty, the limits of cognitive liberty should be taken into consideration. In accordance with the Millian tradition (cf. Mill 1859/2006), Sententia writes that individuals should not be prohibited from using mind-enhancing drugs and technologies "as long as they do not subsequently engage in behavior that harms others" (Sententia 2004, 227). This is a crucial point for it can be expected that within society, when people use effective cognitive enhancers, there will almost certainly be direct or indirect effects on other individuals. Whether these will be positive or negative remains to be established in detail. The first presumption is that often the implications on other individuals will be negative because, in the majority of cases, people who avail themselves of cognitive enhancers aim at increasing performance in some competitive situation – which implies that others will be disadvantaged in some way or another. However, there may also be positive effects on others. For example, it has been argued that cognitive enhancement may increase a society's overall productivity and achievements.

Possible harm to others, in particular negative implications concerning fairness and justice, may limit cognitive liberty and legitimize some sort of formal regulations or policymaking. In order to find a balance between the individual user's point of view and broader implications for society, detailed reflection and research is needed that is not restricted to the individual but that considers the overall situation in society including: common practices in the health care system, analogies and disanalogies with other forms of technology uses, the existing legal framework, social conventions and general assumptions. Such reflections and research are required since the central issues in cognitive enhancement concern the individual's room to maneuver in a modern society.

1.4 Book Outline

The following book is an interdisciplinary approach to cognitive enhancement. The intention of the book is to provide empirical information concerning the various fields of cognitive enhancement and to reflect on its individual, social, ethical and legal implications. The book is a collection of contributions by researchers stemming from different disciplines such as medicine, psychiatry, neuroscience, neurotechnology, philosophy, medical ethics, neuroethics, social sciences and law. The various researchers have very different perspectives with regard to cognitive enhancement, the various strategies, and its implications. What unites these contributions is the ambition to foster a rational and empirically-based debate on cognitive enhancement that serves further development.

The book is divided into three parts. The first part deals with medical and psychosocial aspects of cognitive enhancement. The first four contributions in this part deal with pharmacological strategies for enhancement. Andreas G. Franke and Klaus Lieb give an overview of the various substances for pharmacological neuroenhancement. They distinguish between over-the-counter substances, prescription drugs and illicit drugs. In their chapter, the focus is on "brain doping," a term that refers to the illicit use of a subcategory of these prescription and illicit drugs for neuroenhancement. They draw a very critical picture of the use of these substances for enhancement purposes in that they stress their very limited positive effects, the safety risks and possible side effects. Dimitris Repantis then presents the results of a systematic review of the literature on the available evidence of the risks and benefits of antidepressants, anti-dementia drugs and psychostimulants currently used for pharmacological cognitive enhancement. He concludes that with regard to the use of prescription drugs for enhancement purposes in healthy individuals, there is a considerable lack of evidence both for the effectiveness and for the long-term safety. In the subsequent chapter Brad Partridge critically analyzes currently available data on prevalence rates of non-medical stimulant use by students for cognitive enhancement. In particular, he identifies several examples in the literature in which the prevalence of cognitive enhancement has been uncritically presented and discusses possible implications that may result from such presentation. In order to contribute to a better understanding of the underlying mechanisms by which cognitive neuroenhancers such as modafinil modulate brain functioning, Carlos Trenado and Daniel J. Strauss developed a neural computational model. In their chapter, they present a biologically-inspired, large-scale computational model for studying modafinil effects on electroencephalographic neural correlates of attention.

In the next three chapters, various non-pharmacological strategies for neuroenhancement are presented. Martin Dresler first gives a review of recent research on the various behavioral techniques for the enhancement of cognition in the domains of attention, intelligence, creativity, and memory. His chapter underlines the potential and relevance of these non-technological approaches. Following this, Pavel Dietz summarizes recent research on different kinds of sports and exercise and their potential to enhance cognitive task performance. He then discusses some of the limitations of these studies, particularly focusing on their heterogeneous study design. Subsequently, Colleen Dockery gives an overview of the state of the art of transcranial direct current brain stimulation for neuroenhancement purposes. Her summary of brain stimulation experiments is imbedded in a critical perspective towards neuroenhancement and the current societal context of a growing reliance on high-level cognitive functions for economic competition.

In the second part of this book, philosophical and ethical aspects of cognitive enhancement are discussed. From the point of view of medical ethics, Kirsten Brukamp gives an overview of the ethical issues involved in neuroenhancement. Her focus is on medical risks, the lack of evidence-based medicine, and financial challenges to health care systems, human nature, virtue ethics, liberty, justice and the social value of the purposes behind neuroenhancement. Then, Michael Hauskeller calls into question the purported aim of human enhancement, the making of better human beings. He stresses that what is considered to be a betterment is highly context-dependent for whether some neuroenhancement is desirable for people depends on what they are getting better at, what end the improvement serves, and who benefits from it. In the following contribution, Frederic Gilbert examines the emerging ethical challenges raised by implementation of nanotechnology in brain devices for enhancement purposes in healthy subjects. His particular focus is on the ethics of an adequate informed-consent procedure for invasive nano-bionic brain interventions for neuroenhancement. In the chapter that follows, Emma Peng Chien discusses the mechanism of cognitive enhancement from a philosophical point of view. She argues in favor of mechanistic differences between cognitive-enhancing drugs and behavioral training. In addition, she suggests possible mechanisms for cognitive-enhancing drugs and behavior training and accounts for the characteristic differences of these two strategies. Roland Kipke then examines the question as to whether the ethical problems of cognitive and non-cognitive enhancement are significantly different and, therefore, whether the concentration on cognitive enhancement within the bioethical debate is justified. He concludes that the ethical questions raised by the different forms of pharmacological cognitive and non-cognitive enhancement are in most respects equal or similar, and that the concentration on cognitive enhancement is largely not justified. Following this, Lisa Forsberg discusses three objections to the use of cognitive enhancement based on potential effects that the use of cognitive enhancement is thought to have for the value of achievement. After a detailed examination of a number of counter-arguments against each of these three objections, she concludes that none of the three objections succeeds as an *in principle* objection to the use of cognitive enhancement. Based on an examination of published literature of the perspectives of stakeholders towards cognitive enhancement, Cynthia Forlini and Eric Racine identify three points of contention between the stakeholder perspectives and the ethics debate on cognitive enhancement in academia, which suggest that two separate debates are taking place on parallel tracks. In view of this, they propose that the discipline of bioethics needs to reaffirm its role as a meeting place for the traditional academic ethics debate on cognitive enhancement and the more experientially-based approach of stakeholders to enrich future deliberation.

The third part of the book reflects on social, political and legal aspects of cognition-enhancement. First, Peter Reiner analyzes the current debate on cognitive enhancement and argues that the normative claims at the extremes – by the Transhumanists on the one side and the Bioconservatives on the other side – are driven to a considerable extent by biopolitical intuitions. In order to overcome polemics and to move forward in the discussion on cognitive enhancement, he suggests the adoption of a middle ground position he calls "The view from reasonableness." He also advocates taking empirical data on public attitudes towards cognitive enhancement into consideration. Then, Armin Grunwald's chapter focuses on the question: Are we witnessing a historical change from a performance society to an enhancement society with an inherent and infinite spiral of enhancement including increased self-exploitation and self-instrumentalization? Armin Grunwald's hypothesis is that we can learn from the ongoing debate on human enhancement about our society and contemporary perceptions of ourselves. In the proceeding chapter, Greta Wagner discusses fairness arguments in the cognitive enhancement debate, which focus on the question of whether potential cognitive enhancements would decrease or increase fairness in society. She characterizes both positions as being based on a notion of society as competitive, in which fairness is the purpose of a certain degree of institutional intervention. Within the framework of Michel Foucault's terminology, she argues that the bioethical debate on fairness forms part of a neoliberal governmentality. Jan-Christoph Bublitz then explores some of the legal issues raised by neuroenhancement interventions. He argues that the law will have to recognize a basic human right, namely, cognitive liberty or mental self-determination, which guarantees an individual's sovereignty over her mind and entails the permission to both use and refuse neuroenhancements. The focus of the chapter by Elizabeth Shaw is on a particular group of offenders who appear to have an impaired capacity to appreciate the moral significance of their acts and a limited ability to engage in effective practical reasoning. She discusses whether it could ever be morally permissible to employ certain types of cognitive enhancements to enhance offenders' capacities for practical reasoning and moral communication as part of their rehabilitation. John Danaher then discusses legal issues from a totally different perspective in that he asks whether agents should be held criminally responsible for the consequences of failing to make use of enhancement technologies. He argues that they should, provided such technologies would have allowed them to avoid the risks associated with the state of abnormal agency.

References

Bostrom N, Sandberg A (2009) Cognitive enhancement: methods, ethics, regulatory challenges. Sci Eng Ethics 15:311–341

Bublitz JC, Merkel R (2009) Autonomy and authenticity of enhanced personality traits. Bioethics 23:360–374

Cakic V (2009) Smart drugs for cognitive enhancement: ethical and pragmatic considerations in the era of cosmetic neurology. J Med Ethics 35:611–615

Caplan A (2002) No-brainer: can we cope with the ethical ramifications of new knowledge of the human brain? In: Marcus SJ (ed) Neuroethics: mapping the field. University of Chicago Press, Chicago, pp 95–106

Cole-Turner R (1998) Do means matter? In: Parens E (ed) Enhancing human traits: ethical and social implications. Georgetown University Press, Washington, DC, pp 151–161

de Jongh R, Bolt I, Schermer M, Olivier B (2008) Botox for the brain: enhancement of cognition, mood and pro-social behavior and blunting of unwanted memories. Neurosci Biobehav Rev 32(4):760–776

DeGrazia D (2000) Prozac, enhancement, and self-creation. Hastings Cent Rep 30:34–40

Douglas T (2008) Moral enhancement. J Appl Philos 25:228–245

Dresler M, Sandberg A, Ohla K, Bublitz C, Trenado C, Mroczko-Wąsowicz A, Kühn S, Repantis D (2013) Non-pharmacological cognitive enhancement. Neuropharmacology 64:529–543, Epub 2012 Jul 22. http://dx.doi.org/10.1016/j.neuropharm.2012.07.002

Elliott C (2000) Pursued by happiness and beaten senseless. Prozac and the American dream. Hastings Cent Rep 30:7–12

Faden RR, Beauchamp TL (1986) A history and theory of informed consent. Oxford University Press, Oxford

Farah MJ, Illes J, Cook-Deegan R, Gardner H, Kandel E et al (2004) Neurocognitive enhancement: what can we do and what should we do? Nat Rev Neurosci 5:421–425

Forlini C, Racine E (2009) Disagreements with implications: diverging discourses on the ethics of non-medical use of methylphenidate for performance enhancement. BMC Med Ethics 10:9. doi:10.1186/1472-6939-10-9

Franke AG, Lieb K (2010) Pharmacological neuroenhancement and brain doping: chances and risks. Bundesgesundheitsblatt Gesundheitsforschung Gesundheitsschutz 53:853–860

Franke AG, Bonertz C, Christmann M, Huss M, Fellgiebel A, Hildt E, Lieb K (2011) Non-medical Use of prescription stimulants and illicit use of stimulants for cognitive enhancement in pupils and students in Germany. Pharmacopsychiatry 44:60–66

Franke AG, Lieb K, Hildt E (2012) What users think of the differences between caffeine and stimulants for cognitive enhancement. PLoS One 7(6):e40047

Gordijn B, Chadwick R (eds) (2008) Medical enhancement and posthumanity. Springer, Dordrecht

Greely H, Sahakian B, Harris J, Kessler RC, Gazzaniga M et al (2008) Towards responsible use of cognitive-enhancing drugs by the healthy. Nature 456:702–705

Hamilton R, Messing S, Chatterjee A (2011) Rethinking the thinking cap: ethics of neural enhancement using noninvasive brain stimulation. Neurology 76(2):187–193

Harris J (2011) Moral enhancement and freedom. Bioethics 25(2):102–111

Husain M, Mehta MA (2011) Cognitive enhancement by drugs in health and disease. Trends Cogn Sci 15(1):28–36

Juengst ET (1998) What does enhancement mean? In: Parens E (ed) Enhancing human traits: ethical and social implications. Georgetown University Press, Washington, pp 29–47

Kramer PD (1993) Listening to Prozac. A psychiatrist explores antidepressant drugs and the remaking of the self. Penguin Books, New York

Metzinger TK (2012) Zehn Jahre Neuroethik des pharmazeutischen kognitiven Enhancements – Aktuelle Probleme und Handlungsrichtlinien für die Praxis. Fortschr Neurol Psychiatr 80:36–43

Metzinger T, Hildt E (2011) Cognitive enhancement. In: Illes J, Sahakian B (eds) Oxford handbook of neuroethics. Oxford University Press, Oxford, pp 245–264

Mill JS (1859/2006) On liberty. Adamant Media Corporation, Delaware

Morein-Zamir S, Sahakian BJ (2011) Pharmaceutical cognitive enhancement. In: Illes J, Sahakian B (eds) Oxford handbook of neuroethics. Oxford University Press, Oxford, pp 229–244

Nagel SK (2010) Too much of a good thing? enhancement and the burden of self-determination. Neuroethics 3:109–119

Outram SM (2012) Ethical considerations in the framing of the cognitive enhancement debate. Neuroethics 5:173–184

Outram SM, Racine E (2011) Examining reports and policies on cognitive enhancement: approaches, rationale, and recommendations. Account Res 18:323–341

Parens E (1998) Is better always good? the enhancement project. In: Parens E (ed) Enhancing human traits: ethical and social implications. Georgetown University Press, Washington, pp 1–28

Parens E (2010) The ethics of memory blunting and the narcissism of small differences. Neuroethics 3:99–107

Partridge BJ, Bell SK, Lucke JC, Yeates S, Hall WD (2011) Smart drugs "as common as coffee": media hype about neuroenhancement. PLoS One 6(11):e28416

President's Council on Bioethics (2003) Beyond therapy. Biotechnology and the pursuit of happiness. Dana Press, Washington, DC

Racine E, Forlini C (2010) Cognitive enhancement, lifestyle choice or misuse of prescription drugs? ethics blind spots in current debates. Neuroethics 3:1–4

Repantis D, Laisney O, Heuser I (2010a) Acetylcholinesterase inhibitors and memantine for neuroenhancement in healthy individuals: a systematic review. Pharmacol Res 61:473–481

Repantis D, Schlattmann P, Laisney O, Heuser I (2010b) Modafinil and methylphenidate for neuroenhancement in healthy individuals: a systematic review. Pharmacol Res 62:187–206

Sahakian B, Morein-Zamir S (2007) Professor's little helper. Nature 450:1157–1159

Savulescu J, Bostrom N (eds) (2009) Human enhancement. Oxford University Press, Oxford

Sententia W (2004) Neuroethical considerations: cognitive liberty and converging technologies for improving human cognition. Ann N Y Acad Sci 1013:221–228

Stein DJ (2012) Psychopharmacological enhancement: a conceptual framework. Philos Ethics Humanit Med 7:5

Synofzik M, Schlaepfer TE, Fins JJ (2012) How happy is too happy? euphoria, neuroethics, and deep brain stimulation of the nucleus accumbens. AJOB Neurosci 3(1):30–36

Wilens TE, Adler LA, Adams J, Sgambati S, Rotrosen J et al (2008) Misuse and diversion of stimulants prescribed for ADHD: a systematic review of the literature. J Am Acad Child Adolesc Psychiatry 47:21–31

Part I
Medical and Psychosocial Aspects of Cognitive Enhancement

Chapter 2
Pharmacological Neuroenhancement: Substances and Epidemiology

Andreas G. Franke and Klaus Lieb

Abstract Pharmacological neuroenhancement (PN) refers to the general use of psychoactive substances with the purpose of cognitive enhancement (e.g. enhancement of vigilance, concentration, memory or mood) by healthy subjects. Substances for PN include Over-the-Counter- (OTC-) substances such as coffee, caffeinated drinks/energy drinks, caffeine tablets and Ginkgo biloba as well as prescription drugs and illicit drugs (e.g. (psycho-) stimulants). "Brain doping" refers to the illicit use of a subcategory of these substances. On the one hand, this subcategory includes prescription drugs for the treatment of attention deficit/hyperactivity disorder (ADHD), sleep disorders, Alzheimer's disease and depression. This group consists of (psycho-) stimulants (e.g. amphetamines, methylphenidate), modafinil, antidementives (acetylcholinesterase inhibitors, memantine) and antidepressants (e.g. selective serotonin reuptake inhibitors). On the other hand, this subcategory includes stimulating illicit drugs (e.g. illicit amphetamines, cocaine, ecstasy).

OTC-drugs as well as prescription and illicit drugs have very limited positive effects on cognition and mood. Furthermore, their use leads to safety risks and can cause severe side effects. Therefore, a general use by healthy persons cannot be justified. Claims for liberalization should be critically questioned.

Keywords Brain doping • Cognition • Neuroenhancement • Enhancement • Psychostimulants

---

A.G. Franke (✉) • K. Lieb
Department of Psychiatry and Psychotherapy, University Medical Centre, Johannes-Gutenberg University of Mainz, Mainz, Germany
e-mail: afranke@uni-mainz.de; klaus.lieb@unimedizin-mainz.de

E. Hildt and A.G. Franke (eds.), *Cognitive Enhancement: An Interdisciplinary Perspective*, Trends in Augmentation of Human Performance 1,
DOI 10.1007/978-94-007-6253-4_2,

2.1 Introduction

Enhancement has been an eternal pursuit of humankind. For millennia, humankind tried to improve performance with regard to physical skills by the use of divergent training methods and substances (Baron et al. 2007; Lieb 2010). In the past, physical skills were more important than mental skills. Later, mental skills became increasingly important. For example, famous writers and philosophers (e.g. Balzac, Voltaire) wrote their manuscripts using potential mental performance enhancing substances. For centuries, caffeine was the most important substance for enhancing one's own cognitive abilities. Later, the development of stimulants seemed to be an important step towards academic performance enhancement for wake promoting effects, enhancement of attentiveness, concentration, memory and mood.

Pharmacological neuroenhancement (PN) refers to the use of various substances with the aim of cognitive enhancement via enhancing effects on sub-domains of cognition (e.g. attentiveness, concentration, endurance, memory, mood). Divergent substances belong to the group of pharmacological neuroenhancers. The most famous and widely spread substances are over-the-counter- (OTC-) drugs, which can be bought in supermarkets. Caffeinated beverages (coffee, caffeinated drinks/energy drinks) and caffeine tablets as well as phytopharmaceuticals are pharmacological neuroenhancers. The term "Brain doping" is used infrequently and means the misuse of a subcategory of substances for PN. It refers to the use of prescription and illicit substances by healthy people aiming at academic performance enhancement, mood enhancement and enhancement of pro-social behavior (de Jongh et al. 2008; Franke and Lieb 2010; Lieb 2010). Brain doping is used much like the term of "doping" in sports, which refers to an illicit use of "prohibited substances" for physical performance enhancement. This subcategory of substances consists of prescription substances used without medical indication and illicit drugs. Methylphenidate (MPH), amphetamines (AMPH) and modified amphetamine substances (e.g. amphetamine salts), modafinil, antidementives (acetylcholinesterase inhibitors, memantine) and antidepressants (selective serotonin reuptake inhibitors, SSRIs) for the treatment of attention deficit/hyperactivity disorder (ADHD), sleep disorders, Alzheimer's disease (AD) and depression as well as illicit amphetamines and derivates (3,4-Methylendioxy-N-methylamphetamin, MDMA, ecstasy) and cocaine belong to this subcategory. The use of prescription drugs without medical indication as well as the use of illicit drugs for cognitive enhancing purposes is prohibited and can have serious legal consequences, depending on users' sociocultural background.

Nevertheless, there seems to be a high demand in Western societies for the illicit misuse of brain doping substances. In a non-representative online poll of the journal *Nature*, 80 % of participating graduates who were familiar with *Nature* and read the journal's announcement of the online-poll answered that adults should be allowed to use such substances. Out of these subjects another 20 % answered that they have already used stimulants, modafinil or beta blockers at least once (Maher 2008).

A member poll of a major German health insurance agency (Deutsche Angestelltenkrankenkasse, DAK) found that one-third of the surveyed subjects had already used substances for cognitive enhancing purposes without differentiating prescription, these being illicit and OTC-drugs (DAK 2009). However, this study had severe methodological deficits (response rate, missing differentiation of prescription, illicit and OTC-drugs, etc.) and therefore only has limited statistical validity.

With regard to students in the US, a detailed analysis of the literature has shown that there are only a limited number of studies available that have exclusively studied the use of stimulants for the purpose of cognitive enhancement. In general, there are several studies dealing with the misuse of stimulants in general (Babcock and Byrne 2000; Teter et al. 2003, 2005, 2006; McCabe et al. 2004, 2005, 2006; Hall et al. 2005; Poulin 2007; Wilens et al. 2008). An extensive meta-analysis reported that 5–9 % of grade and high school students, 5–35 % of college aged students misused stimulants during the last year of the survey (Wilens et al. 2008).

This article will present potential substances which are misused for PN, their effects, efficacy, safety risks and side effects. This review is based upon data of randomized controlled trials (RCTs) and meta-analyses of studies on "potential" substances for PN. In a previous study, we surveyed 1,500 pupils and students about their knowledge, attitudes and use of potential pharmacological neuroenhancers (Franke et al. 2011a, b, 2012a). Substances reported by the surveyed subjects correspond to substances cited in international publications regarding substances for PN (Teter et al. 2003, 2005, 2006; McCabe et al. 2004, 2005, 2006; Maher 2008; Wilens et al. 2008; de Jongh et al. 2008; DAK 2009; Franke and Lieb 2010; Lieb 2010; Franke et al. 2011a, b). Presented substances are based on the results of our survey and international literature.

2.2 Prescription and Illicit Drugs for "Brain Doping"

2.2.1 (Psycho-) Stimulants

Methylphenidate (MPH) is approved for the pharmacotherapy of ADHD. In contrast to the US, amphetamines (AMPH) like Adderall® (mixed amphetamine salts) are not approved for ADHD in Germany. Furthermore, illicit drugs contribute to the group of AMPH (e.g. Ecstasy, MDMA). In Germany, MPH is only sold in pharmacies and is only available with a specialized prescription, required for narcotics as specified under the Law of German Medicinal Products (AMG).

The common mode of action of AMPH and MPH is a blockade of pre-synaptic norepinephric and dopaminergic transporters. This means a missing negative feedback regarding a missing blockade of exocytosis of norepinephrine and dopamine which means a higher level of monoaminergic neurotransmission. In contrast to MPH, AMPH additionally leads to vesicular release (exocytosis) of dopamine, which means an increase of action-independent dopaminergic activity (Oades et al. 2005; Prince 2008; Wilens 2008; Franke and Lieb 2010).

Randomized controlled trials (RCTs) show that both substances increase vigilance and attentiveness in healthy subjects, which can lead to shortened reaction time. Effects are even considerably distinctive in healthy subjects who are sleep deprived (Franke and Lieb 2010; Repantis et al. 2010b). Because of the additional action-independent effect of AMPH compared to MPH, AMPH are more effective than MPH. However, effects of previous studies are not very robust and show inconsistent results regarding higher cognitive domains, e.g. memory. These inconsistent effects regarding higher domains may be due to indirect effects mediated via simple cognitive domains, e.g. vigilance and attentiveness.

Furthermore, RCTs show that neither subjects' mood nor subjective self-evaluation undergo changes after using stimulants in doses intended for therapeutic purposes. Higher doses of AMPH and/or intra nasally ingested illicit AMPH have euphoric effects.

Altogether, there are only slightly distinctive pro cognitive effects of stimulants that seem to be indirect and secondary effects of vigilance enhancement. Furthermore, stimulants have a large side effect profile of light to severe side effects and important safety risks (Gould et al. 2009; Arzneimittelfachinformation 2011; Franke et al. 2012b). Side effects include reduced concentration, allergic reactions, divergent gastrointestinal symptoms, headache, dizziness and dyskinesia.

While the pharmacotherapeutic use of stimulants in ADHD patients is well established and effective in symptom reduction and leads to reduced prevalence rates of general drug abuse and addiction in ADHD patients, in healthy subjects it can cause severe addiction (Wilens 2007; Bukstein 2008; Kollins 2008; Franke et al. 2012a). In this respect the use of oral administration is safer than intra nasal ingestion due to a very fast CNS-rise of the stimulants and a subsequent pulsatile dopamine release, which causes addictive behavior (Barkley et al. 2003; Biederman et al. 2008; Bright 2008).

2.2.2 *Modafinil*

Modafinil (Provigil®/Vigil®) is approved for pharmacotherapy of excessive sleepiness in case of narcolepsia. In Europe, modafinil recently lost the approval for excessive sleepiness in cases of sleep apnea syndrome and the so-called "chronic shift worker syndrome."

Modafinil is chemically unrelated to AMPH. Compared to MPH, there is no need for a specialized prescription. Even today, the action profile is not fully understood. It is likely that modafinil acts as a dopamine and norepinephrine reuptake inhibitor and modulates the GABA-ergic and glutamatergic neurotransmission (Minzenberg and Carter 2008).

RCTs show inconsistent positive cognitive effects of modafinil in healthy subjects regarding vigilance, attentiveness, memory, mood and subjective alertness.

However, positive results are consistent for shortened reaction time. In cases of sleep deprivation, effects are more consistent and clearer regarding simple cognitive domains, e.g. vigilance, attentiveness. Higher cognitive domains show less consistent and clear results. This trend is probably due to the fact of secondary effects on higher cognitive domains via vigilance and alertness (Franke and Lieb 2010; Repantis et al. 2010b). RCTs that compare modafinil, MPH and caffeine directly to each other show no relevant differences in divergent neuropsychological tests in healthy subjects between these three substances (Wesensten et al. 2005; Killgore et al. 2009).

Modafinil's side effects and safety risks should not be underestimated (Arzneimittelfachinformation 2011; Franke et al. 2012b). There are many possible infrequent to very frequent side effects of divergent severity, predominantly various gastrointestinal side effects, vision disorders, drowsiness, sleeplessness, etc.

2.2.3 *Acetylcholinesterase Inhibitors*

Acetylcholine esterase inhibitors (donepezil, galantamin and rivastigmin) (ACh-I) are approved for mild to moderate Alzheimer's disease (AD).

They increase concentration of acetylcholine in the synaptic cleft via inhibitory effects at the acetylcholinesterase enzyme, which is thought to increase cognition in patients with AD.

Galantamine solely causes an allosteric modulation of nicotinic acetylcholine receptors. Rivastigmine has additional inhibitory effects on the butyrylesterase enzyme (Franke et al. 2009).

Current RCTs do not have very adequate study designs for evaluating cognitive skills of ACh-I in healthy subjects (Repantis et al. 2010a). Furthermore, there are presently only study results for rivastigmine and donepezil. Existing results are inconsistent; partially there are detrimental effects of ACh-I on reaction time and memory. Altogether, existing results of RCTs indicate that ACh-I have no detectable effects on cognition (Franke and Lieb 2010).

Like MPH and modafinil, ACh-I also have safety risks and side effects, which have to be considered (Franke et al. 2009; Franke and Lieb 2010; Arzneimittelfachinformation 2011). Among these numerous side effects are divergent gastrointestinal symptoms (e.g. diarrhea, nausea, vomitus), loss of appetite and incontinence.

2.2.4 *Memantine*

Memantine, named Axura® or Ebixa®, is approved for moderate to severe AD and has partial antagonistic effects at the glutamatergic N-Methyl-D-Aspartat-(NMDA-) receptor (Franke et al. 2009).

Memantine has no positive effects on healthy subjects: RCTs document neither positive nor negative effects on vigilance, attentiveness or mood but inconsistent effects on particular domains of memory. Again, safety risks and side effects should not be underestimated and can cause divergent gastrointestinal symptoms, hypertonia, headache, dizziness and sleepiness (Franke et al. 2009; Franke and Lieb 2010; Arzneimittelfachinformation 2011).

2.2.5 Antidepressants

Antidepressants are misused by healthy subjects especially in the US for enhanced mood and advanced social skills in general. In particular, users hope to perform better in social interactions with others, to seem more self-confident, reduce social anxiety and to enhance self-presentation in everyday life. Selective serotonin reuptake inhibitors (SSRI) are among the divergent antidepressant pharmaceuticals used for these purposes. The most popular substance is Prozac® (USA) or Fluctin® (Germany), which consists of the active ingredient fluoxetin. The well-known book entitled "*Prozac Nation*" points out the relevance of Prozac® and further SSRI in the US.

While antidepressive pharmacotherapy is effective in depressed patients, there is no effect of SSRI in healthy subjects with regard to vigilance, attentiveness, reaction time and memory. Furthermore, there is discrete evidence, that there are even negative effects on these cognitive domains. Moreover, RCTs show no effects of SSRI on mood or social skills e.g. self-confidence, self-presentation, social anxiety, etc. (Repantis et al. 2009; Franke and Lieb 2010).

Again, safety risks and side effects have to be considered; side effects include not only divergent gastrointestinal symptoms, but also induction of agitation and further severe side effects regarding water and electrolyte balance. These side effects have warranted discouragement of using antidepressants for any purpose except depression (Franke and Lieb 2010; Arzneimittelfachinformation 2011).

2.3 OTC-Drugs for Pharmacological Neuroenhancement

2.3.1 Caffeine

For decades, caffeinated substances have been used for enhancement of vigilance and attentiveness. Today, Coffeinum® is the only approved drug against short time fatigue. Although there is no need for a prescription, Coffeinum® is only sold in pharmacies. It contains 200 mg of caffeine per tablet and is approved up to 400 mg per day – less than the amount of caffeine in a large Starbuck's©-cup of coffee, which contains approximately 500 mg of caffeine.

Caffeine belongs to the group of methylxanthines and has at least three stimulating mechanisms of action in the CNS: (1) Inhibition of depletion of cyclic adenosinmonophosphate (cAMP) via inhibition of the activity of cyclic nucleotidphosphodiesterase, which leads to accumulation of cAMP and its effects.

cAMP as a second messenger which causes a prolongation of epinephrine effects. (2) Blockade of adenosine receptors (especially in the striatum); injections of adenosine in animal experiments lead to reduced vigilance and can be antagonized by caffeine. (3) Mobilization of intra cellular calcium (Franke and Lieb 2010).

Caffeine leads to tachycardia and hypertonia via vasoconstriction as well as bronchia dilatation. Besides these somatic effects, caffeine leads to changed states in cognition. RCTs show that vigilance and attentiveness are increased and reaction is shortened inconsistently. Effects are even more distinct in cases of sleep deprivation. However, there is no direct and clear influence on higher cognitive skills such as memory and mood. As mentioned before, there are some studies that show equal efficacy of caffeine as compared to MPH and modafinil. Thus, caffeine seems an adequate alternative regarding efficacy, safety risks and side effects when compared to those of MPH and modafinil (Wesensten et al. 2005; Killgore et al. 2009).

There are notable inter-individual differences regarding the side effects of caffeine, which leads to a wide spread dose range among users. Toxic effects can first be seen in sensitive subjects at 600 mg; lethal doses are 10,000 mg and higher doses.

2.3.2 *Caffeinated Drinks/Energy Drinks*

The effect of energy drinks consists of a combination of caffeine, the main ingredient, and taurine, which supposedly strengthens the effect of the caffeine. One can of Red Bull® (250 ml) contains caffeine (80 mg), taurine (1,000 mg), glucose (5.25 g), sucrose (21.5 g), glucuronolactone (600 mg) and other flavouring substances. It is supposed that taurine should strengthen glucose utilisation via modifications of insulin level.

The use of energy drinks leads to stronger clinical effects than caffeine itself. Vigilance and attentiveness are increased, reaction time is shortened and single domains of memory show slightly enhancing effects while there is no modulating effect on mood. Safety risks and side effects are not systemically analyzed in the present RCTs. One can assume that energy drinks have the same qualitative side effects but less quantitative side effects than coffee and caffeine tablets because of the same main ingredient which has lower dosages in energy drinks (Franke and Lieb 2010).

2.3.3 *Ginkgo Biloba*

Ginkgo biloba drugs have high concentrations of flavonoids and terpenoids, which have at least antioxidative effects. Beyond that, further mechanisms are assumed that so far are not evidence-based. These include neuroprotection via anti-apoptotic effects, inhibition of β-amyloid-aggregation and changes in gene expression (patterns).

There is a high volume of sales promotion directed to commercials especially for older potential customers; these ads focus on objective, as well as subjective and exploit the fear of cognitive impairment.

Huge RCTs and meta-analyses were able to show that Ginkgo biloba has no clinical effect in younger and older subjects with or without cognitive impairment. While Ginkgo biloba has no consistent pro-cognitive effect, it has nearly no relevant side effects either and can be used almost without any hesitation. However, there are no positive effects on vigilance, attentiveness, reaction time, memory or mood; regardless of whether one used Ginkgo biloba once or during the span of a few months (Solomon et al. 2002; Birks and Grimley Evans 2009; Franke et al. 2009; Franke and Lieb 2010).

2.4 Conclusion

In summary, there are only small pro-cognitive effects of using substances for PN in general and specialized substances for brain doping in particular. Prescription drugs as well as illicit drugs and OTC-drugs have only very limited effects on vigilance, attentiveness, reaction time, subjective self-evaluation, memory and mood. Furthermore, the reported effects are limited to simple cognitive skills, e.g. psychomotor skills and reaction time. In addition, there are nearly no direct effects on higher cognitive skills. One can presume that effects on higher cognitive skills are indirect effects which are mediated via simple cognitive skills, e.g. vigilance. The fact that sleep deprivation leads to clearer results supports this hypothesis.

Regarding prescription and illicit drugs, modafinil and MPH seem to have the strongest effects whereas there are no positive pro-cognitive effects of antidementives and antidepressants at all. Interestingly, caffeine has comparable effects in higher doses (Wesensten et al. 2005). Perhaps energy drinks are not less effective, but unfortunately, at present, there are no comparable studies with prescription stimulants and energy drinks.

In spite of the limited effects, pupils and students seem to "know" which substances are useful for PN (Franke et al. 2012b). Furthermore, there are high prevalence rates regarding the misuse of stimulants in general (5–35 %) (Wilens et al. 2008), but prevalence rates regarding an exclusive use for PN are notably lower (1–4 %) (Franke et al. 2011a). Prevalence rates of caffeine are notably higher (Franke et al. 2011b). In order to properly interpret these different prevalence rates, one should consider the failure to exclude subjects with prescribed stimulants for ADHD in studies with higher prevalence rates and the missing differentiation of the purpose of the use of stimulants. Anyway, the prevalence rate of ADHD is described as 5 % (Goldman et al. 1998; Polanczyk et al. 2007; Huss et al. 2008). Nevertheless, numerous pupils and students seem to know about stimulants and further prescription and illicit drugs for PN and have already misused them.

Thus, it is very important to consider the limited pro-cognitive effects and the safety risks and side effects of these drugs. Furthermore, the misuse of stimulants can cause addiction and further addictive behavior. Also, the misuse of illicit drugs and prescription drugs without prescription is a federal offense. There are also legal alternatives, such as caffeine. However, there are behavioral strategies that can prevent the misuse of these substances for PN. Enough sleep and accurately-timed preparation for exams and sports can enhance higher cognitive functions, which are necessary for good results at school and at the university (Mednick et al. 2003; Korman et al. 2007; Lieb 2010).

References

Arzneimittelfachinformation (2011) Specialized information about drugs. Bundesinstitut für Arzneimittel und Medizinprodukte (Central institute of drugs and medical products) (BfArM), Bonn

Babcock Q, Byrne T (2000) Student perceptions of methylphenidate abuse at a public liberal arts college. J Am Coll Health 49(3):143–145

Barkley RA, Fischer M, Smallish L, Fletcher K (2003) Does the treatment of attention-deficit/hyperactivity disorder with stimulants contribute to drug use/abuse? A 13-year prospective study. Pediatrics 111(1):97–109

Baron DA, Martin DM, Abol Magd S (2007) Doping in sports and its spread to at-risk populations: an international review. World Psychiatry 6(2):118–123

Biederman J, Monuteaux MC, Spencer T, Wilens TE, Macpherson HA, Faraone SV (2008) Stimulant therapy and risk for subsequent substance use disorders in male adults with ADHD: a naturalistic controlled 10-year follow-up study. Am J Psychiatry 165(5):597–603

Birks J, Grimley Evans J (2009) Ginkgo biloba for cognitive impairment and dementia. Cochrane Database Syst Rev (1):CD003120

Bright GM (2008) Abuse of medications employed for the treatment of ADHD: results from a large-scale community survey. Medscape J Med 10(5):111

Bukstein O (2008) Substance abuse in patients with attention-deficit/hyperactivity disorder. Medscape J Med 10(1):24

de Jongh R, Bolt I, Schermer M, Olivier B (2008) Botox for the brain: enhancement of cognition, mood and pro-social behavior and blunting of unwanted memories. Neurosci Biobehav Rev 32(4):760–776

Deutsche Angestelltenkrankenkasse (DAK) (2009) Gesundheitsreport 2009. Analyse der Arbeitsunfähigkeitsdaten. Schwerpunktthema Doping am Arbeitsplatz. IGES Institut GmbH, Berlin

Franke AG, Lieb K (2010) Pharmacological neuroenhancement and brain doping: chances and risks. Bundesgesundheitsblatt Gesundheitsforschung Gesundheitsschutz 53(8):853–859

Franke AG, Lieb K, Fellgiebel A (2009) From symptomatic to disease modifying therapy? Recent developments in the pharmacotherapy of Alzheimer's disease. Fortschr Neurol Psychiatr 77(6):326–333

Franke AG, Bonertz C, Christmann M, Huss M, Fellgiebel A, Hildt E, Lieb K (2011a) Non-medical use of prescription stimulants and illicit use of stimulants for cognitive enhancement in pupils and students in Germany. Pharmacopsychiatry 44(2):60–66

Franke AG, Christmann M, Bonertz C, Fellgiebel A, Huss M, Lieb K (2011b) Use of coffee, caffeinated drinks and caffeine tablets for cognitive enhancement in pupils and students in Germany. Pharmacopsychiatry 44(7):331–338

Franke AG, Bonertz C, Christmann M, Lieb K (2012a) Attitudes toward cognitive enhancement in users and nonusers of stimulants for cognitive enhancement: a pilot study. AJOB Prim Res 3(1):48–57

Franke AG, Konrad A, Lieb K, Huss M (2012b) Stimulant and non-stimulant medication in current and future therapy for ADHD. Fortschr Neurol Psychiatr 80(3):130–140

Goldman LS, Genel M, Bezman RJ, Slanetz PJ (1998) Diagnosis and treatment of attention-deficit/hyperactivity disorder in children and adolescents. Council on Scientific Affairs, American Medical Association. JAMA 279(14):1100–1107
Gould MS, Walsh BT, Munfakh JL, Kleinman M, Duan N, Olfson M, Greenhill L, Cooper T (2009) Sudden death and use of stimulant medications in youths. Am J Psychiatry 166(9):992–1001
Hall KM, Irwin MM, Bowman KA, Frankenberger W, Jewett DC (2005) Illicit use of prescribed stimulant medication among college students. J Am Coll Health 53(4):167–174
Huss M, Holling H, Kurth BM, Schlack R (2008) How often are German children and adolescents diagnosed with ADHD? Prevalence based on the judgment of health care professionals: results of the German health and examination survey (KiGGS). Eur Child Adolesc Psychiatry 17(Suppl):152–158
Killgore WD, Kahn-Greene ET, Grugle NL, Killgore DB, Balkin TJ (2009) Sustaining executive functions during sleep deprivation: a comparison of caffeine, dextroamphetamine, and modafinil. Sleep 32(2):205–216
Kollins SH (2008) A qualitative review of issues arising in the use of psycho-stimulant medications in patients with ADHD and co-morbid substance use disorders. Curr Med Res Opin 24(5):1345–1357
Korman M, Doyon J, Doljansky J, Carrier J, Dagan Y, Karni A (2007) Daytime sleep condenses the time course of motor memory consolidation. Nat Neurosci 10(9):1206–1213
Lieb K (2010) Hirndoping – Warum wir nicht alles schlucken sollten. Verlag Artemis und Winkler, Düsseldorf
Maher B (2008) Poll results: look who's doping. Nature 452(7188):674–675
McCabe SE, Teter CJ, Boyd CJ (2004) The use, misuse and diversion of prescription stimulants among middle and high school students. Subst Use Misuse 39(7):1095–1116
McCabe SE, Knight JR, Teter CJ, Wechsler H (2005) Non-medical use of prescription stimulants among US college students: prevalence and correlates from a national survey. Addiction 100(1):96–106
McCabe SE, Teter CJ, Boyd CJ (2006) Medical use, illicit use and diversion of prescription stimulant medication. J Psychoactive Drugs 38(1):43–56
Mednick S, Nakayama K, Stickgold R (2003) Sleep-dependent learning: a nap is as good as a night. Nat Neurosci 6(7):697–698
Minzenberg MJ, Carter CS (2008) Modafinil: a review of neurochemical actions and effects on cognition. Neuropsychopharmacology 33(7):1477–1502
Oades RD, Sadile AG, Sagvolden T, Viggiano D, Zuddas A, Devoto P, Aase H, Johansen EB, Ruocco LA, Russell VA (2005) The control of responsiveness in ADHD by catecholamines: evidence for dopaminergic, noradrenergic and interactive roles. Dev Sci 8(2):122–131
Polanczyk G, de Lima MS, Horta BL, Biederman J, Rohde LA (2007) The worldwide prevalence of ADHD: a systematic review and metaregression analysis. Am J Psychiatry 164(6):942–948
Poulin C (2007) From attention-deficit/hyperactivity disorder to medical stimulant use to the diversion of prescribed stimulants to non-medical stimulant use: connecting the dots. Addiction 102(5):740–751
Prince J (2008) Catecholamine dysfunction in attention-deficit/hyperactivity disorder: an update. J Clin Psychopharmacol 28(3 Suppl 2):S39–S45
Repantis D, Schlattmann P, Laisney O, Heuser I (2009) Antidepressants for neuroenhancement in healthy individuals: a systematic review. Poiesis Prax 6:139–174
Repantis D, Laisney O, Heuser I (2010a) Acetylcholinesterase inhibitors and memantine for neuroenhancement in healthy individuals: a systematic review. Pharmacol Res 61(6):473–481
Repantis D, Schlattmann P, Laisney O, Heuser I (2010b) Modafinil and methylphenidate for neuroenhancement in healthy individuals: a systematic review. Pharmacol Res 62(3):187–206
Solomon PR, Adams F, Silver A, Zimmer J, DeVeaux R (2002) Ginkgo for memory enhancement: a randomized controlled trial. JAMA 288(7):835–840
Teter CJ, McCabe SE, Boyd CJ, Guthrie SK (2003) Illicit methylphenidate use in an undergraduate student sample: prevalence and risk factors. Pharmacotherapy 23(5):609–617

Teter CJ, McCabe SE, Cranford JA, Boyd CJ, Guthrie SK (2005) Prevalence and motives for illicit use of prescription stimulants in an undergraduate student sample. J Am Coll Health 53(6): 253–262

Teter CJ, McCabe SE, LaGrange K, Cranford JA, Boyd CJ (2006) Illicit use of specific prescription stimulants among college students: prevalence, motives, and routes of administration. Pharmacotherapy 26(10):1501–1510

Wesensten NJ, Killgore WD, Balkin TJ (2005) Performance and alertness effects of caffeine, dextroamphetamine, and modafinil during sleep deprivation. J Sleep Res 14(3):255–266

Wilens TE (2007) The nature of the relationship between attention-deficit/hyperactivity disorder and substance use. J Clin Psychiatry 68(Suppl):114–118

Wilens TE (2008) Effects of methylphenidate on the catecholaminergic system in attention-deficit/hyperactivity disorder. J Clin Psychopharmacol 28(3 Suppl 2):S46–S53

Wilens TE, Adler LA, Adams J, Sgambati S, Rotrosen J, Sawtelle R, Utzinger L, Fusillo S (2008) Misuse and diversion of stimulants prescribed for ADHD: a systematic review of the literature. J Am Acad Child Adolesc Psychiatry 47(1):21–31

Chapter 3
Psychopharmacological Neuroenhancement: Evidence on Safety and Efficacy

Dimitris Repantis

Abstract A number of experts from diverse disciplines have been calling for an evidence-based approach to the evaluation of the risks and benefits of pharmacological cognitive enhancement. If these drugs can be shown to have positive effects in healthy individuals, then this adds urgency to the question of how to regulate their potential use for enhancement purposes. If no evidence of neuroenhancement effects can be found in the existing literature, then this should be made known to healthy individuals who are ready to accept the risk of consuming psychopharmaceuticals.

This paper summarizes the results of a systematic review of the literature that contributes to this quest by collecting and analyzing the available evidence for the most cited neuroenhancement drugs. Based on meta-analyses, it can be shown that expectations regarding the effectiveness of these drugs exceed their actual effects, as has been demonstrated in single- or double-blind randomized controlled trials. According to these data, it seems that the strongest reason not to use prescription drugs for enhancement purposes at the moment is the lack of evidence both for their effectiveness and their long-term safety in healthy people.

Keywords Neuroenhancement • Meta-analysis • Psychostimulants • Antidepressants • Anti-dementia drugs

The primary focus of this paper is not only on cognitive enhancement (Greely et al. 2008), but on *neuroenhancement*, a more general term that refers to the improvement of cognitive performance as well as of emotional well-being and motivation (Schöne-Seifert and Talbot 2011). Of known interventions, psychopharmacology provides readily available options, particularly, the consumption of

D. Repantis (✉)
Department of Psychiatry, Charité-Universitätsmedizin Berlin,
Campus Benjamin Franklin, Berlin, Germany
e-mail: dimitris.repantis@charite.de

E. Hildt and A.G. Franke (eds.), *Cognitive Enhancement: An Interdisciplinary Perspective*, Trends in Augmentation of Human Performance 1,
DOI 10.1007/978-94-007-6253-4_3,

psychopharmaceutical prescription drugs by healthy people, which has given rise to heated debate. Furthermore, these drugs are presumed to be already in widespread use for non-medical reasons as cognitive enhancers, which makes the questions regarding their effects and safety of their use by healthy individuals even more urgent.

A number of experts from diverse disciplines have been calling for an evidence-based approach to the evaluation of the risks and benefits of pharmacological cognitive enhancement (Greely et al. 2008). If these drugs can be shown to have positive effects in healthy individuals, then this adds urgency to the question of how to regulate their potential use for enhancement purposes. If no evidence of neuroenhancement effects can be found in the existing literature, then this should be made known to healthy individuals who are ready to accept the risk of consuming psychopharmaceuticals. Such individuals should be more aware that the presumed benefits of psychopharmaceuticals are not empirically supported.

This paper represents a summary of a systematic review of the literature that contributes to this call for evidence by collecting and analyzing the available evidence for the most cited neuroenhancement drugs. This review provides insight into the empirical research (Repantis et al. 2009, 2010a, b).

The drugs that first sparked this debate were the modern antidepressants. In the last decades, extensive research led to the development of new generations of antidepressants. Surveys indicate that antidepressant use has increased rapidly in most developed countries (McManus et al. 2000; Olie et al. 2002; Raymond et al. 2007). The psychiatrist Peter Kramer noted that Prozac® (fluoxetine) and the other selective serotonin reuptake inhibitors (SSRIs) might have a – possibly positive – effect on the mood and personality of individuals in the absence of mood or personality disorder and coined the term "cosmetic psychopharmacology" (Kramer 1993). However, there is next to no research on the evidence that supports or opposes this assumption. It is this consumption of drugs by normally functioning people for non-therapeutic purposes which can be labeled as neuroenhancement.

Furthermore, psychostimulants are apparently popular among healthy people seeking cognitive enhancement (Talbot 2009). This review concerns the possible neuroenhancement properties of two substances, namely, methylphenidate (MPH) and modafinil. Finally, prescription drugs currently available for the treatment of dementia provide a further possibility for neuroenhancement. Of interest are the drugs used for the treatment of dementia due to Alzheimer's disease, namely, the acetylcholinesterase inhibitors (AChEIs) and memantine. The first category comprises three substances – donepezil, galantamine, rivastigmine – that are recommended for clinical use for the treatment of patients with mild to moderate Alzheimer's disease (Racchi et al. 2004). Memantine is a NMDA receptor antagonist and is registered for the treatment of moderate to severe Alzheimer's disease (Sonkusare et al. 2005).

For this review, all published single- or double-blind randomized controlled clinical trials were considered, including cross-over clinical trials, which compare one of these drugs with a placebo. All interventions in all doses and dosing schedules (single dose or repeated doses) for any duration and by any route of administration were considered. Eligible studies were those involving individuals of any age and either sex who show no evidence of psychiatric disorder, cognitive decline or other diseases. The primary outcomes of interest were emotional, cognitive or motivational parameters. These were measured through objective and subjective psychometric scales and neuropsychological tests. For data processing, these tests were grouped into test clusters according to the predominant neuropsychological domain that they were assessing (Spreen and Strauss 1998; Dumont et al. 2005) and these clusters were aggregated for further analyses into the main factors, the outcomes of the study. These were: mood, wakefulness, motivation, attention, memory and executive functions. Secondary outcomes of interest were adverse effects and acceptability of the medication, measured by numbers of participants dropping out during the trials and post-randomization exclusions due to the drugs' effects.

When numerical data were available, a meta-analysis and a meta-regression for each outcome were performed. For anti-dementia drugs, a statistical analysis was deemed inappropriate and, therefore, not conducted since there were no sufficient data available, and there was large heterogeneity among the studies with respect to outcomes and study interventions.

3.1 Antidepressants

Regarding the research on antidepressants, an unexpected preliminary result was that the majority of the studies (135 studies) examined the effects of a single dose of an antidepressant. This study design did not allow any effect to be found. It is known from clinical populations that the main effects of antidepressants are seen only after several weeks. The main interest here, as well, was obviously in the effect after a drug intake of several days. But even if the drug was given repeatedly, the studies on these healthy individuals had an adequate duration in only very few cases. For instance, only 17 of the 75 repeated dose trials that were found lasted more than 2 weeks. Nevertheless, from the analysis that was performed, a number of conclusions could be drawn. A positive effect on mood was detected that was increasing from the first to the third assessment (14 studies, average duration 16.9 days, $SD = 12$). If the trials had been longer and had had more assessment points, one could speculate that this effect would persist or even become stronger or that a ceiling effect would emerge. For the other outcomes, the small number of studies either did not allow for any effect to emerge or for an analysis to be performed. In summary, no consistent evidence for enhancing effects of antidepressants could be found.

Unfortunately many of the studies did not report their results in numbers, and therefore, although they were formally included in the systematic review, their results were not considered in the analyses. This is a well-known weakness in reporting controlled trials (Egger et al. 2001; Higgins and Green 2006), especially those failing to find any significant result. Consequently, the results that were found through our analyses are to be taken with caution. It is likely that, had the non-significant, not reported results been included in the analyses, the effects might have become less significant.

In a number of studies (79 studies), no comment was made on side effects, while in some studies (20 studies), no adverse effects were observed. In 84 studies there were some sort of side effects. These were mild to moderate and only in few cases led to drop-outs. Adverse reactions of antidepressants, usually observed after the initial administration, wore off with continued intake and primarily consisted of gastrointestinal complaints (e.g. nausea, diarrhea, dry mouth, epigastric pain), sleep disturbances, restlessness, tremor, headache, dizziness, fatigue and drowsiness. Furthermore, sedation was a frequently reported adverse effect.

3.2 Anti-dementia Drugs

Further, studies with anti-dementia drugs were reviewed. However, they were found to be lacking. Only ten trials with donepezil, one with rivastigmine and seven with memantine have been conducted to date. No randomized controlled trials (RCTs) examining the effects of galantamine in healthy individuals were found. Duration of the trials (single dose vs. repeated doses trial) was taken into account in the analysis of available studies. As with antidepressants, anti-dementia drugs show their effect after intake for several weeks. All memantine and the one galantamine trial were, however, single dose trials. Hence, based on these few and insufficient data, no adequate analysis of their potential as neuroenhancement drugs can be performed. Repeated trials have been conducted only with donepezil. These were six small-scale trials, lasting 14–42 days. From these, only two (Beglinger et al. 2005; Fitzgerald et al. 2008) had older persons as participants. The rest of the trials included young healthy participants. This factor complicated the comparison between the results and made it difficult to generalize the results of the latter studies for the main population of interest, namely the growing elderly population.

These few existing studies provide no consistent evidence for a neuroenhancement effect. In one study it was found that donepezil improved the retention of training on complex aviation tasks (Yesavage et al. 2002). In another case, verbal memory for semantically processed words was improved. Donepezil might also improve episodic memory (Gron et al. 2005), but, interestingly, two studies reported transient negative effects on episodic memory (Beglinger et al. 2004, 2005). In a sleep deprivation study, donepezil had no effect when participants were well-rested. Nevertheless, the memory and attention deficits resulting from 24 h of sleep deprivation were attenuated after donepezil intake. This effect however, was seen only in individuals whose performance declined the most after sleep deprivation. Another point that should be made is that in most of the studies a large neuropsychological test battery was applied. However, an effect could be shown in only a few, if not only one, of the tests applied. This could speak either for a selective effect of donepezil or for small effects that in these relatively underpowered studies could be revealed in only one (maybe the most difficult) task. Another possible explanation could be that acetylcholinesterase inhibitors require a pathology of diminished cholinergic transmission to show their effects, and, therefore, it is not possible to optimize performance in healthy individuals who already have an optimal concentration of acetylcholine.

In the majority of the trials, donepezil was well tolerated. In the study of Yesavage et al., no specific adverse effect was reported, but the authors warn that sleep disturbances might become apparent in larger populations (Yesavage et al. 2002). There were some side effects reported in the other five repeated dose trials, but these were benign and only in few cases led to drop-outs. The adverse reactions were mainly gastrointestinal complaints (e.g. nausea), but also headache, dizziness, nightmares and insomnia.

3.3 Psychostimulants

As for the research on the effects of the two stimulants in question here, 46 studies with methylphenidate (MPH) and 45 with modafinil met the inclusion criteria, and their results are considered here. Among the 46 studies of MPH, four were repeated dose trials (Gilbert et al. 1973; Babkoff et al. 1992; Bishop et al. 1997; Gobbi et al. 2003), two of which were with non-sleep deprived volunteers who received MPH once per day for one (Gobbi et al. 2003) and 6 weeks (Gilbert et al. 1973), respectively. The third was a sleep deprivation study where MPH was given every 6 h for a total of 64 h without sleep (Babkoff et al. 1992), while in the fourth, the drug was given twice after a night of either normal sleep or no sleep (Bishop et al. 1997). Regarding the studies on the effect of MPH in sleep deprived individuals, five studies were found. In addition to the two repeated dose studies mentioned above, the other three single-dose studies consist of one 24-h-sleep-deprivation study (Bray et al. 2004) and two studies on partial sleep deprivation, where the drug was given after 4 h of sleep (Roehrs et al. 1999, 2004). The following studies with modafinil were found: of the 45 studies there were 17 on non-sleep deprived individuals in which 100–400 mg of modafinil were administrated. In only two of those, modafinil was given more than once, in one case twice in the evening and in the morning before the testing (Smith et al. 2004) and, in the other case, in a dose of 400 mg/day for 3 consecutive days (Taneja et al. 2007). There were 28 studies with sleep-deprived individuals and in 17 of them, the drug was given more than once with or without napping between the doses. In general, one of two different protocols was used. In the *recovery paradigm*, the volunteers were administered a (typically large) dose of 200–400 mg of modafinil after they had become extremely fatigued by sleep deprivation to determine if, and to what extent the drug could restore cognitive performance to baseline levels. In the maintenance studies (*preventive paradigm*), participants were given smaller (16,7–300 mg), more frequent doses in an attempt to maintain cognitive performance at, or near, baseline levels throughout a period of sleep deprivation (Babkoff and Krueger 1992; Baranski et al. 2004).

The analyses of the results of the trials that gave a single dose of MPH showed a positive effect in one outcome, namely, memory, but no statistically significant effect in the other outcomes: attention, mood and executive functions. In the two repeated drug administration studies, the following results were reported: Gobbi et al. (2003) found that 1 week of MPH significantly increased subjective feelings of energy but did not have any other effects on the other visual analogue scales (VAS) that were used. In another study with a cohort of 27 elderly healthy volunteers, after 6 weeks of daily intake, MPH significantly reduced ratings on a VAS on fatigue but had no effect on five other VAS and no difference on a memory test could be measured (Gilbert et al. 1973). As for the results of the trials giving a single dose of MPH after a night of sleep deprivation, no cognitive enhancing effects were found. In contrast, in one study a negative effect on self-monitoring was observed, with people estimating their performance in a task as better than it actually was (Bray et al. 2004). Repeated intake of MPH during a sustained sleep deprivation period of 64 h (Babkoff et al. 1992) did not effectively reduce sleepiness, while in a study with 36 h of sleep deprivation and in two partial sleep deprivation studies with only 4 h of sleep (Roehrs et al. 1999, 2004), subjective stimulating effects and only a mediocre improvement of attention were found.

As previously mentioned, modafinil was mainly tested in studies with sleep-deprived individuals. However, the fact that trials with non sleep-deprived participants have been performed allowed also for an analysis to be performed. A moderate, positive effect on attention emerged, but no significant effects on mood, memory and motivation could be found. Only two studies with repeated administration of modafinil in well-rested individuals have been performed and, hence, a statistical analysis was not feasible (Smith et al. 2004; Taneja et al. 2007). In the shorter of the two, no effect of an evening and a morning dose of 100 mg on attentional tasks was found (Smith et al. 2004). In the other study, which focused on mood, modafinil was given in a 400 mg dose per day for three consecutive days, and it increased the scores in both the Positive and Negative Affect Scales (Taneja et al. 2007), results which speak for a general mood-elevating effect but with a simultaneous increase of negative affect, namely, anxiety. In sleep-deprived individuals, the effects were more global: a single dose of modafinil after a sleep deprivation of about 36 h (recovery paradigm) had, according to these metaanalyses, a positive effect on wakefulness, executive functions and memory. No evidence of effects on mood and attention was found. Therefore, it can be said that in these paradigms, wakefulness and (to some degree) cognitive functions were maintained. Also, during sustained sleep deprivation over several days and nights (preventive paradigm), repeated intake of modafinil was shown to maintain wakefulness in higher levels than placebo, and this effect lasted for up to 4 days. However, attention and executive functions were not sustained with repeated doses. This fact makes the usability of modafinil as a neuroenhancer in sustained sleep deprivation questionable if one is interested not only in staying awake but also in preserving cognitive performance at high levels at the same time.

Since most of the included papers reported small studies and not large-scale clinical trials, no standardized method of assessing adverse reactions and reporting drop-outs due to adverse effects was used. In a number of studies (26 for MPH and 26 for modafinil), no comment on side effects was made, which leaves us to assume that no severe adverse effects appeared that would deserve a comment in the limited space of a publication. In the majority of the trials, the drugs were well-tolerated. There were some side effects reported, but these were benign and only in few cases lead to drop-outs. For modafinil (Lagarde et al. 1995; Pigeau et al. 1995; Baranski and Pigeau 1997; Caldwell et al. 1999, 2000, 2004; Wesensten et al. 2002; Eddy et al. 2005; Dinges et al. 2006; Gill et al. 2006; Hart et al. 2006; Whitmore et al. 2006), adverse reactions were primarily headache, dizziness, gastrointestinal complaints (e.g. nausea, abdominal pain, dry mouth), increased diuresis, palpitations, nervousness, restlessness, and sleep disturbances and, especially in studies with non-sleep deprived individuals, insomnia. For MPH, a frequently reported side effect (reported in 13 out of 14 trials reporting side effects) (Hink et al. 1978; Wetzel et al. 1981; Strauss et al. 1984; Clark et al. 1986; Peloquin and Klorman 1986; Fitzpatrick et al. 1988; Bray et al. 2004; Brumaghim and Klorman 1998; Rogers et al. 1999; Volkow et al. 1998; Volkow et al. 1999a, b; Mehta et al. 2000) was increased heart rate, while increase in blood pressure was not consistently found. Besides these, typical complaints were headache, anxiety, nervousness, dizziness, drowsiness and insomnia. In total, these drugs seem to be well-tolerated even by this population where the trade-off between side effects and improvement may be less clear. Finally, since the majority of the studies that have been performed were short-term and single-dose studies, no comment can be made on the reinforcing effects, dependence development and drug tolerance of MPH or modafinil in healthy individuals.

3.4 Conclusion

In conclusion, based on meta-analyses, it was shown that expectations regarding the effectiveness of these drugs as neuroenchancers exceed their actual effects, as demonstrated in single- or double-blind randomized controlled trials. According to these data, it seems that the strongest reason not to use prescription drugs for enhancement purposes at the moment is the lack of evidence both for their effectiveness and their long-term safety in healthy people. The – mostly implicit – interpretations of the inconclusive data at hand have often polarized the academic debate. Nonetheless, no evidence of an effect is not paramount to evidence of no effect. Therefore, the question regarding the implications for research is naturally relevant. If there is a societal demand for pharmacological neuroenhancement, then more studies would be necessary to establish efficacy and safety, the latter factor proving to be particularly important for the long run. Moreover, from a pharmacological point of view, it makes quite a difference whether an intervention is supposed to stabilize a disturbed system or to optimize a normally-functioning system. Therefore, it would probably be more promising and safer to develop neuroenhancement compounds by targeting this particular group of healthy people interested in neuroenhancement. One could argue that if safe and effective neuroenhancement drugs were to become available in the future, there would be no sufficient reason to prohibit their use, just like other non-pharmaceutical interventions are also allowed. Still, whatever means may be employed in the long-run, the normative challenges posed by this development pertain to the individual as well as to society. Unequal access to costly enhancement interventions may give rise to issues of distributive justice. The pressure to use pharmacological neuroenhancement might have (implicit) coercion as a result. On an individual level, concerns regarding the authenticity and "personal identity" of enhanced persons seem to play a major role.

References

Babkoff H, Krueger G (1992) Use of stimulants to ameliorate the effects of sleep loss during sustained performance. Mil Psychol 4:191–205

Babkoff H, Kelly T, Matteson L, Gomez S, Lopez A, Hauser S, Naitoh P, Assmus J (1992) Pemoline and methylphenidate: interaction with mood, sleepiness, and cognitive performance during 64 hours of sleep deprivation. Mil Psychol 4:235–265

Baranski J, Pigeau RA (1997) Self-monitoring cognitive performance during sleep deprivation: effects of modafinil, d-amphetamine and placebo. J Sleep Res 6:84–91

Baranski J, Pigeau R, Dinich P, Jacobs I (2004) Effects of modafinil on cognitive and meta-cognitive performance. Hum Psychopharmacol Clin Exp 19:323–332

Beglinger LJ, Gaydos BL, Kareken DA, Tangphao-Daniels O, Siemers ER, Mohs RC (2004) Neuropsychological test performance in healthy volunteers before and after donepezil administration. J Psychopharmacol 18:102–108

Beglinger LJ, Tangphao-Daniels O, Kareken DA, Zhang L, Mohs R, Siemers ER (2005) Neuropsychological test performance in healthy elderly volunteers before and after donepezil administration: a randomized, controlled study. J Clin Psychopharmacol 25:159–165

Bishop C, Roehrs T, Rosenthal L, Roth T (1997) Alerting effects of methylphenidate under basal and sleep-deprived conditions. Exp Clin Psychopharmacol 5:344–352

Bray CL, Cahill KS, Oshier JT, Peden CS, Theriaque DW, Flotte TR, Stacpoole PW (2004) Methylphenidate does not improve cognitive function in healthy sleep-deprived young adults. J Investig Med 52:192–201

Brumaghim JT, Klorman R (1998) Methylphenidate's effects on paired-associate learning and event-related potentials of young adults. Psychophysiology 35:73–85

Caldwell JA, Smythe NK, Caldwell JL, Hall KK, Norman DN et al (1999) The effects of modafinil on aviator performance during 40 hours of continuous wakefulness: a UH-60 helicopter simulator study. USAARL Report No. 99-17. United States Air Force Research Laboratory, Fort Rucker

Caldwell JA, Caldwell JL, Smythe NK, Hall KK (2000) A double-blind, placebo-controlled investigation of the efficacy of modafinil for sustaining the alertness and performance of aviators: a helicopter simulator study. Psychopharmacology (Berl) 150:272–282

Caldwell JA, Caldwell JL, Smith JK, Brown DL (2004) Modafinil's effects on simulator performance and mood in pilots during 37 h without sleep. Aviat Space Environ Med 75:777–784

Clark CR, Geffen GM, Geffen LB (1986) Role of monoamine pathways in attention and effort: effects of clonidine and methylphenidate in normal adult humans. Psychopharmacology (Berl) 90:35–39

Dinges DF, Arora S, Darwish M, Niebler GE (2006) Pharmacodynamic effects on alertness of single doses of armodafinil in healthy subjects during a nocturnal period of acute sleep loss. Curr Med Res Opin 22:159–167

Dumont GJ, de-Visser SJ, Cohen AF, van-Gerven JM (2005) Biomarkers for the effects of selective serotonin reuptake inhibitors (SSRIs) in healthy subjects. Br J Clin Pharmacol 59:495–510

Eddy D, Storm W, French J, Barton E, Cardenas R (2005) An assessment of modafinil for vestibular and aviation-related effects. Report AFRL-HE-BR-TR-2005-0129, United States Air Force Research Laboratory, Brooks City-Base

Egger M, Smith GD, Altman DG et al (2001) Systematic reviews in health care: meta-analysis in context. BMJ Books, London

FitzGerald DB, Crucian GP, Mielke JB, Shenal BV, Burks D, Womack KB, Ghacibeh G, Drago V, Foster PS, Valenstein E et al (2008) Effects of donepezil on verbal memory after semantic processing in healthy older adults. Cogn Behav Neurol 21:57–64

Fitzpatrick P, Klorman R, Brumaghim JT, Keefover RW (1988) Effects of methylphenidate on stimulus evaluation and response processes: evidence from performance and event-related potentials. Psychophysiology 25:292–304

Gilbert JG, Donnelly KJ, Zimmer LE, Kubis JF (1973) Effect of magnesium pemoline and methylphenidate on memory improvement and mood in normal aging subjects. Int J Aging Hum Dev 4:35–51

Gill M, Haerich P, Westcott K, Godenick KL, Tucker JA (2006) Cognitive performance following modafinil versus placebo in sleep-deprived emergency physicians: a double-blind randomized crossover study. Acad Emerg Med 13:158–165

Gobbi G, Slater S, Boucher N, Debonnel G, Blier P (2003) Neurochemical and psychotropic effects of bupropion in healthy male subjects. J Clin Psychopharmacol 23:233–239

Greely H, Sahakian B, Harris J, Kessler RC, Gazzaniga M, Campbell P, Farah MJ (2008) Towards responsible use of cognitive-enhancing drugs by the healthy. Nature 456:702–705

Gron G, Kirstein M, Thielscher A, Riepe MW, Spitzer M (2005) Cholinergic enhancement of episodic memory in healthy young adults. Psychopharmacology 182:170–179

Hart CL, Haney M, Vosburg SK, Comer SD, Gunderson E, Foltin RW (2006) Modafinil attenuates disruptions in cognitive performance during simulated night-shift work. Neuropsychopharmacology 31:1526–1536

Higgins J, Green S (2006) Cochrane handbook for systematic reviews of interventions 4.2.6. Wiley, Chichester

Hink RF, Fenton WH, Pfefferbaum A, Tinklenberg JR, Kopell BS (1978) The distribution of attention across auditory input channels: an assessment using the human evoked potential. Psychophysiology 15:466–473

Kramer P (1993) Listening to Prozac. Penguin, New York

Lagarde D, Batejat D, Van-Beers P, Sarafian D, Pradella S (1995) Interest of modafinil, a new psychostimulant, during a sixty-hour sleep deprivation experiment. Fund Clin Pharmacol 9:271–279

McManus P, Mant A, Mitchell PB, Montgomery WS, Marley J, Auland ME (2000) Recent trends in the use of antidepressant drugs in Australia, 1990–1998. Med J Aust 173:458–461

Mehta MA, Owen AM, Sahakian BJ, Mavaddat N, Pickard JD, Robbins TW (2000) Methylphenidate enhances working memory by modulating discrete frontal and parietal lobe regions in the human brain. J Neurosci 20:RC65

Olie JP, Elomari F, Spadone C, Lepine JP (2002) Antidepressants consumption in the global population in France. Encéphale 28:411–417

Peloquin LJ, Klorman R (1986) Effects of methylphenidate on normal children's mood, event-related potentials, and performance in memory scanning and vigilance. J Abnorm Psychol 95:88–98

Pigeau R, Naitoh P, Buguet A, McCann C, Baranski J, Taylor M, Thompson M, Mack I (1995) Modafinil, d-amphetamine and placebo during 64 hours of sustained mental work. I. Effects on mood, fatigue, cognitive performance and body temperature. J Sleep Res 4:212–228

Racchi M, Mazzucchelli M, Porrello E, Lanni C, Govoni S (2004) Acetylcholinesterase inhibitors: novel activities of old molecules. Pharmacol Res 50:441–451

Raymond CB, Morgan SG, Caetano PA (2007) Antidepressant utilization in British Columbia from 1996 to 2004: increasing prevalence but not incidence. Psychiatr Serv 58:79–84

Repantis D, Schlattmann P, Laisney O, Heuser I (2009) Antidepressants for neuoenhancement in healthy individuals: a systemic review. Poiesis Prax 6:139–174
Repantis D, Laisney O, Heuser I (2010a) Acetylcholinesterase inhibitors and memantine for neuroenhancement in healthy individuals: a systemtic review. Pharmacol Res 61:473–481
Repantis D, Schlattmann P, Laisney O, Heuser I (2010b) Modafinil and methylphenidate for neuroenhancement in healthy individuals: a systematic review. Pharmacol Res 62:187–206
Roehrs T, Papineau K, Rosenthal L, Roth T (1999) Sleepiness and the reinforcing and subjective effects of methylphenidate. Exp Clin Psychopharmacol 7:145–150
Roehrs T, Meixner R, Turner L, Johanson CE, Roth T (2004) Reinforcing and subjective effects of methylphenidate: dose and time in bed. Exp Clin Psychopharmacol 12:180–189
Rogers RD, Blackshaw AJ, Middleton HC, Matthews K, Hawtin K, Crowley C, Hopwood A, Wallace C, Deakin JF, Sahakian BJ et al (1999) Tryptophan depletion impairs stimulus-reward learning while methylphenidate disrupts attentional control in healthy young adults: implications for the monoaminergic basis of impulsive behaviour. Psychopharmacology 146:482–491
Schöne-Seifert B, Talbot D (2011) (Neuro-)enhancement. In: Helmchen H, Sartorius N (eds) Textbook of psychiatry. Springer, New York
Smith D, Pernet A, Rosenthal JM, Bingham EM, Reid H, Macdonald IA, Amiel SA (2004) The effect of modafinil on counter-regulatory and cognitive responses to hypoglycaemia. Diabetologia 47:1704–1711
Sonkusare SK, Kaul CL, Ramarao P (2005) Dementia of Alzheimer's disease and other neurodegenerative disorders – memantine, a new hope. Pharmacol Res 51:1–17
Spreen O, Strauss E (1998) A compendium of neuropsychological tests: administration, norms, and commentary. Oxford University Press, Inc., New York
Strauss J, Lewis JL, Klorman R, Peloquin LJ, Perlmutter RA, Salzman LF (1984) Effects of methylphenidate on young adults' performance and event-related potentials in a vigilance and a paired-associates learning test. Psychophysiology 21:609–621
Talbot M (2009) A reporter at large brain gain. The underground world of neuroenhancing drugs. The New Yorker, April 27, 2009
Taneja I, Haman K, Shelton RC, Robertson D (2007) A randomized, double-blind, crossover trial of modafinil on mood. J Clin Psychopharmacol 27:76–79
Volkow ND, Wang GJ, Fowler JS, Gatley SJ, Logan J, Ding YS, Hitzemann R, Pappas N (1998) Dopamine transporter occupancies in the human brain induced by therapeutic doses of oral methylphenidate. Am J Psychiatry 155:1325–1331
Volkow ND, Wang GJ, Fowler JS, Logan J, Gatley SJ, Wong C, Hitzemann R, Pappas NR (1999a) Reinforcing effects of psychostimulants in humans are associated with increases in brain dopamine and occupancy of D(2) receptors. J Pharmacol Exp Ther 291:409–415
Volkow ND, Wang GJ, Fowler JS, Gatley SJ, Logan J, Ding Y, Dewey SL, Hitzemann R, Gifford AN, Pappas NR (1999b) Blockade of striatal dopamine transporters by intravenous methylphenidate is not sufficient to induce self-reports of 'high'. J Pharmacol Exp Therbreak 288:14–20
Wesensten NJ, Belenky G, Kautz MA, Thorne DR, Reichardt RM, Balkin TJ (2002) Maintaining alertness and performance during sleep deprivation: modafinil versus caffeine. Psychopharmacology 159:238–247
Wetzel CD, Squire LR, Janowsky DS (1981) Methylphenidate impairs learning and memory in normal adults. Behav Neural Biol 31:413–424
Whitmore J, Hickey P, Doan B, Harrison R, Kisner J, Beltran T, McQuade J, Fischer J, Marks F (2006) A double-blind placebo controlled investigation of the efficacy of modafinil for maintaining alertness and performance in sustained military ground operations. Report AFRL-HE-BR-TR-2006-0005. United States Air Force Research Laboratory, Brooks-City-Base
Yesavage JA, Mumenthaler MS, Taylor JL, Friedman L, O'Hara R, Sheikh J, Tinklenberg J, Whitehouse PJ (2002) Donepezil and flight simulator performance: effects on retention of complex skills. Neurology 59:123–125

Chapter 4
A Bubble of Enthusiasm: How Prevalent Is the Use of Prescription Stimulants for Cognitive Enhancement?

Bradley Partridge

Abstract This chapter focuses on the evidence for the prevalence of non-medical stimulant use by students for cognitive enhancement. Some of what is cited as apparent evidence for widespread cognitive enhancement (also known as neuroenhancement) has a number of weaknesses that we should be aware of. Here, I expand upon several examples whereby the prevalence of cognitive enhancement has been uncritically presented. Caution needs to be exercised to avoid whipping up hype about neuroenhancement by overextending what the currently available data on prevalence really says.

Keywords Cognitive enhancement • Neuroenhancement • Prevalence • Psychostimulants • Prescription drugs • Students

A number of proponents argue that cognitive enhancement has much to offer individuals, and perhaps society, and so ought to be facilitated. If we're to understand the phenomenon, we need to look critically at who is engaging in this kind of neuroenhancement, how many people, and for what for reasons? University students are often considered to be among the highest users of drugs for cognitive enhancement. US studies have reported that stimulants typically indicated for the treatment of Attention Deficit Hyperactivity Disorder (ADHD) are used without prescription by some healthy university students as a "study aid" or for other cognitive enhancement purposes (often interchangeably termed "neuroenhancement") (Teter et al. 2006). Good social science research can help us understand this practice, thereby informing bioethical and policy discussions about cognitive enhancement. In this chapter, I focus on the evidence for the

B. Partridge (✉)
The University of Queensland, UQ Centre for Clinical Research (UQCCR), Herston, Queensland, 4029, Australia
e-mail: b.partridge@uq.edu.au

E. Hildt and A.G. Franke (eds.), *Cognitive Enhancement: An Interdisciplinary Perspective*, Trends in Augmentation of Human Performance 1,
DOI 10.1007/978-94-007-6253-4_4,

prevalence of non-medical stimulant use by students for cognitive enhancement and how this has been portrayed in some prominent articles from the bioethics literature. In particular, I examine whether some articles have exaggerated the prevalence of cognitive enhancement. Several authors have recently pointed out that enthusiasm about cognitive enhancement among students has often relied on weak evidence for the assertion that cognitive enhancement is widespread (Forlini and Racine 2009; Outram 2010; Hall and Lucke 2010; Lucke et al. 2011). In this paper, I expand upon some of these points. We need better evidence about how widespread neuroenhancement is and more in depth investigations about attitudes and motivations related to the practice.

4.1 How Prevalent Is the Use of Prescription Stimulants for Cognitive Enhancement?

In the bioethics and neuroscience literature, some have claimed that the use of prescription stimulant medications for neuroenhancement is widespread and increasing in prevalence. "Neuroenhancement" *per se* can encompass a wide range of functions and methods of enhancement, but this paper focuses particularly on the use of prescription drugs such as methylphenidate and modafinil to improve functions such as alertness, memory or concentration. These are the drugs that are the most routinely discussed in the bioethics literature, and empirical evidence about their use by US college students as a "study aid" is often cited in support of these claims about prevalence.

When it comes to questions about how widespread this form of neuroenhancement is, there is some uncertainty. Caution needs to be exercised to avoid whipping up hype about neuroenhancement by overextending what the currently available data on prevalence really says (Hall and Lucke 2010; Lucke et al. 2011; Partridge et al. 2011). This should not be construed as a call to dismiss neuroenhancement as a real phenomenon – there are important public health, ethical, and social issues that require our attention. However, some of what is cited as apparent evidence for *widespread* neuroenhancement has a number of weaknesses that we should be aware of.

One concern is that exaggerated claims can inflate an uncritical bubble of enthusiasm about neuroenhancement. Uncritical claims that many students are using Ritalin as a study aid, for example, can be perpetuated throughout the literature and media (Partridge et al. 2011), perhaps resulting in policy recommendations that are based on this uncritical belief. For example, there have been recommendations that we should relax laws that prohibit psychostimulant use without a prescription if it is for cognitive enhancement purposes and that pharmaceutical companies ought to be allowed to market "cognitive enhancement drugs" to healthy people (Greely et al. 2008). These recommendations are largely based on the assumption that many people (especially students) are seeking and engaging in cognitive enhancement already.

More troubling to some is the possibility that uncritically favorable portrayals of using stimulants for neuroenhancement might entice people to engage in the practice before we know exactly what the potential long term harms might be. This enticement might also occur as a result of inflated claims about the prevalence of neuroenhancement, not just inflated claims about the potential benefits. Let's look at some examples from the literature that may give the impression that neuroenhancement is widespread or becoming increasingly common.

4.2 Responding to Requests from Healthy Patients for "Cognitive Enhancers"

Larriviere and colleagues have offered guidance to neurologists on how to respond to healthy patient requests for "neuroenhancing" drugs (Larriviere et al. 2009). This article was written on behalf of the Ethics Law and Humanities Committee of the American Academy of Neurology (AAN) and defends the ethical and legal acceptability of prescribing stimulants to healthy people for the purposes of enhancement. It is essentially premised on the following assertion that neuroenhancement is widespread:

> In the last decade, persons with no diagnosed medical or mental health condition have been increasingly seeking and utilizing, for the purpose of enhancing their memory or cognitive skills, prescription drugs originally developed to improve executive function or memory in persons with disorders such as attention deficit hyperactivity disorder or Alzheimer disease (McCabe et al. 2004, 2005a, b; Farah 2005). This practice, now known as neuroenhancement, is gathering momentum (Farah et al. 2004; Maher 2008).

The impression given to the reader is that a lot of people desire psychostimulants for cognitive enhancement, that the practice has become more common over the last 10 years and that there will only be more to come – so much so, that the AAN sees the need to advise their physicians about how to navigate the growing tide of people wanting neuroenhancers. But let's look closer at this assumption and assess whether it is well founded on the basis of the evidence cited by the AAN committee.

Of the five articles Larriviere et al. (2009) cite to support their claims about the demand and prevalence of people seeking neuroenhancement, two are commentary articles on the ethics of neuroenhancement by Farah and colleagues (Farah et al. 2004; Farah 2005). These commentaries don't report direct evidence of prevalence – for example, they didn't conduct an empirical study about the prevalence of neuroenhancement or systematically review such studies. This is not a failing of these articles of course – they weren't intended to perform such a function – but the problem is that Larriviere et al. (2009) cite them as evidence to support their own claim that stimulants are being increasingly sought for neuroenhancement. Also cited by Larriviere et al. (2009) is an often cited online survey about neuroenhancement among 1,400 readers of *Nature* (Maher 2008). However, this poll was an exercise intended to stimulate debate about neuroenhancement rather than an attempt to conduct a rigorous scientific study – it surveyed a self-selected sample of academics and there are few details of the methodology available, casting doubts about how representative this data really is. This online poll has also often been referred to in the media as evidence for widespread and increasing neuroenhancement (Partridge et al. 2011).

Also cited as evidence of increasing neuroenhancement in the AAN paper, is a survey of the non-medical use of prescription *opioids* by college students (McCabe et al. 2005a) – not ADHD drugs or Alzheimer's drugs as indicated by Larriviere et al. (2009). It is unclear why this paper was referred to given that there has been little suggestion elsewhere of the possibility that prescription opioids are commonly used by healthy people to improve memory or other cognitive skills, and in any case, the survey referred to did not ask about prescription opioid use for neuroenhancement. In fact, of the sources cited by Larriviere et al. (2009) to support the claim that stimulants are being increasingly sought for neuroenhancement, only one is an empirical study estimating the prevalence of illicit Ritalin use (McCabe et al. 2004). However, this survey was conducted with high school students only, and there is little information about why Ritalin was taken illicitly. That is, the results of that study don't allow us to say exactly how much of this non-medical use was for neuroenhancement purposes.

In summary, on the basis of the evidence cited by Larriviere et al. (2009), we don't see a robust case that healthy people are increasingly seeking and utilizing prescription drugs specifically for the purposes of neuroenhancement – let alone that the practice has increased over the last decade. That doesn't necessarily mean that there isn't any evidence that some people are engaging in cognitive enhancement, but this example shows that the assumption of widespread cognitive enhancement is sometimes made uncritically. The clinical implications of the AAN's guidance on this issue have attracted some criticism (Boot et al. 2012), and we might have reason to also be concerned that the impetus for providing such guidance in the first place has been ill-informed.

4.3 Surveying "Non-medical Use of Stimulants"

Surveys of the non-medical use of prescription stimulants are useful to the extent that they may give us an indication of how many people have used the drugs most often thought to be enhancers. But estimates of prevalence can vary because some studies have small sample sizes, or, despite having larger samples, they survey from a particular pool (e.g. students from only one university). Some ask only about lifetime prevalence, rather than recent use or frequency of use. Some only ask about specific stimulants. Most surveys have been with US samples only.

One of the largest nationwide surveys of US college students about their non-medical use of prescription stimulants involved over 10,000 students at 119 colleges (McCabe et al. 2005b). This data was collected in 2001 and found that the lifetime prevalence of non-medical prescription stimulant use was 6.9 %, past year prevalence was 4.1 %, and past month prevalence was 2.1 %. As seen in the Larriviere et al. (2009) example, findings about the non-medical use of stimulants are often cited as *de facto* evidence of cognitive enhancement. The problem is that such surveys do not typically ask students specifically whether they use these drugs for enhancement – perhaps because these surveys are conducted by public health researchers primarily interested in prescription drug abuse and issues of addiction, rather than cognitive enhancement. Many of these surveys have asked about the use of a stimulant "without a prescription" or for "non-medical reasons." As well as for cognitive enhancement, non-medical use of prescription stimulants might potentially be for a number of other reasons – for fun, to get high, possibly to alleviate the effects of a hangover, to lose weight, or as a form of self-treatment of symptoms in people who believe they have ADHD. And yet, when the prevalence of non-medical prescription stimulant use is reported simply as the prevalence of "neuroenhancement," then the prevalence of neuroenhancement may be overestimated when we take into account all the potential reasons for non-medical stimulant use.

Greely et al. (2008), begin their *Nature* article by saying that:

> Today, on university campuses around the world, students are striking deals to buy and sell prescription drugs such as Adderall and Ritalin – not to get high, but to get higher grades, to provide an edge over their fellow students or to increase in some measurable way their capacity for learning.

What they are describing is the non-medical use of stimulants specifically for cognitive enhancement. They go on to say:

> One survey estimated that almost 7 % of students in US universities have used prescription stimulants **in this way**, and that on some campuses, up to 25 % of students had used them in the past year. These students are early adopters of a trend that is likely to grow.

The survey they refer to is the McCabe et al. (2005b) survey of over 10,000 students, just mentioned. However, although that survey didn't ask *why* students had used prescription stimulants illicitly, Greely et al. (2008) seem to assume it was for neuroenhancement. This assumption misconstrues the findings. It is also interesting to note that highlighting the *lifetime* prevalence (which is higher than the past year or past month prevalence) may give the reader the impression that the behavior is more widespread. Reporting an outlier may have the same effect. The overwhelming majority of colleges that were surveyed by McCabe et al. (2005b) had a past year prevalence rate of between 0 and 4 %. Only one college out of 119 had a prevalence of 25 %, but it is this one outlier that is mentioned in the Greely et al. (2008) article.

To avoid some of these problems there is a clear need for better estimates of how many people are using prescription stimulants specifically for cognitive enhancement, and the discussion about cognitive enhancement would benefit from more international evidence. In countries such as Australia and New Zealand, there have been no studies exploring the prevalence of prescription stimulant use for cognitive enhancement. Although the prescription rates of stimulants have increased significantly in Australia over the last decade (Hollingworth et al. 2011), whether or not these drugs are being increasingly used by healthy people to improve cognition is unknown. In Germany, a recent study surveyed 1,547 pupils and students about their use of prescription stimulants specifically for cognitive enhancement (Franke et al. 2011). The lifetime prevalence for cognitive enhancement was 1.29 %, past year prevalence 0.26 %, and past month prevalence only 0.06 % (one student). These results indicate a relatively low prevalence, particularly when put in the context of US data (although these results should be generalized to other contexts with some caution). While it would not be surprising if the prevalence of cognitive enhancement varied across countries, it is interesting to contrast the low prevalence of cognitive enhancement in the German survey with some claims that the use of prescription stimulants as a study aid (particularly Ritalin) is much more widespread in the USA.

4.4 Perpetuating the Assumption that Neuroenhancement Is Widespread

Let us look at some examples of how the neuroenhancement "bubble of enthusiasm" can be inflated when a survey's findings are misconstrued by others as evidence of widespread cognitive enhancement. A good example of this is a survey titled "Student perceptions of methylphenidate abuse at a public liberal arts college" (Babcock and Byrne 2000). Babcock and Byrne surveyed 283 students at one US college (MCLA) using a simple 10 item, "yes/no" questionnaire – this included no items asking specifically about the use of Ritalin (or any other substances) as a study aid or for cognitive enhancement. It did, however, include the item "*Have you ever taken Ritalin for fun (non-medical purposes)?*" – to which 16.6 % of respondents indicated "yes."

However, since their publication over a decade ago, these findings about the lifetime prevalence of *recreational* Ritalin use have been inaccurately reported on a number of occasions by other authors as evidence for the widespread use of stimulants for *neuroenhancement.* For example, Farah et al. (2004) in the journal *Nature Reviews Neuroscience*, cite the results of the Babcock and Byrne survey when they say that:

> The use of prescription stimulants (such as methylphenidate and dextroamphetamine) as study aids by high school and college students who do not have ADHD has recently drawn attention, and might include as many as 16 % of the students on some campuses (Babcock and Byrne 2000).

Although 16.6 % of participants had used Ritalin for fun, Babcock and Byrne's survey included no items assessing the prevalence of using methylphenidate or dextroamphetamine as a study aid (and did not survey any high school students). In fact, Babcock and Byrne make no actual mention of neuroenhancement apart from one sentence in the discussion that says, "Personal communications with students at MCLA suggest that methylphenidate is sometimes used as a study aid for 'pulling all-nighters." This is worded as a tentative aside by Babcock and Byrne, not an actual finding of the survey – we aren't told how many students were personally contacted; how they were contacted; exactly how frequently Ritalin was used this way; whether this was a rumor they had heard or whether they had actually engaged in the practice; and it is not even clear whether the students who were personally communicated with were even participants in the original study. With this in mind, it is surprising to read that Farah et al. (2004) and other articles in the bioethical literature cite this survey as good evidence that many students at some universities are engaging in neuroenhancement, by using stimulants (particularly Ritalin) as a study aid. In discussing the potential use of stimulants for cognitive enhancement by surgeons, Warren et al. (2009) perpetuate the claim by saying that, "an estimated 16 % of students at some United States universities take prescription medication as study aids (Babcock and Byrne 2000)."

Again, the finding that 16.6 % of students had used Ritalin for fun has been incorrectly reported as evidence that 16 % had used prescription medications as a study aid – two very different purposes! Other examples uncritically extend the Babcock and Byrne data even further. In his 2006 paper in the *Journal of Medical Ethics*, Chatterjee says: "Based on the belief that these drugs improve test performance, the use of stimulant medications among college students in the US is widespread (Babcock and Byrne 2000)."

In the above example, the Babcock and Byrne data is embellished as evidence for widespread use of stimulants across US colleges – no longer even using the caveat "at some colleges". Riis et al. (2008) make a similar claim:

> ... many college students are aware of Ritalin's effectiveness. One study estimated that as many as 16 % of college students have used it as a study aid, often illegally using pills prescribed for someone else (Babcock and Byrne 2000).

In their article discussing ethical issues related to psychopharmacology and adolescents, Koelch et al. (2008) claim, "A growing number of publications show that especially stimulants are being used for improving task performance during examinations."

Babcock and Byrne (2000) is among the five articles cited to support this claim, although it is unclear why. Mehlman (2004) also cites Babcock and Byrne when discussing the ethics of cognition enhancing drugs, saying, "Students have long used amphetamines as a study aid with methylphenidate (Ritalin®) being the current cognitive enhancement drug of choice on U.S. college campuses."

Aside from giving readers the impression that cognitive enhancement is more widespread than the evidence indicates, there are other important consequences. For example, we have seen that Farah et al. (2004) cite the Babcock and Byrne survey as evidence that up to 16 % of students at some colleges use stimulants for cognitive enhancement – as mentioned earlier in this paper, the AAN guidelines by Larriviere et al. (2009) on prescribing neuroenhancers to healthy people is essentially based on the assumption that neuroenhancement is widespread and becoming increasingly common. Interestingly, Farah et al. (2004) is among the evidence cited by Larriviere et al. (2009) for this claim.

Babcock and Byrne were interested in investigating the abuse of Ritalin by students, and prescription stimulants do carry a risk of addiction. Studies have found that students who said they had used prescription stimulants non-medically were more likely to be white, male, and live in a fraternity, and they were also much more likely to have used other illicit drugs (McCabe et al. 2005b). This trend might indicate a number of different things that we need to explore in order to conduct good social science research. Perhaps college students who use illicit drugs are also inclined to seek out pharmacological solutions to their study pressures. Or we might find that within a university there are "hot spots" of neuroenhancement – groups (such as fraternities) where stimulant use for neuroenhancement is common, but outside these groups the behavior may be rare. A person's social proximity to these "hot spots" may then influence whether/how they are exposed to pharmaceutical cognitive enhancement. When it comes to estimating how prevalent neuroenhancement is, if a person is close to these "hot spots," then neuroenhancement may seem more prevalent than it really is.

Bioethicists are probably not the only ones who may have overestimated the prevalence of neuroenhancement. A 2008 survey, querying more than 3,000 students, asked students whether they had actually used prescription stimulants for non-medical reasons in the past year – 6 % had (the details of the survey mean that we don't know how much of this was for cognitive enhancement) (McCabe 2008). But participants were also asked to estimate how prevalent they thought stimulant use was on campus, and the average estimate was 20 %. Around 70 % of participants overestimated the prevalence of stimulant use by their peers. In particular, those who had used stimulants thought the prevalence was much higher than those who hadn't used stimulants. If this is true of non-medical stimulant use in general, then it is worth exploring whether it is true for neuroenhancement. One of the potential problems is that a social norm about this kind of substance use may be created. If people think that many others are taking prescription stimulants for neuroenhancement, then perhaps it seems more acceptable, and maybe more people will be enticed to try it.

We need to understand the potential short and long term health risks of stimulant use by healthy people, before we accept (or promote) the practice. Well conducted social science research using both qualitative and quantitative methods will help to inform us about the prevalence, motivations and attitudes of students using prescription stimulants for cognitive enhancement purposes. This empirical evidence will then help to better inform ethical and policy discussions about neuroenhancement.

References

Babcock Q, Byrne T (2000) Student perceptions of methylphenidate abuse at a public liberal arts college. J Am Coll Health 49:143–145

Boot BP, Partridge B, Hall W (2012) Better evidence for safety and efficacy is needed before neurologists prescribe drugs for neuroenhancement to healthy people. Neurocase 18(3):181–184

Chatterjee A (2006) The promise and predicament of cosmetic neurology. J Med Ethics 32:110–113

Farah MJ (2005) Neuroethics: the practical and the philosophical. Trends Cogn Sci 9:34–40

Farah MJ, Illes J, Cook-Deegan R et al (2004) Neurocognitive enhancement: what can we do and what should we do? Nat Rev Neurosci 5:421–425

Forlini C, Racine E (2009) Disagreements with implications: diverging discourses on the ethics of non-medical use of methylphenidate for performance enhancement. BMC Med Ethics 10:9

Franke AG, Bonertz C, Christmann M et al (2011) Non-medical use of prescription stimulants and illicit use of stimulants for cognitive enhancement in pupils and students in Germany. Pharmacopsychiatry 44:60–66

Greely H, Sahakian B, Harris J et al (2008) Towards responsible use of cognitive-enhancing drugs by the healthy. Nature 456:702–705

Hall WD, Lucke JC (2010) The enhancement use of neuropharmaceuticals: more scepticism and caution needed. Addiction 105:2041–2043

Hollingworth SA, Nissen LM, Stathis SS et al (2011) Australian national trends in stimulant dispensing: 2002–2009. Aust N Z J Psychiatry 45:332–336

Koelch M, Schnoor K, Fegert JM (2008) Ethical issues in psychopharmacology of children and adolescents. Curr Opin Psychiatry 21:598–605

Larriviere D, Williams MA, Rizzo M et al (2009) Responding to requests from adult patients for neuroenhancements Guidance of the Ethics, Law and Humanities Committee. Neurology 73:1406–1412

Lucke JC, Bell S, Partridge B et al (2011) Deflating the neuroenhancement bubble. AJOB Neurosci 2:38–43

Maher B (2008) Poll results: look who's doping. Nature 452:674–675

McCabe SE (2008) Misperceptions of non-medical prescription drug use: a web survey of college students. Addict Behav 33:713–724

McCabe SE, Teter CJ, Boyd CJ et al (2004) Prevalence and correlates of illicit methylphenidate use among 8th, 10th, and 12th grade students in the United States, 2001. J Adolesc Health 35:501–504

McCabe SE, Boyd CJ, Teter CJ (2005a) Illicit use of opioid analgesics by high school seniors. J Subst Abuse Treat 28:225–230

McCabe SE, Knight JR, Teter CJ et al (2005b) Non-medical use of prescription stimulants among US college students: prevalence and correlates from a national survey. Addiction 100:96–106

Mehlman MJ (2004) Cognition-enhancing drugs. Milbank Q 82:483–506

Outram SM (2010) The use of methylphenidate among students: the future of enhancement? J Med Ethics 36:198–202

Partridge BJ, Bell SK, Lucke JC et al (2011) Smart drugs "as common as coffee": media hype about neuroenhancement. PLoS One 6:e28416

Riis J, Simmons J, Goodwin G (2008) Preferences for enhancement pharmaceuticals: the reluctance to enhance fundamental traits. J Consum Res 35(3):495–508

Teter CJ, McCabe SE, LaGrange K et al (2006) Illicit use of specific prescription stimulants among college students: prevalence, motives, and routes of administration. Pharmacotherapy 26:1501–1510

Warren OJ, Leff DR, Athanasiou T et al (2009) The neurocognitive enhancement of surgeons: an ethical perspective. J Surg Res 152:167–172

Chapter 5
Modeling the Effects of Modafinil on Selective Attention Electroencephalographic Neural Correlates

Carlos Trenado

Abstract Nowadays, the development of pharmaceutical neuroenhancers attracts considerable attention due to their applicability and controversial implications within human society. Modafinil is one such neurorenhancer that has gained notoriety due to its widespread use in military and college environments. Several studies in the field of cognitive neuroscience, psychophysics and pharmacology have reported on the human brain differential processing under the influence of modafinil. Unfortunately, the underlying mechanisms by which modafinil modulates excitatory and inhibitory neural populations on cortical and subcortical neural networks as well as its positive or negative effects are not well understood. To cope with this, we propose a biologically-inspired, large-scale computational model for studying modafinil effects on electroencephalographic neural correlates of attention. Numerical simulations are performed and reconciled with recent experimental findings on the influence of modafinil on the locus coeruleus (LC) and the corticothalamic pathways. It is concluded that neural computational models represent a promising approach to gain insight into the neurodynamics of cognitive neuroenhancers.

Keywords Neuroenhancement • Modafinil • Mathematical modeling • Neurodynamics • Electroencephalography

---

C. Trenado (✉)
Systems Neurosciences and Neurotechnology Unit, Saarland University Hospital, Saarbrücken, Germany

Department of Kinesiology, School of Public Health, University of Maryland College Park, College Park, MD, USA
e-mail: trenado@cdb-unit.de

E. Hildt and A.G. Franke (eds.), *Cognitive Enhancement: An Interdisciplinary Perspective*, Trends in Augmentation of Human Performance 1,
DOI 10.1007/978-94-007-6253-4_5,

5.1 Outline

The present chapter introduces for the first time computational modeling results on the effect of modafinil on electroencephalographic attention neural correlates based on the hypothesis that modafinil modulates neural activity in the locus coeruleus. Notably, such a modulation has been implicated with an enhancement in connectivity between the prefrontal cortex and subcortical structures as in the case of the thalamus and the limbic system. The organization of this chapter is as follows: Sect. 5.2 introduces the basics about modafinil and the emerging field of neural enhancement modeling; Sect. 5.3 presents our biologically-inspired model of modafinil effects on electroencephalographic correlates of selective attention; Sect. 5.4 introduces our numerical results; Sect. 5.5 presents our conclusions.

5.2 Introduction

5.2.1 Modafinil

Modafinil is a psychostimulant that has been utilized to enhance wakefulness, attention capacity, and vigilance (Franke and Lieb 2010). It has attracted considerable interest due to its pharmacological profile that differs from traditional neuroenhancers such as amphetamines and methylphenidate. Several research studies support that modafinil is a safe, well-tolerated drug, with long-lasting effects (Sugarman et al. 2011). In fact, modafinil has been tested in the treatment of Alzheimer's disease (Bransfield 2004); depression (Frye et al. 2007); attention-deficit hyperactivity disorder (ADHD) (Keen and Hadijikoumi 2011); multiple sclerosis-induced fatigue (Téllez and Montalbán 2004; Nicholas and Chataway 2007); post-cognitive impairment in schizophrenia (Ballon and Feifel 2006; Henderson et al. 2005); spasticity associated with cerebral palsy (Ballon and Feifel 2006); age-related memory decline as a neuroprotective agent against the "dopamine-deficiency disorder" Parkinson's disease (Knie et al. 2011; van Vliet et al. 2006) and the treatment of cancer-related fatigue (Morrow et al. 2005). The use of modafinil in the treatment of shift work sleep disorder and obstructive sleep apnea has also been emphasized (Ballon and Feifel 2006). Focusing on arousal enhancement, modafinil has gained widespread attention in popular press articles because of its increasing popularity among students at leading American colleges and military personnel.

In spite of the aforementioned potential curative benefits, concerns about possible side effects related to modafinil consumption have also been expressed by several authors, who advocate limitations on its daily use; these authors' concerns are based on the fact that there is a lack of knowledge regarding modafinil's neural mechanisms and the way it influences brain structures (Joos et al. 2010).

With regard to neural mechanisms, some authors hypothesized that modafinil promotes wakefulness via its effects in the anterior hypothalamus, whereas a dopamine-related action in the nucleus accumbens has been considered moderate (Scamell et al. 2000). With respect to alertness and partial wakefulness, it has been assumed that modafinil inhibits the reuptake of noradrenaline by the noradrenergic terminals on sleep-promoting neurons of the ventrolateral preoptic nucleus (VLPO) (Gallopin et al. 2004). More significant, perhaps, has been the report of modafinil's ability to increase excitatory glutamatergic transmission in thalamic and limbic structures. This mechanism reduces local GABAergic transmission, thereby diminishing GABA (A) receptor signaling on the mesolimbic and thalamic dopamine terminals (Ferraro et al. 1997). With regard to attentional control, recent studies have stressed the importance of the locus coeruleus (LC), a portion of the brainstem. It has been hypothesized that the LC is involved in shifts from distractible to attentive states. By visualizing activity from functional magnetic resonance imaging (fMRI), researchers have shown that modafinil indeed alters the state of the LC (Minzenberg et al. 2008). Moreover, these authors were able to shift subjects into a more attentive state by administering modafinil, as reflected by enhanced coordinated brain activity and better performance on an attentional test. In particular, the reported experimental findings suggest that modafinil decreases activity in the LC while favoring subcortical connectivity with the prefrontal cortex.

5.2.2 Neural Enhancement Modeling

Modeling, in the most abstract sense, provides a conceptual structure by which assumptions are incorporated and hypotheses articulated. A mathematical model differs from a theoretical, descriptive model in that the assumptions are articulated via mathematical rules. In particular, mathematical modeling is intended not only to describe, explain and test hypotheses about real-world phenomena arising from fields such as engineering, physics, physiology ecology, neuroscience, wildlife management, chemistry, economics, etc. but, more importantly, to make predictions. In particular, neural modeling addresses the proposal and development of computational and mathematical schemes aimed to study the human brain function. Due to the brain's high complexity (the brain involves a large number of interconnected neurons underlying responses at different temporal and spatial scales), a modeling approach represents a plausible methodology for describing and extracting relevant features of the brain's functionality in a tractable and reduced manner.

In order to successfully model and elucidate the neurodynamics of prospective neuroenhancers such as modafinil, one should consider the multiscale nature of the human brain. Theoretically, neuroenhancement modeling is achieved by integrating responses of neural populations into large scale neural responses that are comparable to data normally obtained by way of experimental techniques. Such modeling formulations have already proven their capability for capturing details ranging from single neurons to neural networks and cortical column maps. Such models, might in principle, provide deeper insights into the influence of neuroenhancers at neural network and cortical levels. In order to test prospective functional mechanisms of pharmaceutical neuroenhancers, the multiscale modeling approaches enable the integration of spatial (fMRI data, optogenetics, voltage sensitive dye imaging) and temporal data (behavioral studies, electroencephalographic data and intracranial recordings) from current brain imaging technologies.

In addition, implementing an efficient and validated model for studying the dynamics of brain structures under the influence of a specific psychostimulant, could be useful not only for developing better cognitive enhancers, but also for supporting the development of diagnostic tools and treatments for people suffering from a variety of neurodevelopmental disorders, such as attention deficit hyperactivity disorder (ADHD), autism, schizophrenia, etc.

5.3 Computational Model

The proposed model is a biologically plausible, large-scale, computational model, whose architecture and connectivity adopt principles from brain cortical and subcortical structures (Trenado et al. 2009). In particular, the model incorporates excitatory and inhibitory neural population effects that correspond to neural networks at the cortex, the thalamus, and the limbic system. Modulatory effects of modafinil are thus incorporated by a "modafinil protocol" block that regulates activity in the locus coeruleus (LC) (see Fig. 5.1). The critical role of the interaction between the cortex, the thalamus and the limbic system in relation to control and gating

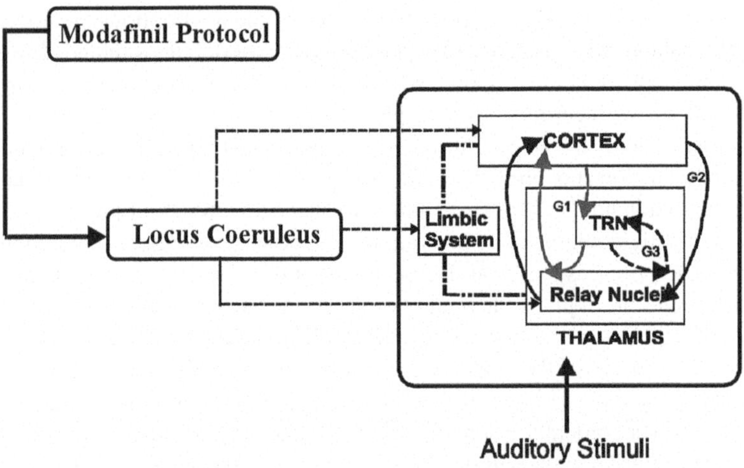

Fig. 5.1 Scheme of the brain structures considered in the model for studying modafinil effects on neural correlates of selective attention. The modafinil protocol block refers to a dose of modafinil being administered to a patient or subject; the locus coeruleus block (*LC*) represents the mean neural activity at that structure, which in turn influences the limbic system and corticothalamic connectivity

of selective attention has already been addressed by several authors (Ledoux et al. 1991; Destexhe 2000); whereas, the involvement of the LC in relation to modafinil has just recently been addressed (Minzenberg et al. 2008).

Interestingly, the LC, a small nucleus located in the pons, represents the main source of noradrenaline in the forebrain. In addition, the LC is involved in what is known as the ascending reticular activating system, a neural path critical for arousal and wakefulness. Specifically, LC neurons possess wide projections and their activity varies according to the degree of arousal and specific cognitive processes. A crucial observation is that the variability in LC neural responses results in a release of noradrenaline that targets numerous brain areas including the human prefrontal cortex (Bouret and Sara 2005).

As suggested by some authors (Minzenberg et al. 2008), we assume that modafinil acts by decreasing the neural activity in the LC, while enhancing neural activity between cortical and prominent subcortical structures. In this regard, we propose a relationship between modafinil dose (D) (typical range of dose 0–400 mg) and the LC neural activity (LCN) as provided by

$$LCN = \exp(-\theta D + K) \tag{5.1}$$

where θ denotes a parameter governing the degree of decay of LCN activity and K is a threshold determined by the minimum level of LCN activity.

With respect to thalamic modulation, we propose a relationship between corticothalamic gains G_1 and G_2 and the LCN activity as provided by,

$$\begin{aligned} \Psi &= \exp(-(LCN)^2) \\ \Psi &= G_1 + G_2 \end{aligned} \tag{5.2}$$

Here, the values of parameters G_1 and G_2 are selected in accordance with Trenado et al. (2009).

In our model, consideration of excitatory and inhibitory neural activities from both cortical and subcortical populations (Fig. 5.2) is achieved by means of a mean field equation of the form,

$$P_a = N_{ae} s_e \phi_e + N_{ai} s_i \phi_i + N_{as} s_s \phi_s$$

where ϕ_e, ϕ_i and ϕ_s denote mean field excitatory, inhibitory and subcortical neural activities, and N_{ae}, N_{ai}, N_{as} denote parameters related to excitatory, inhibitory and subcortical connectivity of neural populations.

The computation of the mean soma potential V_a is performed by integrating neural activities from different neural populations as provided by the following equation,

$$V_a(\vec{r}, t) = \int_{-\infty}^{\infty} L(t - t') P_a(\vec{r}, t') dt'$$

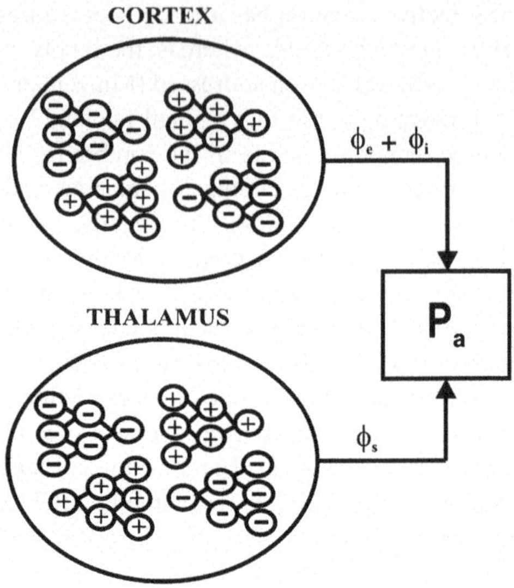

Fig. 5.2 The figure shows schematically the interaction and integration of cortical and subcortical neural populations which are summarized by the term (P_a)

The term L accounts for low-pass dendritic effects. Here, the mean firing rate is defined by a sigmoidal type function; the propagation of action potentials is accounted for by a damped wave equation; whereas, corticothalamic modulation is modeled by means of a transfer function involving gains G_1, G_2 and G_3 (see Trenado et al. 2009 for details).

5.4 Numerical Results

By specifying a modafinil protocol (namely, by defining a relationship between modafinil doses administered to a subject and the degree of LC activity, Eq. (5.1)), we are able to carry out the modulation of mean neural activity of LC neurons. Subsequently, cortical and subcortical modulation is achieved by varying gains G_1 and G_2 under the influence of LC neuron activity as specified by Eq. (5.2). In particular, numerical simulation of neural correlates of selective attention as reflected by the degree of synchronization of relevant evoked response components N1 (occurring at $\sim$100 ms) and P2 (occurring at $\sim$200 ms), was performed by following the approach in Trenado et al. (2009). Figure 5.3 shows numerical results of the N1 component synchronization by varying LC neural activity. In agreement with our simulations and the results reported in Minzenberg et al. (2008), it was observed that a high synchronization of the N1 component corresponded to a low LC neural activity; whereas, a high activity of the LC led to a decrease in the synchronization of the N1 component. As suggested by Trenado et al. (2009) and Bollimunta (2011), varying parameters G_1 and G_2 reflects a modulation of corticothalamic circuits that has implications on a mechanism for selective attention.

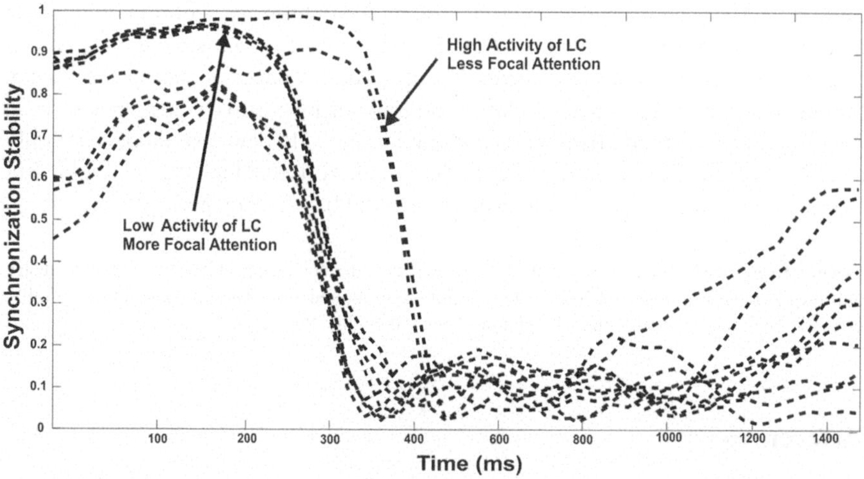

Fig. 5.3 Shows synchronization results of relevant evoked response components by varying LC neural activity. Here, a coherent synchronization corresponds to a state of focal attention, while a decrease in synchronization corresponds to a state of non-focal attention

Thus, our simulations provide support for the hypothesis that an increase (decrease) in LC neural activity leads to a non-focal (focal) attention state. In light of previous studies on neural correlates of selective attention, an observed synchronization (desynchronization) of the N1 component has been interpreted as intensification (relaxation) of the activity of the thalamic relay nuclei that leads to an increasing (decreasing) allocation of brain resources towards a target stimulus.

5.5 Conclusions

While the present study is preliminary, the obtained results showed the potential of applying neural computational models to test different hypotheses regarding the influence of neuroenhancers on humans. The present chapter addresses, for the first time, the effects of modafinil on a corticothalamic large-scale model of neural correlates of selective attention. In particular, the model is useful to test recent experimental findings regarding the involvement of the LC neural activity and its role in facilitating bottom-up projections thus enabling improvement in focal attention (Fig. 5.3). An interesting feature that is yet to be explored is the transition between states of focal and non-focal attention under the influence of modafinil. In particular, since fMRI possesses a good spatial resolution but lacks good temporal resolution, the applicability of a well-calibrated model promises to be instrumental in providing deeper insight into the influence of modafinil at time scales of milliseconds. In addition, the possibility of understanding the effect of modafinil on different cortical and subcortical brain structures might prove useful in elucidating treatments for disorders such as ADHD, schizophrenia, autism, among others. Needless to say, a deeper understanding of modafinil mechanisms in the human brain is instrumental in order to clarify the issue of its positive or negative effects. As a final note, it should be pointed out that biologically plausible models not only offer the possibility of incorporating findings from different brain technologies, but also the opportunity to test mechanisms that could be experimentally unfeasible.

Acknowledgement The author gratefully acknowledges financial support by the German Ministry of Education and Research (BMBF) and the Volkswagen Foundation. The kindness and hospitality from PD Dr. Elisabeth Hildt and Ms. Sheila Madary is highly appreciated.

References

Ballon JS, Feifel D (2006) A systematic review of modafinil: potential clinical uses and mechanisms of action. J Clin Psychiatry 67:554–566

Bollimunta A (2011) Neuronal mechanisms and attentional modulation of corticothalamic alpha oscillations. J Neurosci 31(13):4935–4943

Bouret S, Sara SJ (2005) Network reset: a simplified overarching theory of locus coeruleus noradrenaline function. Trends Neurosci 28(11):574–582

Bransfield RC (2004) Potential uses of modafinil in psychiatric disorders. J Appl Res 4(2):198–207

Destexhe A (2000) Modeling corticothalamic feedback and gating of the thalamus by the cerebral cortex. J Physiol Paris 94(5–6):391–410

Ferraro L, Antonelli T, O'Connor WT, Tanganelli S, Rambert F, Fuxe K (1997) The antinarcoleptic drug modafinil increases gluatamate release in thalamic areas and hippocampus. Neuroreport 8(13):2883–2887

Franke AG, Lieb K (2010) Pharmacological neuroenhancement and brain doping: chances and risks. Bundesgesundheitsblatt Gesundheitsforschung Gesundheitsschutz 53(8):853–859

Frye MA, Grunze H, Suppes T, McElroy SL, Keck PE Jr, Walden J, Leverich GS, Altshuler LL, Nakelsky S, Hwang S, Mintz J, Post RM (2007) A placebo-controlled evaluation of adjunctive modafinil in the treatment of bipolar depression. Am J Psychiatry 164(8):1242–1249

Gallopin T, Luppi PH, Rambert FA, Frydman A, Fort P (2004) Effect of the wake-promoting agent modafinil on sleep-promoting neurons from the ventrolateral preoptic nucleus: an in vitro pharmacologic study. Sleep 27(1):19–25

Henderson DC, Louie PM, Koul P, Namey L, Daley TB, Nguyen DD (2005) Modafinil-associated weight loss in a clozapine-treated schizoaffective disorder patient. Ann Clin Psychiatry 17(2):95–97

Joos L, Docx L, Schmaal L, Sabbe BG, Dom G (2010) Modafinil in psychiatric disorders: the promising state reconsidered. Tijdschr Psychiatr 52(11):763–773

Keen D, Hadijikoumi I (2011) ADHD in children and adolescents. Clin Evid (Online) Feb 4, pii: 0312. http://www.ncbi.nlm.nih.gov/pmc/articles/PMC3217800/

Knie B, Mitra MT, Logishetty K, Chauhuri KR (2011) Excessive daytime sleepiness in patients with Parkinson's disease. CNS Drugs 25(3):203–212

Ledoux J, Farb R, Romanski L (1991) Overlaping projections to the amygdala and stiatum from auditory processing areas of the thalamus and cortex. Neurosci Lett 134:139–144

Minzenberg MJ, Watrous AJ, Yoon JH, Ursu S, Carter CS (2008) Modafinil shifts human locus coeruleus to low-tonic, high-phasic activity during functional MRI. Science 322(5908):1700–1702

Morrow GR, Shelke AR, Roscoe JA, Hickok JT, Mustian K (2005) Management of cancer-related fatigue. Cancer Invest 23(3):229–239

Nicholas R, Chataway J (2007) Multiple sclerosis. Clin Evid (Online), Aug 15, doi:pii: 1202. http://www.ncbi.nlm.nih.gov/pmc/articles/PMC2907805/

Scamell TE, Estabrooke IV, McCarthy MT, Chemelli RM, Yanagisawa M, Miller MS, Spacer CB (2000) Hypothalamic arousal regions are activated during modafinil-induced wakefulness. J Neurosci 20(22):8620–8628

Sugarman DE, Poling J, Sofuoglu M (2011) The safety of modafinil in combination with oral Δ9-tetrahydrocannabinol in humans. Pharmacol Biochem Behav 98(1):94–100

Téllez N, Montalbán X (2004) Modafinil and fatigue in multiple sclerosis. Neurologia 19(8):434–437

Trenado C, Haab L, Strauss DJ (2009) Corticothalamic feedback dynamics for neural correlates of attention. IEEE Trans Neural Syst Rehabil Eng 17(1):46–52

van Vliet SA, Vanwersch RA, Jongsma MJ, van der Gugten J, Olivier B, Philippens IH (2006) Neuroprotective effects of modafinil in a marmoset Parkinson model: behavioral and neurochemical aspects. Behav Pharmacol 17(5–6):453–462

Chapter 6
Behavioral Neuroenhancement

Martin Dresler

Abstract The term neuroenhancement usually characterizes interventions in humans that aim to improve mental functioning beyond what is necessary to sustain or restore good health. While the current bioethical debate on neuroenhancement typically concentrates on pharmacological methods of enhancement, according to the given characterization, enhancement of mental capabilities also by technological, nutritional, or behavioral means has to be seen as neuroenhancement proper. In this chapter, research on behavioral techniques for the enhancement of cognition in the domains of attention, intelligence, creativity, and memory will be summarized.

Keywords Neuroenhancement • Behavioral techniques • Cognition • Attention • Intelligence • Creativity • Memory

6.1 Introduction

The term *neuroenhancement* usually characterizes interventions in humans that aim to improve mental functioning beyond what is necessary to sustain or restore good health (Juengst 1998). While the current bioethical debate on neuroenhancement typically concentrates on pharmacological methods of enhancement, according to the given characterization, enhancement of mental capabilities also by technological, nutritional, or behavioral means has to be seen as neuroenhancement proper. While the scientific literature provides a plethora of publications, studying for example 'brain training' methods in pathological conditions or age-related cognitive decline, only few studies examine healthy subjects. In the following, I will

M. Dresler (✉)
Department of Endocrinology of Sleep, Max Planck Institute of Psychiatry, Munich, Germany
e-mail: dresler@mpipsykl.mpg.de

E. Hildt and A.G. Franke (eds.), *Cognitive Enhancement: An Interdisciplinary Perspective*, Trends in Augmentation of Human Performance 1,
DOI 10.1007/978-94-007-6253-4_6,

summarize research on behavioral techniques for the enhancement of cognition in the domains of attention, intelligence, creativity, and memory. I will discuss briefly why these empirical data have important implications for the ethical debate on neuroenhancement.

6.2 Attention

The ability to focus one's attention underlies success in many everyday tasks, however voluntarily sustaining attention for extended periods of time is not easy. Very diverse training methods have been shown to improve several aspects of attention. Studies have focused on two different groups of attention training methods: computerized training, developed in Europe and in the USA and meditation methods, arising from Asian traditions. While computerized attention training has been shown to be successful both in children and adults (Cho et al. 2002; Tang and Posner 2009), a growing number of recent studies concentrates on attention-enhancing effects of meditation. Meditation can be conceptualized as a family of complex emotional and attentional regulatory training methods developed for various ends, including the cultivation of well-being and emotional balance. The two probably most common meditation styles are focused attention meditation, entailing the voluntary focusing of attention on a chosen object, and open monitoring meditation, involving non-reactive monitoring of the content of experience from moment to moment (Lutz et al. 2008). Regular meditation was shown to be associated with more accurate, efficient and flexible attentional processes across diverse tasks (Lutz et al. 2009; Moore and Malinowski 2009; Hodgins and Adair 2010; MacLean et al. 2010), with the resulting enhancement in behavioral measures being related to changes in neural activity (Lutz et al. 2004, 2008, 2009). While most studies work with experienced meditators or utilize intensive meditation training of several weeks' length, even a short meditation training of 4 days was shown to enhance attentional processes (Zeidan et al. 2010).

6.3 Intelligence

Intelligence is one of the most fundamental mental domains: Since people who perform well in one cognitive domain typically also perform well in others, a general factor of intelligence is thought to strongly influence all fields of intellectual ability. General intelligence is one of the most validly measurable human traits and a reliable predictor for educational and professional success (Gottfredson 2002). It is thought to gradually stabilize during childhood and change little thereafter, being only negligibly trainable in adulthood (Jensen 1969; Gottfredson 1997). Hence, few studies of intelligence training in adults exist (Buschkuehl and Jaeggi 2010). In particular, elderly subjects and subjects with low intelligence profit from intelligence training: playing video games or participating in cognitively stimulating activities seems to enhance intelligence in the elderly (Basak et al. 2008; Tranter and Koutstaal 2008), and even thinking aloud while performing a classical IQ test was shown to improve the result by almost one standard deviation (Fox and Charness 2010). While practicing intelligence tests leads to performance increases also in younger subjects (Bors and Vigneau 2003), such training gains probably do not lead to gains in general intelligence proper (te Nijenhuis et al. 2007).

General intelligence and working memory are considered to be highly interrelated constructs from an individual-differences perspective, highly reliably predicting each other (Colom et al. 2004; Ackermann et al. 2005). Hence, working memory training can be considered to imply intelligence training. Despite also being considered a constant trait, in several recent studies, working memory capacity has been successfully trained (Dahlin et al. 2009; Klingberg 2010), a process in which changes in fronto-parietal brain activity, increased myelination and dopamine receptor density are involved (Olesen et al. 2004; McNab et al. 2009; Takeuchi et al. 2010). These training-induced improvements in working memory seem to be transferred to intelligence proper, resulting in enhanced performance in tests of general intelligence (Jaeggi et al. 2008).

6.4 Creativity

Creativity is the ability to intentionally produce work that is novel, valuable and appropriate (Sternberg and Lubart 1999; Dresler 2008). Creativity is often thought of as the human capacity most pivotal for progress in science, art, engineering and economics; it is a non-trainable gift of rare geniuses. However, several studies have shown that creativity can be enhanced both by training (Fink 2008) and sleep (Dresler 2011).

Classical group-based methods like the brain storming technique, i.e. cognitive stimulation via the exposure to ideas of other people, have been shown to be effective under certain conditions (Dugosh et al. 2000; Fink et al. 2010). Several cognitive techniques are focused on enhancing creativity also on the individual level. A large meta-analysis by Scott et al. (2004a) found that creativity training programs induce gains in performance especially if they focus on the development of cognitive skills and heuristics involved in skill application by using realistic exercises appropriate to the domain at hand. In contrast, some commonly applied training strategies like imagery training proved less effective (Scott et al. 2004b). In recent years, computer-based training programs have been developed that seem to enhance ideational fluency, i.e. the quantity of ideas produced, but not necessarily the originality of these ideas (Benedek et al. 2006).

The classical creativity model of Helmholtz (1896) and Wallas (1926) proposes incubation, a process of unconscious recombination of thought elements that were stimulated through previous conscious work, as a pivotal phase in the creative process. On a theoretical level, sleep provides an ideal neural environment for incubation (Dresler 2011), and indeed sleep and dreaming have been shown to enhance several creative processes. For example in a study on creative insight, more than twice as many subjects gained insight into a hidden task rule if they had the opportunity to sleep after first task exposure compared to a wake control group (Wagner et al. 2004). In particular, REM sleep, the sleep stage most strongly associated with intense dreaming, enhances the formation of associative networks in creative problem solving (Cai et al. 2009). Selective deprivation of REM sleep but not of other sleep stages impairs post-sleep performance in creativity tasks that are presented to the subjects before sleep (Cartwright 1972; Glaubman et al. 1978). Interestingly, subjects show greater cognitive flexibility in creativity tasks immediately after awakenings from REM sleep compared to awakenings from other sleep stages (Walker et al. 2002a).

6.5 Memory

More than half a century ago, Miller (1956) published his observation of a severe limitation of our capacity to encode and later remember information: the number of objects an average human can hold in short-term memory is seven, plus or minus two. Since then, the term *memory span* refers to the longest list of items (e.g. digits, letters, words) that a person can repeat back immediately after presentation in correct order on 50 % of trials – even subjects with extraordinary memory typically can't exceed Miller's magical number seven by far. By contrast, already a century ago some case reports mention exceptional memorizers with memory spans of several dozens (Brown and Deffenbacher 1975), and since the early 1990s, memory athletes regularly prove memory spans of several hundreds of digits (Konrad and Dresler 2010). However, these superior memorizers typically don't claim to have a good memory by birth but have acquired it by intensive training in memory techniques. Most prominent is the so called *method of loci*, an ancient technique utilizing routes, visualizing to-be-remembered items at salient points along the routes, and then mentally retracing those routes during recall (Yates 1966). Parallel to its success in memory sports, training in the method of loci has also been shown to enhance memory capacity in the laboratory (Verhaeghen et al. 1992; Nyberg et al. 2003; Kondo et al. 2005). Interestingly, enhanced memory skills are related to a change in neural activity during encoding, shifting from language- to space-processing areas (Maguire et al. 2003; Nyberg et al. 2003; Kondo et al. 2005).

Besides learning and training mnemonics, a very simple behavioral technique has been shown to reliably enhance memory: sleep. A rapidly growing number of studies show that different memory systems and processes benefit from sleep (Diekelmann and Born 2010) – even a nap as short as 6 min is sufficient to promote memory performance (Lahl et al. 2008). Besides its stabilizing function, sleep boosts certain kinds of memories even above the level of initial acquisition: Procedural memories like motor skills typically reach a plateau after some time of training – however after a night of sleep motor performance starts from a higher level despite the absence of further training (Walker et al. 2002b). Interestingly, the sleep-memory relationship is specifically influenced by personal factors like gender, hormonal status or mental health (Dresler et al. 2010; Genzel et al. 2012). While for several years it was thought that REM sleep supports the consolidation of procedural memories while non-REM sleep supports declarative memories like verbal information, recent studies suggest that this model was too simplistic (Genzel et al. 2009; Rasch et al. 2009).

6.6 Differential Neuroenhancement

Behavioral neuroenhancement techniques have proven useful in a wide selection of cognitive domains. However, a recent study with a huge sample size questioned the transfer of training gains even to closely-related untrained tasks (Owen et al. 2010). By contrast, memory athletes often report that a small number of mnemonics help them to profoundly enhance their performance in almost any memory task and in academic and everyday life (Karsten 2011); working memory training was shown to improve general intelligence and thereby the most influential cognitive trait (Jaeggi et al. 2008). Obviously, a more differential approach is needed to elucidate which techniques are successful in which cognitive domains and how their enhancing effects are transferred to other domains. It might turn out that only very few behavioral techniques, e.g. working memory training, meditation or sleep, affect a wider range of cognitive capabilities. However this constraint would not reduce the value of other techniques or of behavioral neuroenhancement in general. Also, in sports training only few methods exist that benefit all sports; however, just because one wouldn't expect to become a better swimmer by tennis training or a better runner by golf training, this expectation wouldn't make tennis or golf training any less valuable as an effective means to its intended end.

So far, the public debate on neuroenhancement concentrates mainly on pharmacological interventions. The impressive success of some behavioral techniques, however, suggests that this focus might be misleading: techniques like the method of loci might exceed any pharmacological intervention by far (Verhaeghen et al. 1992; Konrad and Dresler 2010; Repantis et al. 2010). Such success would put discussions about unfair advantages of pharmacologically enhanced individuals in competitive situations in a different light. Behavioral neuroenhancement might actually differ from pharmacological neuroenhancement in another ethically important aspect: several drugs enhance cognition only in pathological cases; whereas, in healthy subjects, these drugs are effective to a weaker degree or not at all. This phenomenon might be different for certain behavioral neuroenhancement techniques: mnemonics seem to benefit young and healthy subjects to a larger extent than older or cognitively impaired subjects (Yesavage et al. 1990; Verhaeghen et al. 1992), creating an even larger gap between these groups. Again, a differential approach is needed that compares the effects of different methods of neuroenhancement in different populations.

In conclusion, despite outperforming pharmacological neuroenhancement, behavioral methods are a yet under-recognized form of neuroenhancement. The aforementioned comparisons show that empirical studies on behavioral neuroenhancement touch on important issues in the ethics of neuroenhancement. While ethical debates can in principle discuss purely theoretical or counterfactual scenarios, their impact on society and real life necessarily needs to be informed by empirical data. A differential approach, comparing the effects of different neuroenhancers on different cognitive domains in different populations, is therefore required to provide an empirical foundation for the debate on neuroenhancement.

References

Ackermann PL, Beier ME, Boyle MO (2005) Working memory and intelligence: the same or different constructs? Psychol Bull 131:30–60

Basak C, Boot WR, Voss MW, Kramer AF (2008) Can training in a real-time strategy video game attenuate cognitive decline in older adults? Psychol Aging 23:765–777

Benedek M, Fink A, Neubauer A (2006) Enhancement of ideational fluency by means of computer-based training. Creat Res J 18:317–328

Bors DA, Vigneau F (2003) The effect of practice on Raven's advanced progressive matrices. Learn Individ Diff 13:291–312

Brown E, Deffenbacher K (1975) Forgotten mnemonists. J Hist Behav Sci 11:342–349

Buschkuehl M, Jaeggi SM (2010) Improving intelligence: a literature review. Swiss Med Weekly 140:266–272

Cai DJ, Mednick SA, Harrison EM et al (2009) REM, not incubation, improves creativity by priming associative networks. PNAS 106:10130–10134

Cartwright RD (1972) Problem solving in REM, NREM, and waking. Psychophysiology 9:108

Cho BH, Ku J, Jang DP et al (2002) The effect of virtual reality cognitive training for attention enhancement. Cyberpsychol Behav 5:129–138

Colom R, Rebollo I, Palacios A et al (2004) Working memory is (almost) perfectly predicted by g. Intelligence 32:277–296

Dahlin E, Bäckman L, Neely AS et al (2009) Training of the executive component of working memory: subcortical areas mediate transfer effects. Restor Neurol Neurosci 27:405–419

Diekelmann S, Born J (2010) The memory function of sleep. Nat Rev Neurosci 11:114–126

Dresler M (2008) Kreativität als offenes Konzept. In: Dresler M (ed) Kreativität. Hirzel, Stuttgart

Dresler M (2011) Kreativität, Schlaf und Traum – Neurobiologische Zusammenhänge. In: Herrmann K (ed) Neuroästhetik. University Press, Kassel

Dresler M, Kluge M, Genzel L, Schüssler P, Steiger A (2010) Impaired off-line memory consolidation in depression. Eur Neuropsychopharmacol 20:553–561

Dugosh KL, Paulus PB, Roland EJ, Yang HC (2000) Cognitive stimulation in brainstorming. J Pers Soc Psychol 79:722–735

Fink A (2008) Möglichkeiten zur Förderung des kreativen Denkens. In: Dresler M, Baudson TG (eds) Kreativität. Hirzel, Stuttgart

Fink A, Grabner RH, Gebauer D et al (2010) Enhancing creativity by means of cognitive stimulation: evidence from an fMRI study. Neuroimage 52:1687–1695

Fox MC, Charness N (2010) How to gain eleven IQ points in ten minutes: thinking aloud improves Raven's matrices performance in older adults. Aging Neuropsychol Cogn 17:191–204

Genzel L, Dresler M, Wehrle R, Grözinger M, Steiger A (2009) Slow wave sleep and REM sleep awakenings do not affect sleep dependent memory consolidation. Sleep 32:302–310

Genzel L, Kiefer T, Renner L, Wehrle R, Kluge M, Grözinger M, Steiger A, Dresler M (2012) Sex and modulatory menstrual cycle effects on sleep related memory consolidation. Psychoneuroendocrinology 37(7):987–998

Glaubman H, Orbach I, Aviram O et al (1978) REM deprivation and divergent thinking. Psychophysiology 15:75–79

Gottfredson LS (1997) Mainstream science on intelligence: an editorial with 52 signatories, history, and bibliography. Intelligence 24:13–23

Gottfredson LS (2002) Where and why g matters: not a mystery. Hum Perform 15:25–46

Helmholtz HV (1896) Vorträge und Reden. Vieweg, Braunschweig

Hodgins HS, Adair KC (2010) Attentional processes and meditation. Conscious Cogn 19:872–878

Jaeggi SM, Buschkuehl M, Jonides J et al (2008) Improving fluid intelligence with training on working memory. PNAS 105:6829–6833

Jensen AR (1969) How much can we boost IQ and scholastic achievement? Harv Educ Rev 39:1–123

Juengst ET (1998) What does enhancement mean? In: Parens E (ed) Enhancing human traits: ethical and social implications. Georgetown University Press, Washington, DC

Karsten G (2011) Mnemotechniken – Strategien für außergewöhnliche Gedächtnisleistungen. In: Dresler M (ed) Kognitive Leistungen. Spektrum, Heidelberg

Klingberg T (2010) Training and plasticity of working memory. Trends Cogn Sci 14:317–324

Kondo Y, Suzuki M, Mugikura S et al (2005) Changes in brain activation associated with use of a memory strategy: a functional MRI study. Neuroimage 24:1154–1163

Konrad BN, Dresler M (2010) Grenzen menschlicher Gedächtnisleistungen. In: Baudson TG, Seemüller A, Dresler M (eds) Grenzen unseres Geistes. Hirzel, Stuttgart

Lahl O, Wispel C, Willigens B, Pietrowsky R (2008) An ultra short episode of sleep is sufficient to promote declarative memory performance. J Sleep Res 17:3–10

Lutz A, Greischar LL, Rawlings NB, Ricard M, Davidson RJ (2004) Long-term meditators self-induce high-amplitude gamma synchrony during mental practice. PNAS 101:16369–16373

Lutz A, Slagter HA, Dunne JD, Davidson RJ (2008) Attention regulation and monitoring in meditation. Trends Cogn Sci 12:163–169

Lutz A, Slagter HA, Rawlings NB et al (2009) Mental training enhances attentional stability: neural and behavioral evidence. J Neurosci 29:13418–13427

MacLean KA, Ferrer E, Aichele SR et al (2010) Intensive meditation training improves perceptual discrimination and sustained attention. Psychol Sci 21:829–839

Maguire EA, Valentine ER, Wilding JM et al (2003) Routes to remembering: the brains behind superior memory. Nat Neurosci 6:90–95

McNab F, Varrone A, Farde L et al (2009) Changes in cortical dopamine D1 receptor binding associated with cognitive training. Science 323:800–802

Miller GA (1956) The magical number seven, plus or minus two: some limits on our capacity for processing information. Psychol Rev 63:81–97

Moore A, Malinowski P (2009) Meditation, mindfulness and cognitive flexibility. Conscious Cogn 18:176–186

Nyberg L, Sandblom J, Jones S et al (2003) Neural correlates of training-related memory improvement in adulthood and aging. PNAS 100:13728–13733

Olesen PJ, Westerberg H, Klingberg T (2004) Increased prefrontal and parietal activity after training of working memory. Nat Neurosci 7:75–79

Owen AM, Hampshire A, Grahn JA et al (2010) Putting brain training to the test. Nature 465:775–778

Rasch B, Pommer J, Diekelmann S, Born J (2009) Pharmacological REM sleep suppression paradoxically improves rather than impairs skill memory. Nat Neurosci 12:396–397

Repantis D, Schlattmann P, Laisney O, Heuser I (2010) Modafinil and methylphenidate for neuroenhancement in healthy individuals: a systematic review. Pharm Res 62:187–206

Scott G, Leritz LE, Mumford MD (2004a) The effectiveness of creativity training: a quantitative review. Creat Res J 16:361–388

Scott G, Leritz LE, Mumford MD (2004b) Types of creativity training: approaches and their effectiveness. J Creat Behav 38:149–179
Sternberg RJ, Lubart TI (1999) The concept of creativity: prospects and paradigms. In: Sternberg RJ (ed) Handbook of creativity. Cambridge University Press, Cambridge
Takeuchi H, Sekiguchi A, Taki Y et al (2010) Training of working memory impacts structural connectivity. J Neurosci 30:3297–3303
Tang YY, Posner MI (2009) Attention training and attention state training. Trends Cogn Sci 13:222–227
te Nijenhuis J, van Vianen AEM, van der Flier H (2007) Score gains on g-loaded tests: no g. Intelligence 35:283–300
Tranter LJ, Koutstaal W (2008) Age and flexible thinking: an experimental demonstration of the beneficial effects of increased cognitively simulating activity on fluid intelligence in healthy older adults. Neuropsychol Dev Cogn B 15:184–207
Verhaeghen P, Marcoen A, Goossens L (1992) Improving memory performance in the aged through mnemonic training: a meta-analytic study. Psychol Aging 7:242–251
Wagner U, Gais S, Haider H et al (2004) Sleep inspires insight. Nature 427:352–355
Walker MP, Liston C, Hobson JA, Stickgold R (2002a) Cognitive flexibility across the sleep-wake cycle: REM-sleep enhancement of anagram problemsolving. Brain Res Cogn Brain Res 14:317–324
Walker MP, Brakefield T, Morgan A et al (2002b) Practice with sleep makes perfect: sleep-dependent motor skill learning. Neuron 35:205–211
Wallas G (1926) The art of thought. Harcourt, New York
Yates FA (1966) The art of memory. Routledge, London
Yesavage JA, Sheikh JI, Friedman L, Tanke E (1990) Learning mnemonics: roles of aging and subtle cognitive impairment. Psychol Aging 5:133–137
Zeidan F, Johnson SK, Diamond BJ et al (2010) Mindfulness meditation improves cognition: evidence of brief mental training. Conscious Cogn 19:597–605

Chapter 7
The Influence of Sports on Cognitive Task Performance – A Critical Overview

Pavel Dietz

Abstract The potential effect of different methods on cognitive performance is of great interest to scientific researchers. In addition to drugs, nutrition and sleeping behavior, extensive research has focused on different kinds of sports and exercise and their potential to enhance cognitive task performance. In this chapter, I will give the reader an overview of studies dealing with athletic activities of healthy people and their effects on cognitive task performance. Furthermore, I present and discuss limitations and problems found in these studies. One problem that I point out that makes the comparison of the study results difficult is the heterogeneous study design regarding mode, duration and intensity of the exercise as well as the type and the setting in which a specific cognitive task is applied among the different studies.

Keywords Sports • Cognitive enhancement • Cognitive performance • Training

7.1 Introduction

Cognitive Enhancement or Neuroenhancement is an issue of recent scientific interest in the fields of medicine, psychology, philosophy, politics and ethics and is defined as the use of diverse methods that have the potential of increasing cognitive functions such as attention, memory, learning performance or mood in healthy people (Lieb 2010). Extensive research has focused on pharmaceuticals and illicit drugs such as stimulants (e.g. methylphenidate and amphetamines), antidepressants or modafinil and mephedrone, all of which are implemented primarily for the treatment of diseases, but have been – and still are – abused by healthy individuals

P. Dietz (✉)
Department of Sports Medicine, Rehabilitation and Disease Prevention, Faculty of Social Science, Media and Sports, Johannes Gutenberg-University of Mainz, Mainz, Germany
e-mail: pdietz@uni-mainz.de

E. Hildt and A.G. Franke (eds.), *Cognitive Enhancement: An Interdisciplinary Perspective*, Trends in Augmentation of Human Performance 1,
DOI 10.1007/978-94-007-6253-4_7,

as cognitive enhancers (Maher 2008; Cakic 2009). The effects of such substances on cognitive task performance are mainly dependent on dosage and timing of intake (Caldwell et al. 2000; Wesensten et al. 2005; Minzenberg and Carter 2008; Repantis et al. 2009; Tye et al. 2010).

In addition to the use of pharmaceuticals and illegal drugs, other methods are used in relation to cognitive enhancement. For example, Dockery and colleagues (2009) have shown positive effects of transcranial direct current stimulation, a brain stimulation technique, on planning performance which includes working memory and skill performance. Additionally, the effects of sleep on cognitive performance, especially on memory and learning performance have been investigated and discussed in several studies. Significant sleep benefits on memory are observed after short naps of 1–2 h, as well as after an 8-h night of sleep (Mednick et al. 2003; Tucker et al. 2006; Korman et al. 2007; Nishida and Walker 2007).

In this overview, I will focus on intervention studies dealing with athletic activities of healthy people and their effects on cognitive task performance. I consider the possibility of using sports as a potential method for cognitive enhancement. Clinical studies in patients that apply sports for treatment of cognitive diseases, such as Alzheimer's disease (Coelho et al. 2009), are not included in this overview because they do not fall within the definition of cognitive enhancement given above. The central question at the onset of this literature overview is: Can we use sports for the purpose of cognitive enhancement and if so, which sports should be considered for such a purpose?

7.2 Overview of the Literature

The methodological structure of intervention studies that measure the effects of sports on cognitive task performance is mostly homogeneous regarding the following general conditions: a cognitive pre-test or test battery is followed by a defined exercise intervention and a second cognitive test is performed afterwards to measure the changes in cognitive task performance from pre- to post-exercise intervention (Tomporowski 2003; Lambourne and Tomporowski 2010).

In a current review article, Lambourne and Tomporowski (2010) have brought together studies based on endurance exercise, including studies using treadmills or bicycles. The levels of intensity were differentiated by the following categories: fatiguing, steady-state and inverted-U-exercises.

Fifty studies were included in the analysis. Twenty-one studies provided cognitive measures prior to and during exercise, with a mean sample size of 14 (8–41) participants. Eighteen of these twenty-one studies used an uncontrolled study design. The preferred exercise mode was cycling (16 studies); the exercise duration, ranging from 6 to 90 min, was very heterogeneous also, as well as the demands of the exercises and the types of cognitive tasks. One hundred and twenty-six effects were derived from these studies; the preferred cognitive measurement was response time, which was measured in 13 cases.

Twenty-nine studies, with a total sum of 545 participants (mean: 15; 6–100), were included in the analysis that provided pre- and post-exercise cognitive measures. Most studies ($n = 23$) used an uncontrolled trial, with a cycling treatment in 19 cases. The duration of the exercises ranged from 3 to 120 min. One hundred and nine effects from pre- to- post-exercise were derived from these studies with a high number of different cognitive tasks used in each study.

During exercise, they found a significant but very heterogeneous mean sample size ($\Delta = -0.14$; 95 % CI = −0.26 to −0.01, $p = 0.04$) and concluded that acute exercise led to an impairment of cognitive task performance. Effect size was dependent on exercise mode with larger effect sizes in studies that involved running or treadmill exercise, compared with cycling exercise ($p < 0.001$), the type of task used ($p < 0.001$), as well as when the cognitive tasks were performed. Negative effects were measured during the first 20 min of exercise, while positive effects were measured after the first 20 min ($p < 0.001$). The interaction of task type and exercise demand was also significant ($p = 0.02$), with positive effects in processing speed during steady-state-exercise, in contrast to negative effects during fatiguing- and inverted-U-exercise.

In contrast to measuring cognitive task performance during exercise, measuring following the intervention showed that acute exercise led to a small improvement in cognitive task performance ($\Delta = 0.20$; 95 % CI = 0.14–0.25, $p < 0.001$). Here, effect sizes were also dependent on exercise mode ($p = 0.04$) and the type of task used ($p = 0.002$), as well as study design, with larger effect sizes in uncontrolled trials compared with controlled trials ($p = 0.004$).

Contrary to studies dealing with endurance training and cognitive task performance, the number of studies dealing with resistance training and its effect on cognitive performance is much lower. Chang and Etnier (2009) randomly assigned 68 participants to a control group and three resistance training groups of 40, 70 and 100 % of the 10 repetition maximum. By using the Stroop Test (Stroop 1935) and Paced Auditory Serial Addition Task (Deary et al. 1991), they were able to show a significant linear effect of resistance exercise intensity on information processing speed and a significant quadratic trend for resistance exercise intensity on executive functions (Chang and Etnier 2009). This linear dose-response relationship of exercise on cognitive task performance has also been shown in several other studies (Mc Morris and Graydon 2000; Davranche and Audiffren 2004). By contrast, other studies have shown that there is not a linear but rather an inverted-U relationship (Aks 1998; Arent and Landers 2003), and, in turn, others again found no dose-response relationship at all (Cote et al. 1992). These findings, however, underline the heterogeneity of the results in studies dealing with cognitive task performance due to an exercise intervention as shown by Lambourne and Tomporowski (2010).

Pontifex and colleagues (2009) compared the effects of aerobic exercise and resistance exercise on cognitive task performance in one trial by measuring reaction time. They indicated a greater decrease in reaction time immediately after and 30 min after acute aerobic exercise, compared to the resistance training and seated control group. They concluded that exercise-induced changes in cognition are specific to the aerobic exercise domain (Pontifex et al. 2009), but since only a few studies are concerned with resistance training, more scientific research is needed.

7.3 Conclusion and Critique

These studies show that there are positive effects of sports and exercise on cognitive task performance. These effects are generally regarded to be based on the increase of cerebral blood flow (Herholz et al. 1987; Secher et al. 2008; Ogoh and Ainslie 2009) and on an increase of growth factors, like the brain-derived neurotrophic factor (BDNF) (Knaepen et al. 2010), or an increase of neurotransmitters like dopamine and norepinephrine (Winter et al. 2007) during and after exercise and sports. Many studies show negative or no effects as well (Lambourne and Tomporowski 2010); however, the results of these studies and the feasible effects of exercise on cognition quality are certainly influenced by mode, duration and intensity of the exercise, the type and the setting in which a specific cognitive task is applied, as well as the general study design. The timing of when cognitive tasks are carried out (e.g. during or following exercise) is a very important factor, but Lambourne and Tomporowski (2010) as well as Tomporowski (2003) do not give any advice on which point of time is the most beneficial to apply a cognitive task. Perhaps most importantly in such intervention studies, each result of a cognitive task is affected by a learning effect from one repetition to the next. To calculate how much the obtained results are influenced by this learning effect, a control group has to be used to assess how large this learning effect really is. In the review article of Lambourne and Tomporowski (2010), mainly uncontrolled trials were found. This situation, however, heavily reduces the quality and explanatory power of the presented effects of exercise on cognitive performance in this review. A separate calculation of the controlled trials alone is needed in order to achieve more reliable data that factors in the learning effect.

Additionally, there are no methodological standards (duration, intensity, mode, type of cognitive task), which makes the summarization of the existing literature on exercise and its effects on cognitive task performance difficult. As a result, each study has its own specific study design. This heterogeneity of methods and the consequent heterogeneity of results, combined with missing analytical results of controlled trials, renders the question posed at the beginning of this paper unanswerable at this point in time. Aerobic exercise seems to be the most beneficial athletic activity for cognitive task performance, but more scientific research of randomized controlled trials is needed, especially in studies dealing with resistance training, as well as standardized methods and study designs.

In consideration of these factors that influence the results of studies dealing with exercise and cognitive task performance, it would be helpful to assess data from a population of athletes who report about their subjective feelings regarding cognitive performance during or following exercise. The results of such surveys in athletes, which inquire about the preferred mode, intensity and duration of sports, the perceived effects on cognition and how long these effects persist, could very well be relevant to creating more suitable intervention studies in the future. A first pilot study in endurance athletes and ball sports athletes is already in progress.

Acknowledgement The project is funded by the Initiative PRO-Humanities and Social Sciences 2015 (Initiative PRO Geistes- und Sozialwissenschaften 2015), University of Mainz.

References

Aks DJ (1998) Influence of exercise on visual search: implications for mediating cognitive mechanisms. Percept Mot Skills 87:771–783

Arent SM, Landers DM (2003) Arousal, anxiety, and performance: a reexamination of the inverted-u hypothesis. Res Q Exerc 74:436–444

Cakic V (2009) Smart drugs for cognitive enhancement: ethical and pragmatic considerations in the era of cosmetic neurology. J Med Ethics 35:611–615

Caldwell JA Jr, Caldwell JL, Smythe NK III, Hall KK (2000) A double-blind, placebo-controlled investigation of the efficacy of modafinil for sustaining the alertness and performance of aviators: a helicopter simulator study. Psychopharmacology 150:272–282

Chang YK, Etnier JL (2009) Exploring the dose-response relationship between resistance exercise intensity and cognitive function. J Sport Exerc Psychol 31:640–656

Coelho FG, Santos-Galduroz RF, Gobbi S, Stella F (2009) Systematized physical activity and cognitive performance in elderly with Alzheimer's dementia: a systematic review. Rev Bras Psiquiatr 31:163–170

Cote J, Salmela J, Papathanasopoulu KP (1992) Effects of progressive exercise on attentional focus. Percept Mot Skills 75:351–354

Davranche K, Audiffren M (2004) Facilitating effects of exercise on information processing. J Sports Sci 22:419–428

Deary IJ, Langen SJ, Hepburn DA, Frier BM (1991) Which abilities does the PASAT test. Pers Individ Diff 12:983–987

Dockery CA, Hueckel-Weng R, Birbaumer N, Plewina C (2009) Enhancement of planning ability by transcranial direct current stimulation. J Neurosci 29:7271–7277

Herholz K, Buskies W, Rist M, Pawlik G, Hollmann W, Heiss WD (1987) Regional cerebral blood flow in man at rest and during exercise. J Neurol 234:9–13

Knaepen K, Goekint M, Heyman EM, Meeusen R (2010) Neuroplasticity – exercise-induced response of peripheral brain-derived neurotrophic factor. Sports Med 40:765–801

Korman M, Doyon J, Doljansky J, Carrier J, Dagan Y, Karni A (2007) Daytime sleep condenses the time course of motor memory consolidation. Nat Neurosci 10:1206–1213

Lambourne K, Tomporowski PD (2010) The effect of exercise-induced arousal on cognitive task performance: a meta-regression analysis. Brain Res 41:12–24

Lieb K (2010) Hirndoping – Warum wir nicht alles schlucken sollten. Artemis & Winkler, Mannheim

Maher B (2008) Poll results: look who's doping. Nature 452:674–675

Mc Morris T, Graydon J (2000) The effect of incremental exercise on cognitive performance. Int J Sport Psychol 31:66–81

Mednick S, Nakayama K, Stickgold R (2003) Sleep-dependent learning: a nap is as good as a night. Nat Neurosci 6:697–698

Minzenberg MJ, Carter SC (2008) Modafinil: a review of neurochemical actions and effects on cognition. Neuropsychopharmacology 33:1477–1502

Nishida M, Walker MP (2007) Daytime naps, motor memory consolidation and regionally specific sleep spindles. PLoS One 4:e341

Ogoh S, Ainslie PN (2009) Cerebral blood flow during exercise: mechanisms of regulation. J Appl Physiol 107:1370–1380

Pontifex MB, Hillman CH, Fernhall B, Thompson KM, Valentini TA (2009) The effect of acute aerobic and resistance exercise on working memory. Med Sci Sports Exerc 41:927–934

Repantis D, Schlattmann P, Laisney O, Heuser I (2009) Antidepressants for neuroenhancement in healthy individuals: a systematic review. Poiesis Prax 6:139–174

Secher NH, Seifert T, Van Lieshout JJ (2008) Cerebral blood flow and metabolism during exercise: implications for fatigue. J Appl Physiol 104:306–314

Stroop JR (1935) Studies of interference in serial verbal reactions. J Exp Psychol 18:643–662

Tomporowski PD (2003) Effects of acute bouts of exercise on cognition. Acta Psychol 112: 297–324

Tucker MA, Hirota Y, Wamsley EJ, Lau H, Chaklader A, Fishbein W (2006) A daytime nap containing solely non-REM sleep enhances declarative bit not procedural memory. Neurobiol Learn Mem 86:241–247

Tye KM, Tye LD, Cone JJ, Hekkelman EF, Janak PH, Bonci A (2010) Methylphenidate facilitates learning-induced amygdale plasticity. Nat Neurosci 13:475–480

Wesensten NJ, Killgore WDS, Balkin TJ (2005) Performance and alertness of caffeine, dextroamphetamine, and modafinil during sleep deprivation. J Sleep Res 14:255–266

Winter B, Breitenstein C, Mooren FC, Voelker K, Fobker M, Lechtermann A, Krueger K, Fromme A, Korsukewitz C, Floel A, Knecht S (2007) High impact running improves learning. Neurobiol Learn Mem 87:597–609

Chapter 8
The Human Experiment: How We Won't Win the Rat Race. What Can We Learn from Brain Stimulation in Humans and Rats About Enhancing the Functional Neurobiology of Higher Cognitive Functions?

Colleen A. Dockery

Abstract This chapter addresses neuroenhancement and is divided into three parts. Firstly, neuroenhancement is considered in terms of the current societal context of a growing reliance on high level cognitive functions for economic competition. Then, specific research examples involving an increasingly popular neuroenhancement method, transcranial direct current brain stimulation, are discussed regarding what contributions enhancement technologies can make to these higher level cognitive functions. Speculations are made about the dynamics of relationships between brain structures and functions. The complexity of the involved brain mechanisms is discussed to highlight the intricacy of neural engagement to support these functions. And finally, the indications from empirical research are re-applied to the current state of the systems that employ higher level cognitive functions. Questions are presented about the viability of the so-called "More is Better" (MiB) model, in relation to neuroenhancement and for supporting cognitive functions.

Keywords Neuroenhancement • Executive function • Optimization • Inverted-U dose-response curve • State-dependence • Task load

Abbreviations

MiB	"More is Better" (model)
BOLD	Blood Oxygenation Level Dependent
NE	neuroenhancement

C.A. Dockery (✉)
Institute of Medical Psychology and Behavioral Neurobiology, University of Tübingen, Tübingen, Germany

Hochschule Albstadt-Sigmaringen, Sigmaringen, Germany
e-mail: dockery@hs-albsig.de

E. Hildt and A.G. Franke (eds.), *Cognitive Enhancement: An Interdisciplinary Perspective*, Trends in Augmentation of Human Performance 1,
DOI 10.1007/978-94-007-6253-4_8,

ADHD	Attention Deficit Hyper Activity Disorder
tDCS	Transcranial direct current stimulation
EF	Executive functions
PFC	prefrontal cortex
TOL	Tower of London test
DLPFC	dorsolateral prefrontal cortex
RT	reaction times
ACC	accuracy
rTMS	transcranial magnetic stimulation
PET	positron-emission tomography
DA	dopamine
PISA	Programme for International Student Assessment

8.1 Introduction

In the context of a growing global competition for resources (EEA 2010), there is a common perception that growth will support higher rates of employment and resource productivity, in addition to better living standards (Merkel 2007). At the governmental level, greater competition through technological innovations and economic integration is encouraged as a palliative measure to reduce unemployment rates and stimulate economic growth. A drive to overcome a perceived deficit of workforce to fulfill economic needs may contribute to the public and institutional impetus towards working harder or longer, with the aim to produce better conditions. The idea that "more is better" is a model that may have been conceptually popularized during the Industrial Revolution in which large scale production at low cost solved supply and demand. The rise in productivity due to technological sophistication is a laudable human achievement that is associated with an improved quality of life. Also, thanks to technological advances, humans rely increasingly less on physical prowess in their daily work lives and more on their mental faculties, which creates an increasingly sedentary lifestyle (Paffenbarger et al. 1986). A transition from the "industrial society" to the "knowledge-based" society is important for sustainable development (Merkel 1998). As a result, greater reliance upon cognitive functions, such as the ability to monitor and manipulate information, plan, form strategies, solve problems and make decisions has emerged.

Enhancement technologies have been long available and are stimulated by economic competition. Currently a surge in interest in "neuroenhancement" (NE) has been spurred by attention in the media, science and medicine (Farah et al. 2004; Galert et al. 2009; Larriviere et al. 2009). The definition has generally been taken as the use of pharmacological substances to improve neural function, which can include cognition and mood (Greely et al. 2008). The word "enhancement" itself implies an increase of quantity, value or power, though it does not always mean an improvement. A meta-analysis of misuse of stimulants related to Attention Deficit Hyper Activity Disorder (ADHD) among students suggests that neuroenhancement is an existing issue (Wilens et al. 2008). Despite concern and assertions about the imminent trend for increased substance misuse, empirical data from population sampling revealed that the annual prevalence rates for non-medical use of prescribed pharmacological substances in pupils and students is as low as 4.1–5.4 % in America (Sussman et al. 2006) and 0.3 % in Germany (Franke et al. 2011). Users' tendency to misuse other substances also suggests that addiction disorder may be as relevant as substance misuse for study aids. Furthermore, literature reviews of the efficacy of memantine, anti-dementia drugs, methylphenidate and modafinil on cognitive performance do not provide firm conclusions and indicate weak if negligible effects on cognition (Normann et al. 2010; Repantis et al. 2010a, b).

This evidence suggests that with pharmacological enhancement, more (e.g. substance) is not necessarily better. This could simply be due to poor experimental designs, since studies often included only single-dose trials with testing over a short time period, and without concurrent training or learning. When found, positive effects were often associated with deficient states, such as with sleep deprivation or in disorders such as ADHD (see reviews above). A focus on enhancement itself can eclipse the implied underlying aim, which presumably is optimization of a given function. If a technological enhancement confers improvement, the underlying mechanisms should be evaluated in order to broaden their applicability and ensure safety. Apart from efficacy and prevalence, the safety and ethics are far from being established, which makes this form of "enhancement" seemingly less viable than presumed (Flower et al. 2010; Quednow 2010). Neuroenhancement is not limited to the misuse of pharmacological substances. Other forms of NE include cognitive training (Brenes 2003; Olesen et al. 2004; Klingberg 2010; La Rue 2010), meditation (Chiesa et al. 2011), exercise (Pereira et al. 2007; Lambourne and Tomporowski 2010; Yanagisawa et al. 2010) and brain stimulation (Siebner et al. 2009; Zimerman and Hummel 2010). The following chapter will consider a particular technology used for brain stimulation.

Transcranial direct current stimulation (tDCS) is a method of brain stimulation used in humans and animals to transiently alter neuronal excitability via weak direct currents with the aim to alter functions associated with the underlying cortical areas (Stagg and Nitsche 2011). Depending on the brain region being stimulated and the polarity used, the excitability can be increased or decreased (anodal and cathodal respectively), which generally makes a cell more or less likely to spontaneously fire. The duration of stimulation and current strength also influence the duration and intensity of the after-effects, which can last up to 1 h (Nitsche et al. 2003, 2007). Due to these effects, tDCS shows promise for use in clinical applications to treat neurological (e.g. stroke) and neuropsychiatric (e.g. depression) disorders (Schlaug and Renga 2008; Nitsche et al. 2009; Utz et al. 2010). It is unclear which electrode montage, treatment frequency and current intensity will result in the most optimal effects for any disorder; however, recently for healthy participants, a dose-response curve was reported in regard to learning (identification of concealed objects) in which a higher current intensity lead to greater performance benefits (Clark et al. 2010). Whether the MiB model always applies to tDCS effects on structure-function deserves further attention, though data suggest that the relationship is more complex.

More recently tDCS is being studied in rat models to evaluate the brain effects at the cellular and molecular level, which may elucidate why tDCS results in functional improvements (Liebetanz et al. 2006a, b, 2009b; Takano et al. 2011). The efficacy of prospective treatments developed for humans can be enhanced by animal models, which emulate in-tact and pathological functions to study the underlying neurobiology of higher-order cognitive processes.

8.2 Consideration of Scientific Results

Executive functions (EF), defined as the cognitive capacity to regulate and control behavior, include goal formation, planning, execution of goal-directed plans and effective performance (Jurado and Rosselli 2007). EF is a component of higher cognitive function that is essential for successful daily living and a high quality of life. Human evolution is associated with increased brain size and metabolism (especially in the prefrontal cortex (PFC)), which are concurrent with expanded capabilities in cognitive function (Fu et al. 2011). Executive functions, particularly those depending on working memory and planning ability, degrade with age (Penner et al. 2010). The functions that are generally targeted for neuroenhancement encompass EF, which are associated with the neocortex of the human brain and specifically the PFC (Leh et al. 2010). EF deficits are prevalent in frontal lobe associated disorders such as depression and schizophrenia (Martinez-Aran et al. 2002; Ottowitz et al. 2002) or frontal lesions (Jacobs et al. 2007), emphasizing the integrity of the prefrontal cortex as essential for intact performance. For this reason, in this chapter the prefrontal cortex is used to serve as a model system to consider plasticity-related changes in function.

Though limited in scope, the following research examples address the use of one particular NE technology and its impact on higher cognitive functions. In the following studies, brain stimulation by tDCS was used to manipulate working memory and skill learning, which support EF. Possible underlying mechanisms to explain the results will then be considered. Since successful visuospatial working memory is associated with increased prefrontal activity and supports problem solving and planning (Newman et al. 2003; Olesen et al. 2004), it is reasonable to posit that altered activity in these areas can lead to altered functions. tDCS of the frontal cortex in rats has been found to affect the hemodynamic activity in the frontal cortex and in more distal regions (Takano et al. 2011). In the following studies, tDCS was applied to the PFC in humans and the frontal cortex of rats to test for performance changes on PFC-related tasks of working memory and skill learning.

8.2.1 Human Study

The Tower of London test (TOL) is a neuropsychological test to evaluate executive function (Fig. 8.1) and is sensitive in revealing impairments in patient performance

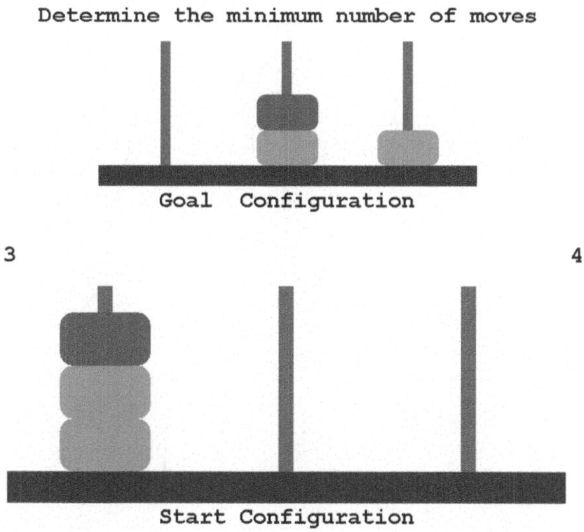

Fig. 8.1 The TOL test is a puzzle involving beads that "slide" on pegs which must be moved to match a designated goal configuration. The "start" and "goal" configurations are presented on the computer screen with two answers at the sides of the screen. In this version, subjects must mentally determine the minimum number of moves necessary to reach the target configuration and press the button corresponding to the side of the screen with the correct answer

relative to healthy subjects (Shallice 1982). Neuroimaging and lesion studies reveal that TOL performance involves a distributed network of integrated brain activity to produce planning, including bilateral dorsolateral prefrontal cortex (DLPFC) activation (Unterrainer and Owen 2006). Planning performance is supported by working memory, and the left DLPFC has been thought to be the critical structure for solving working memory problems with higher task loads (Olesen et al. 2004) and relational complexity (Kroger et al. 2002). In the TOL, graded activation increases have been found in relation to increased task difficulty (a higher number of moves) (Baker et al. 1996; Newman et al. 2003; Unterrainer and Owen 2006). A number of patient populations show deficits in TOL performance including those with frontal lobe damage (Shallice and Burgess 1991), schizophrenia (Tyson et al. 2004; Rasser et al. 2005; Ungvari et al. 2008), depression (Elliott et al. 1997; Goethals et al. 2005), dementia (Rainville et al. 2002), and Parkinson's Disease (Beauchamp et al. 2008; Rektorova et al. 2008). Patient performance, particularly at higher levels of task load, is distinguishable from healthy controls, and neuroimaging studies support the DLPFC as the most critical structure for solving the TOL task (Owen et al. 1990; Grafman et al. 1992; Owen 1997; Rainville et al. 2002; Lazeron et al. 2004; Rasser et al. 2005; van den Heuvel et al. 2005). Cognitive skill learning depends on the prefrontal cortex; it is the ability to acquire how to solve complex problems in intellectual tasks through practice, which, in turn, improves performance (Cerella et al. 2006). Repetition of the TOL test also leads to cognitive skill learning (Ouellet et al. 2004).

To determine the efficacy of brain stimulation-induced activity changes, a study in healthy students was designed to investigate the effects of tDCS over the left DLPFC on planning performance in the TOL test over multiple sessions (Dockery et al. 2009). Due to the known task-related Blood Oxygenation Level Dependent (BOLD) activation increases (Unterrainer and Owen 2006), it was proposed that anodal tDCS (known to increase excitability and BOLD) would lead to improved

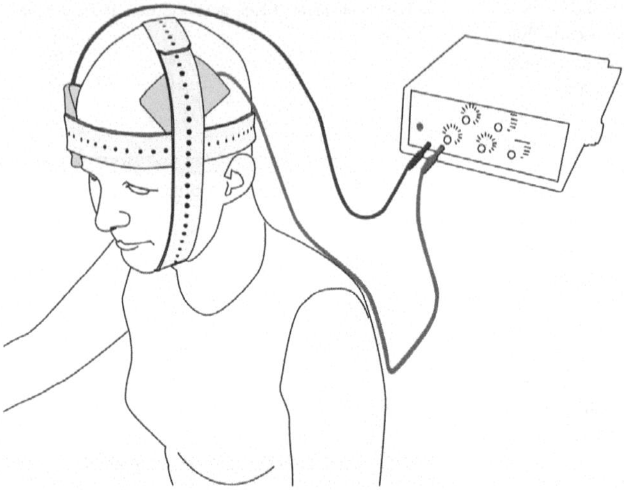

Fig. 8.2 Transcranial direct current stimulation of the left DLPFC and the contralateral right orbit of a human participant. The electrodes, enveloped in wet sponges, are fixed to the head with adjustable latex bands. The stimulator is located out of view of the participant

performance, particularly at high task load levels, while cathodal (known to decrease excitability and often associated with negligible performance effects) and sham tDCS would not. This hypothesis, based on existing literature, was in support of the MiB model. In a cross-over design, 24 healthy participants (5 men, 19 women) performed the TOL test during and after 15 min of active anodal, cathodal and sham tDCS of the DLPFC over three sessions with a long-term follow-up session (after 6 months or 1 year). The 1 mA current was delivered between a pair of water-soaked sponge electrodes (35 cm^2) with electrodes fixed over F3 (International 10–20 system of electrode placement) and contralaterally above the right orbit (Fig. 8.2).

Brain stimulation by tDCS boosted TOL performance for both anodal and cathodal stimulation, causing a significant improvement in planning performance compared to sham tDCS. Anodal tDCS resulted in improvements, particularly in later sessions as indexed by faster reaction times (RT) with equal to higher accuracy (ACC), while cathodal tDCS showed benefits in early sessions leading to a flattened learning curve across sessions due to better initial performance. Retrospectively, the participants were grouped according to the order in which they received the different types of tDCS as defined by: A/C = Anodal before Cathodal, C/A = Cathodal before Anodal. These results were indicated by significant order effects of the stimulation (RT: [$F(1,22) = 8.935$, $p = 0.007$]; borderline for ACC: [$F(1,22)$ 3.494, $P = 0.075$]). In Fig. 8.3 the order effects relating to tDCS sequence are apparent across sessions. A significant interaction for stimulation order and task load (high, low) (RT only: [$F(1,22) = 7.749$, $P = 0.011$]) showed a distinct advantage for tDCS C/A at high task loads (Fig. 8.4). The behavioral results reflect phase-specific performance gains particularly at higher levels of task demand by

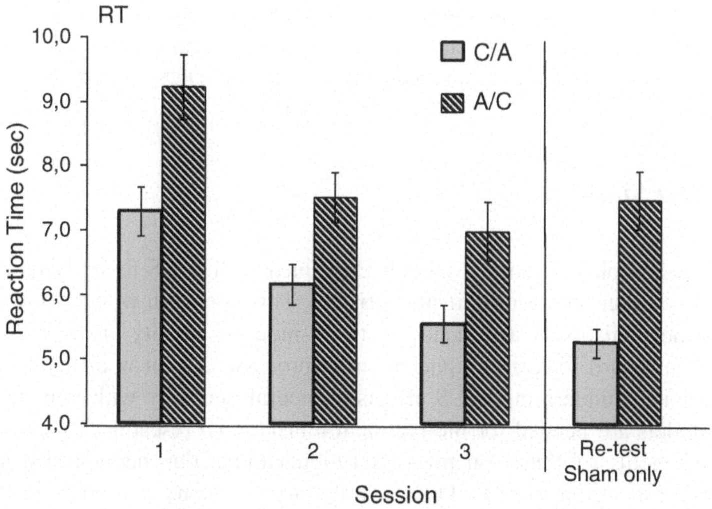

Fig. 8.3 The mean reaction times (seconds) for the TOL task for each order of tDCS sequence (A/C: Anodal before Cathodal, C/A: Cathodal before Anodal) across all four sessions. The order of stimulation conditions was counterbalanced across participants. Error bars indicate ± SEM (standard error of the mean)

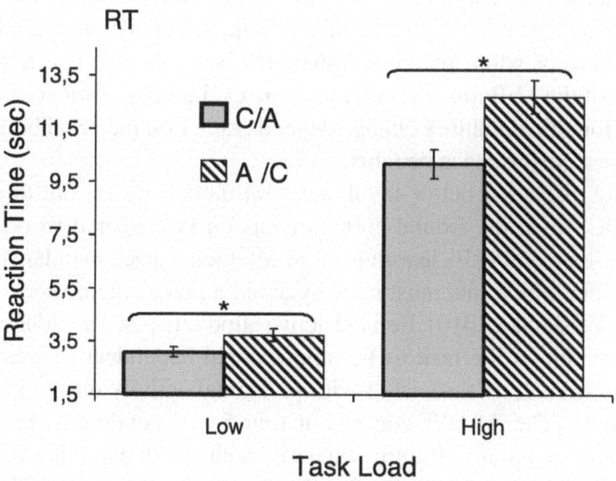

Fig. 8.4 The mean reaction times (seconds) for the TOL task for each order of tDCS sequence (A/C: Anodal before Cathodal, C/A: Cathodal before Anodal) according to task load levels (Low: 1–2-, High: 4–5-move problems). Error bars indicate ± SEM (standard error of the mean)

acute brain stimulation. The results from the re-test session (n = 19) under sham stimulation (at 6 months or 1 year) show that these phase and polarity-specific benefits persist well beyond the acute application (RT only: [F(1,16) = 17.357,

P = 0.001]) by which the pretreatment with tDCS C/A during training, yielded a 42 % faster RT than tDCS A/C at follow-up. The long-term cognitive benefits may result as a function of learning mechanisms paired with tDCS-altered brain activity.

8.2.2 *Rat Study*

Experiments employing animal models combined with tDCS to study its efficacy on cognitive function are rare, if non-existent. This approach would allow for the study of mechanisms of action due to tDCS-induced activity changes and their relevance in regard to learning and memory processes. As it is difficult to study the mechanisms underlying tDCS effects on neural plasticity with humans alone, animal models are needed for bidirectional translational research. To evaluate the potential benefits of tDCS on prefrontal-hippocampal dependent tasks, a novel paradigm for assessing emulated human cognitive functions in a rodent model was developed (Dockery and Wesierska 2010), as was the methodology for transcranial direct current stimulation in rats (Dockery et al. 2011). This may help to increase knowledge about the mode of action of beneficial tDCS effects on cognitive tasks by establishing an animal model that bridges the human studies and supports testing of the neurobiological basis of induced changes. Due to the findings in the previously reported human study, it was proposed that during early learning, cathodal tDCS, known to decrease excitability, would lead to improved performance particularly at high task load levels, while anodal and sham tDCS would not. This hypothesis was not in support of the MiB model, such that more excitability is not necessarily better and the direction of excitability change depends rather on the basal brain activity in order to produce performance benefits.

In the study discussed below involving a rat model, we set out to examine the efficacy of tDCS over the frontal cortex of rats on visuospatial working memory, long-term memory and skill learning in an allothetic place avoidance alternation task (APAAT), in which rats must actively avoid a place where shock is presented (Dockery and Wesierska 2010). Related active allothetic place avoidance paradigms (Fig. 8.5) have shown the task to be hippocampal dependent (Cimadevilla et al. 2001) and the APAAT is associated with prefrontal activity due to its demand on working memory. The APAAT consists of four 5 min conditions: habituation (no shock), two place avoidance training intervals with shock and, after a 5 min delay, a retrieval test (shock inactivated). Over three consecutive days (D1, 2, 3), prior to behavioral training, freely behaving rats received 30 min of 200 μA of tDCS over the frontal cortex (Fig. 8.6), which is thought to increase (Anodal n = 15) or decrease (Cathodal n = 13) neuronal excitability (Liebetanz et al. 2009a) relative to control rats (n = 12). For each training day, the location of the shock sector was alternated. The long term effect of stimulation and training on behavior was tested without tDCS on D21.

Performance improved with place avoidance training, within daily sessions as indexed by a decreased number of entrances ($F_{6,228} = 4.17$; $P = 0.0004$) (Fig. 8.7).

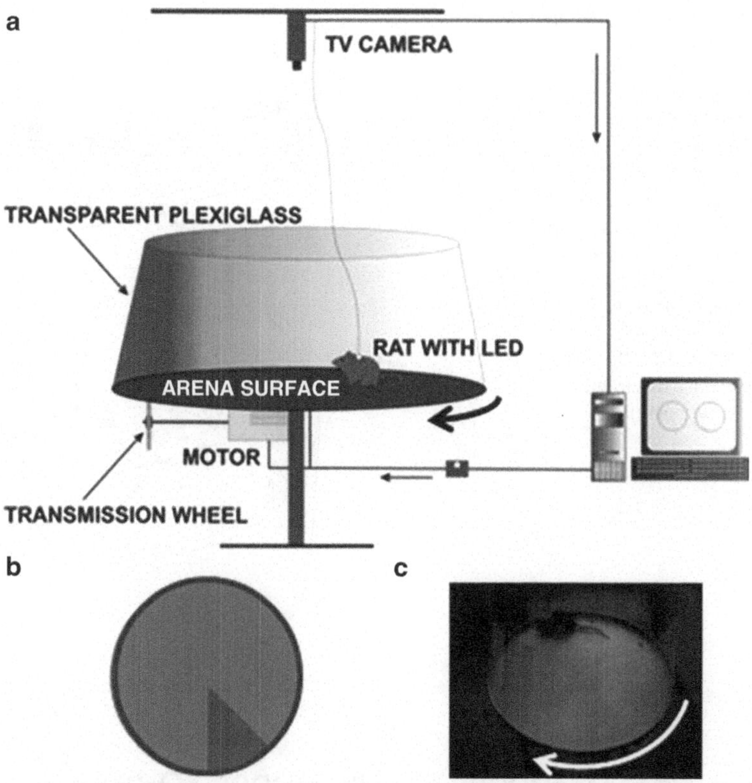

Fig. 8.5 A figure modified from Bubenikova-Valesova et al. (2008) shows a schematic representation of the place avoidance set-up. In (**a**), the arena is depicted located in a room in which salient room-based cues are presented. Via a diode on the periphery of the arena and another on the rat's back, the rat's trajectory can be recorded and monitored by a camera mounted on the ceiling. Thereby coordinates in a 2-D frame, both according to the arena and the room frames can be registered by a computer program (Track Analysis, Bio-signal Group, Brooklyn, New York) and a monitor located in another room. In (**b**), a schema depicts the aerial view of the to-be-avoided sector (e.g. 45° sector) and in (**c**), a photo depicting the scale of the rat on the 80 cm diameter arena relative to the to-be-avoided sector. The *arrows*, depicted in segments (**a**) and (**c**), represent the movement of the arena when the active place avoidance task is being employed

The complexity of the task was ensured by a continuously altered shock sector location each day and was supported by the results, which show that they avoided better on D2 than on D1 and D3 (ENTR: $F\ 2{,}76 = 5.41$; $P < 0.006$; $D1 > D2 < D3$, $P < 0.004$). These results likely express poor performance in the naïve state (D1) and higher load on D3 from exposure to previous shock sectors on D1 and D2. Here, D2 represents the optimum for having advantages from task experience and still a moderate task load. Improved performance was also found within sessions in which skill learning (not depicted) occurred as shown by a low number of shocks per entrance during the second training interval ($F\ 3{,}114 = 39.39$; $P < 1 \times 10^{-16}$;

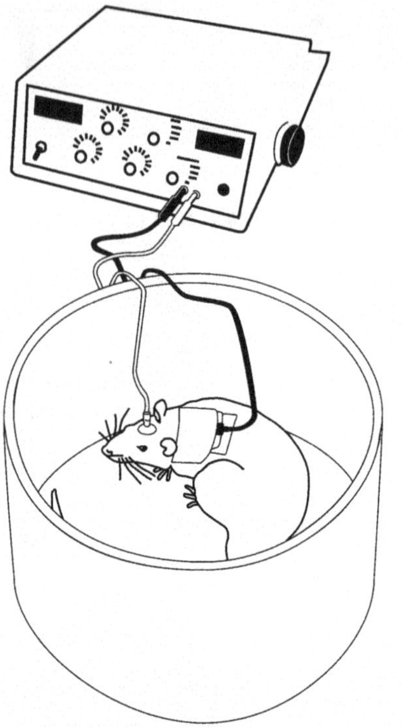

Fig. 8.6 Experimental set-up with transcranial direct current stimulation (figure from the supplementary material in Dockery et al. 2011). The epicranial electrode (target) was plugged into the cannula fixed over the frontal cortex, and the second electrode was strapped to the rat's back by a latex jacket. The constant current was supplied by a portable stimulator (model: CX 6650, Rolf Schneider Electronics, Gleichen, Germany)

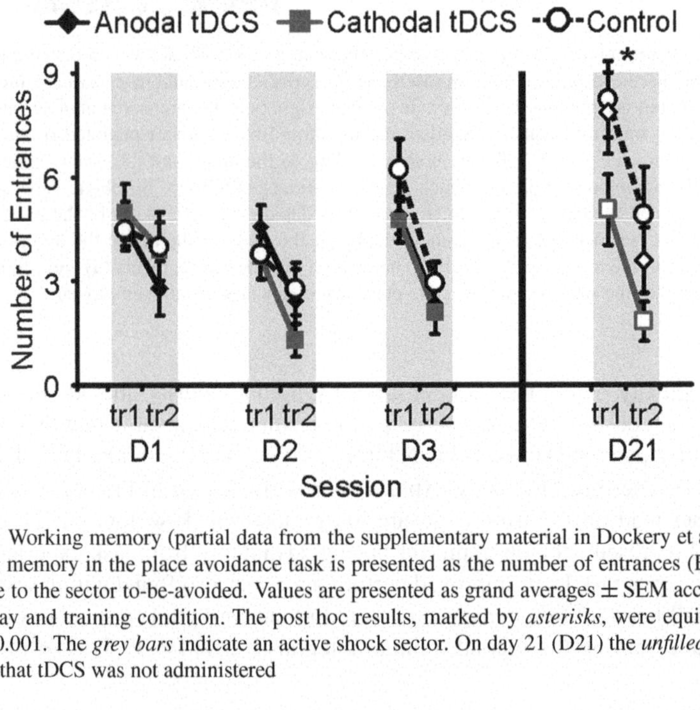

Fig. 8.7 Working memory (partial data from the supplementary material in Dockery et al. 2011). Working memory in the place avoidance task is presented as the number of entrances (ENTR) in reference to the sector to-be-avoided. Values are presented as grand averages ± SEM according to group, day and training condition. The post hoc results, marked by *asterisks*, were equivalent to: ***$P < 0.001$. The *grey bars* indicate an active shock sector. On day 21 (D21) the *unfilled symbols* indicate that tDCS was not administered

ha > t1 > t2 < ts = t1; $P < 0.01$). This means that even though rats must newly acquire the location of the shock sector for each day, their ability to avoid the sector (and not just escape), once they know where it is, improved with training for each session. This ability is referred to as cognitive skill learning.

There is an effect of day on skill learning in which performance is optimal on D2 ($F_{2,76} = 7.50$; $P < 0.001$; D1 > D2 < D3; $P < 0.002$). The long-term benefits of brain stimulation by tDCS on early learning were found on D21 (no tDCS) in rats after cathodal stimulation; the results indicate that these rats performed better with less entrances ($F_{2,37} = 3.61$; $P < 0.036$; $tDCS_a = tDCS_c <$ Contr, $P < 0.07$) and fewer shocks per entrances ($F_{2,38} = 4.67$; $P < 0.015$; $tDCS_a = tDCS_c$, $tDCS_c <$ Contr, $P < 0.02$) during the training intervals than control rats. This suggests that by pairing a highly cognitively demanding task with cathodal frontal tDCS (thought to decrease cortical excitability) during training/early acquisition, later performance (without stimulation) will show advantages. The appearance of latent effects even while no significant differences were found during the training is unusual. It is intriguing in light of the course of long-term plasticity changes with learning and, conversely, with regard to the delayed effects of insult manifesting in neurodegenerative disorders.

Our results indicate that complex cognitive functions, which are frequently associated with pathology in various human diseases and disorders, are captured in our rodent paradigm and that these functions can be altered by tDCS with long-term benefits to performance. This supports the plausibility of neuroenhancement in healthy humans and rats. The cumulative effects of tDCS on visuospatial working memory and skill learning in rats suggest that, as in humans, they are phase-dependent (requiring time and experience) and polarity-specific. Further, since cathodal tDCS conferred benefits, especially under highly novel and highly demanding conditions, there may be a role for exogenously decreased frontal excitability, which may temper the high arousal associated with the task novelty, high load, brain stimulation itself or foot shock. Perhaps the nootropic potential of tDCS of the frontal cortex on spatial memory and learning operates via a kind of anxiolytic effect of inhibitory stimulation on the PFC activity. Based on current literature, this would be particularly advantageous during early acquisition/novelty and under high task demand (Salehi et al. 2010).

8.3 Discussion and Summary

Two major findings emerge from the results: (1) the phase of learning/memory and (2) the level of task load (at least in humans) are important determinants of the efficacy of tDCS effects on frontal cortex function. These findings indicate that the parameters of tDCS (current strength, duration of stimulation, polarity) alone do not determine how the current will alter function due to neuronal excitability changes in the (pre)frontal cortex in healthy young adults and rats. The results also suggest that the existing state of the dynamic system that is targeted must be determined in order to establish which direction of tDCS-induced changes constitutes an enhancement leading to optimization. Otherwise, the risk in driving the system in the wrong direction is plausible, which is likely due to homeostatic mechanisms. Here, the phase of learning was determined across sessions for humans (same test, 1 week inter-trial interval) and also within sessions for rats (novel task condition each day, 1 day inter-trial interval). In human planning performance, the phase- and polarity-dependent effects were most apparent for the more difficult problems.

The idea that the existing activity state of the PFC (novelty/stress vs. learned state; high arousal vs. low arousal) determines which direction of activity changes affect performance benefits, is in contrast to commonly held ideas that tDCS-induced excitability increase would statically lead to performance benefits, while decreases would lead to null or negative effects. These assumptions are based on the MiB model, that is, that more excitability would result in higher performance gains when the stimulated area is needed for the given task. The results reported here suggest that the MiB model does not always apply, at least not to brain stimulation of the (pre)frontal cortex and function in novel PFC-related tasks.

In rats, phases of learning and memory for a spatial learning task (i.e., acquisition, delay, and retrieval) have been associated with changes in the amount of extracellular dopamine in the mPFC (Phillips et al. 2004). The PFC may differentially modulate distinct phases of visuospatial learning (Rinaldi et al. 2007). The DLPFC is the most crucial site for dopaminergic effects on cognitive functions (Braver and Cohen 2000; Cools et al. 2002) that are associated with endogenous DA release (Aalto et al. 2005; Phillips et al. 2004). This seems appropriate since dopamine (DA) is a neuromodulator implicated in synaptic mechanisms mediating cognitive functions such as attention, learning, memory formation and reward behavior. The plasticity of corticostriatal circuitry and dopamine levels are differentially modulated during different learning phases, with the activity of DA neurons decreased after extensive training compared to the early stages of learning a novel action (Costa 2007). Dopamine release increases during acquisition of novel information (Goto and Grace 2005; Lemon and Manahan-Vaughan 2006), while, conversely, subsequent presentations of a novel stimulus lead to its down regulation in the PFC (Wilkinson et al. 1998).

In humans, working memory capacity increases with training, which yields plasticity of dopamine (D1) receptor densities and brain activity pattern changes (McNab et al. 2009; Klingberg 2010). Concerning the phase-dependent tDCS results, there is much experimental evidence to support the relevance of previous experience of a particular cortical region in constraining the subsequent response to tDCS in a homeostatic manner (Ridding and Ziemann 2010). Our findings stand in contrast to tDCS of the motor cortex, in which increased excitability by anodal tDCS enhances motor performance, while cathodal tDCS reduces improvement in skill acquisition (Vines et al. 2006); in our studies, the behavioral effects of the excitability changes seem to depend on the pre-existing state of the cortex in relation to previous experience and, possibly, arousal. State-dependent effects of tDCS on motor cortex excitability have been shown when pharmacological substances were introduced. For example, anodal tDCS was found to have reverse effects, inhibiting motor cortex excitability, with L-dopa administration (Kuo et al. 2008). This suggests that more excitability does not necessarily facilitate enhanced plasticity. Furthermore, dose-dependent impairment by a DA D2-like agonist on tDCS-induced motor cortex excitability changes was found, in which "impairment" referred to both blunted plasticity and to a reversal of excitability changes, such as inhibition induced by anodal tDCS at 0.125 mg or excitation by cathodal tDCS at 1.0 mg (Monte-Silva et al. 2009). Besides pharmacological modulation, seemingly paradoxical effects of tDCS on the occipital lobe were found on visual-evoked potentials and attributed to the duration of the polarization and the stimuli used (Accornero et al. 2007). These findings suggest that tDCS stimulation can yield different effects, which are not always enhancements that lead to improved function.

According to the Yerkes-Dodson law, performance improves with increases in arousal level; however, beyond a certain optimal medium point it has deleterious effects (Yerkes and Dodson 1908). This inverted U-shaped relationship exists between task performance and the beneficial effects of dopamine agonists on cognition (Kimberg et al. 2001). Seamans et al. (1998) showed that in rats, working memory depends on the maintenance of an optimal range of DA activity in the medial PFC and that there is phase specificity by selective disruption of behavior with DA receptor blockade. In a PET study, Parkinson's patients, with known DA disturbance, showed altered activity and predominant use of explicit memory strategies to acquire the cognitive skill underlying TOL planning performance at a lower rate of accuracy (Beauchamp et al. 2008). Single-photon emission computed tomography imaging of striatal dopaminergic deficits have been linked to both poor TOL performance and depressive mood (Rektorova et al. 2008). Furthermore, dopamine has been found to modulate task-related fronto-striatal activation and default mode network deactivation in TOL performance (Nagano-Saito et al. 2009).

Not only was learning phase found to be important, but in these studies, task difficulty was also a factor influencing the effects of brain stimulation on cognitive functions. Interestingly, both new situations and harder tasks tend to increase an individual's arousal levels, which relate to activity in the PFC. The level of difficulty of a task influences the connectivity of the brain areas involved in working memory (Rissman et al. 2008). The importance of task load (and associated brain structures) on the MiB model is not trivial, as the inverted U-shape function is thought to be representative for difficult, but not easy tasks since those follow a linear relationship of brain activity and performance (Salehi et al. 2010). This, then, could clearly influence the impact of any stimuli (or brain stimulation) that alter the excitability of the (pre)frontal cortex on cognition. These can include brain stimulation, stressors, mood or prior experience. The impact of these can be greater when task load is higher, as shown by the greater effects of stress on spatial learning and memory in rats when reversal learning was introduced (Salehi et al. 2010).

Apart from learning and memory (Baldi and Bucherelli 2005), there are other factors which modulate an organism's physiology according to an inverted-U shaped dose response curve. This means that both too much and too little are associated with poor performance and in common parlance this is referred to as homeostasis (Chrousos 2009). The body's response to stress can influence the

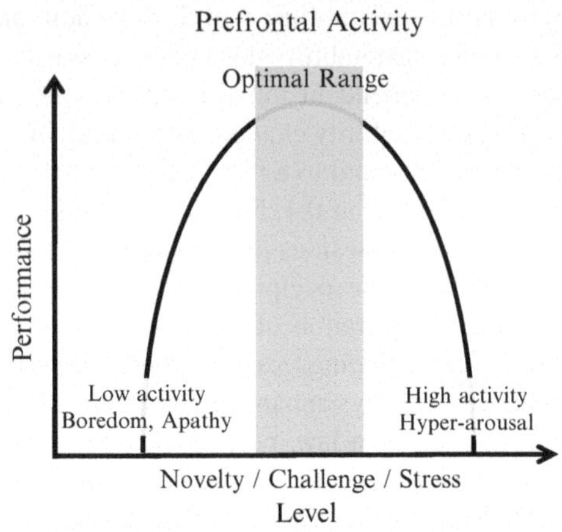

Fig. 8.8 The relationship between prefrontal activity levels and higher level cognitive function follows an 'inverted U-shape' in which only moderate activity results in optimal performance. Data suggest that the likelihood of tDCS of the prefrontal cortex to affect gains or detriments in cognitive performance is related to the existing activity level of the prefrontal cortex, which is known to relate to factors such as novelty, challenge and stress

activity of the prefrontal cortex. The study by Salehi et al. showed that in rats the inverted-U shaped relationship between stress and cognitive functions is most evident during early acquisition as opposed to in over-trained conditions (Salehi et al. 2010). This suggests that the inverted-U shaped relationship is phase-specific. According to Fig. 8.8, it is possible to see how knowing the state of a system, such as prefrontal circuitry, can help determine which changes (direction, amount) can drive it toward more optimal function.

Concerning tDCS effects, to date, no studies have been performed in order to more directly determine extrastriatal DA modulation following acute tDCS of the DLPFC with a specific focus on the prefrontal cortex. Another type of brain stimulation, repetitive transcranial magnetic stimulation (rTMS), showed endogenous dopamine changes (release) in the ipsilateral medial PFC, which were specific to left DLPFC stimulation; these changes were determined by positron-emission tomography (PET) (Cho and Strafella 2009). In the motor cortex, measurement of DA modulation by tDCS using PET has also not been performed. However, DA receptor activation by L-dopa showed nonlinear dosage effects on neuroplasticity for non-focal (Monte-Silva et al. 2010) and focal (Thirugnanasambandam et al. 2011) brain stimulation; the effects indicate the need for an optimal DA level for functional plasticity. Reversed or abolished effects of tDCS effects on excitability were found depending on the L-dopa dosages. This suggests that the MiB model does not apply to the interaction of the dopaminergic system and tDCS effects on neural activity. Experimental data rather support the need for homeostasis of that system. This could be more thoroughly tested in a rat model in which frontal cortex dopamine release could be monitored throughout different learning phases, arousal states and task difficulty levels, and in association with frontal tDCS.

Dopamine has a neuroplasticity-modifying influence on tDCS effects (Kuo et al. 2008), which indicates that while the effects require a physiological concentration of DA, a reversal of effects on excitability can occur depending on the DA level. More is not simply always better. Extremely high or low DA concentrations, for example, due to periods of stress, can alter the DLPFC network balance to incommensurate inhibitory interneuron activation (Kroner et al. 2007). In the PFC, both deficient and exorbitant levels of DA receptor stimulation, expressing an inverted U-shaped function, impair working memory (Cools et al. 2008). Brain stimulation-induced changes in excitability or learning processes themselves can cause inverse or preventative effects on proceeding manipulations of neuronal excitability and synaptic plasticity (Lang et al. 2004; Siebner et al. 2004; Ziemann et al. 2004; Stefan et al. 2006). This model coincides with homeostatic plasticity, whereby low background activity (e.g. pre-treatment with cathodal tDCS in an earlier session) would enhance the associative plasticity related to learning (Nitsche et al. 2007), whereas high excitability (e.g. anodal tDCS) would inhibit it. Likewise, the strength of this homeostatic effect would diminish as learning reached asymptotic levels. This representation of homeostatic plasticity stands in contrast to the MiB model.

In the described human study, for participants who were naïve to tDCS and the TOL test, cathodal tDCS of the DLPFC facilitated acquisition of executive functions as decreased excitability by exogenous stimulation paired with the endogenously increased activation due to a novel task lead to benefits (Dockery et al. 2009). This may be most important during initial learning because less experience allows for a greater number of possible paths to reach the goal and, therefore, a higher likelihood for error. It is possible that cathodal tDCS mediates its early beneficial effect through noise reduction of neuronal activity by metaplastic regulatory dopaminergic activity. Anodal tDCS may confer benefits by exogenously increasing DLPFC activity in the later training phase when basal dopamine (DA) levels have receded, which causes enhanced efficacy of active receptors. An exogenously-induced excitability increase in healthy, naïve participants may not be beneficial during initial exposure to the novel task because a kind of "overstimulation" could lead to excitotoxicity. Though not determined for the PFC in a single session, anodal tDCS of the motor cortex was found in association with decreased GABA while cathodal tDCS was associated with both decreased GABA and glutamate (Stagg et al. 2009).

tDCS provides a tool to induce lasting improvements in cognitive function (e.g. frontal lobe pathologies) and skill acquisition (e.g. learning disabilities) by optimizing neural activity and thereby strengthening the connections associated with compromised prefrontal-hippocampal circuitry or sub-optimal activity levels. The results and theories presented here are not in favor of a general, More is Better approach to brain stimulation of prefrontal-based functions. Translational research aims to elucidate the nature of executive functions by using animal models to investigate generally how they might be enhanced and later apply the findings to humans. With these models it is possible to directly test for an inverted-U-shaped dose response curve between PFC activity changes and dependent cognitive functions. Future work should aim to increase knowledge about the mode of action of beneficial tDCS effects on cognitive tasks by establishing the neurobiological basis of tDCS-induced changes.

8.4 Conclusions

The aim of neural enhancement is to improve cognitive function or mood. To ensure that enhancement equates to improvement, valid tests are needed and must be employed. Otherwise, the efficacy of the NE is unclear, as are the potential safety hazards. With only the research examples given, it is clear that altering brain activity in order to reach performance gains is not trivial or straight-forward. When gains are achieved, the underlying mechanisms are not necessarily easy to estimate or understand. While idealism drives technological advancement, in the case of brain stimulation, it must be moderated by objectivity via empirical evidence from scientific inquiry. Though the studies discussed here did not directly test the physiological basis for the functional changes due to stimulation of the (pre)frontal cortex, they laid the groundwork to directly test such theories. The results reported suggest that the MiB model is not always the most fitting to achieve an optimal performance level, and rather moderate activity is needed to support these high level cognitive functions governed by the PFC. It is clear that rather than make assumptions, direct testing, which requires time and resources, is necessary in order to gain greater understanding.

One method currently engaged with the aim to test the fitness of a society's individuals, is the use of standardized assessment. Such types of assessment aim to measure the capacities of students, in terms of abilities and skills, by a constructed basis of comparison for students of different backgrounds. The Programme for International Student Assessment (PISA), organized by the Organisation for Economic Co-operation and Development, presents PISA test results, which are meant to reflect students' knowledge and capabilities (OECD 2010). The test has been developed to measure the extent to which education systems prepare students for life. This then can help policy makers make informed decisions about how best to "enable citizens to take advantage of a globalised world economy" (OECD 2010). With the PISA results, much speculation has been observable in the media about a country's educational system falling behind, with the implied notion that high scores on standardized tests correlate to students' employability. The notion of MIB is reflected by the Secretary General of the OECD, with the statement, "stay ambitious; work harder to reach your full potential, no matter how you come out in the picture" (Gurría 2010).

If working harder means studying longer and more intensely, the question is whether it helps to enable citizens, in this case, by promoting effective learning and memory in students. Is it necessary to determine the current state of the students in order to determine which conditions will produce the best results? If so, this is an alternative to focusing primarily on the quantity of work load. Furthermore, is it important to verify whether a higher load translates to better learning or better eventual employability? If we empirically know that more work hours or a higher load do not necessarily lead to more desirable results, then what should change about the expectations and also the structures of reward? It is also constructive to determine which institutions are suited to make contributions and assert structural changes to support reform.

Responsibility for development in any evidence-based reform movement does not lie solely on the government or social institutions. Individuals and small collectives also make choices that determine behavioral outcomes. In the research studies addressed here and for other NE methodologies, a question concerning the role of autonomy remains. It is unclear whether NE is even possible without effort. It is plausible that engagement may be a criterion for NE related to improved plasticity in the executive functions, learning and memory. It is interesting that task difficulty plays a role in the relevance of induced changes in PFC-dependent functions since task load influences the connectivity of the brain areas involved (Rissman et al. 2008). Working memory is associated with fluid intelligence through common neural circuitry, and processes related to attentional control, the ability to manipulate abstract relations and maintain possible paths to reach goals (Jaeggi et al. 2008). Previously, it was thought that intelligence was a fixed trait. However, training of working memory, possibly through constant engagement of EF components, has been found to transfer to fluid intelligence (Jaeggi et al. 2008). This training itself requires attention and engagement. If the benefits on working memory are viable only through active engagement of particular functions paired with some form of NE, then NE may simply boost natural mechanisms. This is very relevant considering the importance of brain plasticity for learning and memory. On the other hand, if the mechanisms are in place, why are "healthy" participants not already optimized? If one knows that an optimal state for EF, learning and memory exists, then is it possible to train oneself to recognize and maintain such a state?

Considering the influence of government, institutions and media on work ethic, the question is whether this More is Better model actually results in attainment of goals. Does longer duration and higher intensity in work performance result in better employment rates, more resources and better living standards? Does performance depend on the type of work? This chapter addresses some consequences of applying this model to a complex system such as the structure-function relationship between the prefrontal cortex and executive function. Empirical data suggest that the MiB model is inappropriate for higher level cognitive function. For human evolution it is important to recognize that the MiB model is supported by rhetoric and policy that in the short term drive norms towards skills being upgraded that do not necessarily match the demand, and in the long term lead to an effete labor force. More generally, applying the wrong model to such a homeostatic system as high level cognitive function drives individuals, particularly those most vulnerable, towards possibly irreversible structural damage and pathological function. This then changes the demand for skills. Instead of promoting a model that may not be the best fit to achieve aims of higher cognitive function, government bodies and social institutions can constructively influence economic outcomes through incentives and disincentives that support changes. These reforms would take into account the current state of individuals in society and their workload in order to garner optimal changes.

The popularity of neuroenhancement in the media, science and medicine should not be discounted as it indicates affinity to the concept. Increased attention to the topic is also of concern because it may correlate with a rise in prevalence rates since people can be influenced by the presentation of reasons to change their behavior despite lack of supporting evidence (Larriviere and Williams 2010). The popularity may indicate a perceived need for the (presumed) advance that neural enhancement may provide. In social systems, due to natural selection, with increased demand on higher cognitive functions, the capacity for such functions would be favored and therefore behavior would be under pressure to follow suit. Both the lay public and the professionals involved in the topic are subject to pressures of increased performance demand. As the renowned developmental psychologist Piaget pointed out, we are unable to recognize the stage of development that we are currently in. While a biological system such as the human brain has natural breaks (e.g. GABAergic system), it seems humans are poor at estimating the impact of when too much is too much, when more in not necessarily better. It appears that despite our evolving (pre)frontal cortex, we are not always making decisions that lead to improvement by employing enhancement.

Despite the enthusiasm for an "enhancement society", as the current Chancellor of Germany once said, "In the long term, 'progress' works against us if it continues to be detrimental to nature" (Merkel 1998). This, of course, applies not only to the earth upon which we live but also to the body and mind that make us who we are.

Acknowledgement I thank Andrzej Wichrowski, Linda van der Heiden and Malgorzata Wesierska for critical reading of this manuscript. Thanks to Petr Cempirek for the illustrations in Figs. 8.2 and 8.6.

References

Aalto S, Bruck A, Laine M, Nagren K, Rinne JO (2005) Frontal and temporal dopamine release during working memory and attention tasks in healthy humans: a positron emission tomography study using the high-affinity dopamine D2 receptor ligand [11C]FLB 457. J Neurosci 25: 2471–2477

Accornero N, Li Voti P, La Riccia M, Gregori B (2007) Visual evoked potentials modulation during direct current cortical polarization. Exp Brain Res 178:261–266

Baker SC, Rogers RD, Owen AM, Frith CD, Dolan RJ, Frackowiak RS, Robbins TW (1996) Neural systems engaged by planning: a PET study of the Tower of London task. Neuropsychologia 34:515–526

Baldi E, Bucherelli C (2005) The inverted "u-shaped" dose-effect relationships in learning and memory: modulation of arousal and consolidation. Nonlinear Biol Toxicol Med 3:9–21

Beauchamp MH, Dagher A, Panisset M, Doyon J (2008) Neural substrates of cognitive skill learning in Parkinson's disease. Brain Cogn 68:134–143

Braver TS, Cohen JD (2000) On the control of control: the role of dopamine in regulating prefrontal function and working memory. In: Monsell S, Driver J (eds) Attention and performance XVIII; control of cognitive processes. MIT Press, Cambridge, pp 713–737

Brenes GA (2003) Cognitive training may improve targeted cognitive functions in older adults. Evid Based Ment Health 6:54

Bubenikova-Valesova V, Stuchlik A, Svoboda J, Bures J, Vales K (2008) Risperidone and ritanserin but not haloperidol block effect of dizocilpine on the active allothetic place avoidance task. Proc Natl Acad Sci USA 105:1061–1066

Cerella J, Onyper SV, Hoyer WJ (2006) The associative-memory basis of cognitive skill learning: adult age differences. Psychol Aging 21:483–498

Chiesa A, Calati R, Serretti A (2011) Does mindfulness training improve cognitive abilities? A systematic review of neuropsychological findings. Clin Psychol Rev 31:449–464

Cho SS, Strafella AP (2009) rTMS of the left dorsolateral prefrontal cortex modulates dopamine release in the ipsilateral anterior cingulate cortex and orbitofrontal cortex. PLoS One 4:e6725

Chrousos GP (2009) Stress and disorders of the stress system. Nat Rev Endocrinol 5:374–381

Cimadevilla JM, Wesierska M, Fenton AA, Bures J (2001) Inactivating one hippocampus impairs avoidance of a stable room-defined place during dissociation of arena cues from room cues by rotation of the arena. Proc Natl Acad Sci USA 98:3531–3536

Clark VP, Coffman BA, Mayer AR, Weisend MP, Lane TD, Calhoun VD, Raybourn EM, Garcia CM, Wassermann EM (2010) TDCS guided using fMRI significantly accelerates learning to identify concealed objects. Neuroimage 59(1):117–128

Cools R, Stefanova E, Barker RA, Robbins TW, Owen AM (2002) Dopaminergic modulation of high-level cognition in Parkinson's disease: the role of the prefrontal cortex revealed by PET. Brain 125:584–594

Cools R, Gibbs SE, Miyakawa A, Jagust W, D'Esposito M (2008) Working memory capacity predicts dopamine synthesis capacity in the human striatum. J Neurosci 28:1208–1212

Costa RM (2007) Plastic corticostriatal circuits for action learning: what's dopamine got to do with it? Ann N Y Acad Sci 1104:172–191

Dockery CA, Wesierska MJ (2010) A spatial paradigm, the allothetic place avoidance alternation task, for testing visuospatial working memory and skill learning in rats. J Neurosci Methods 191:215–221

Dockery CA, Hueckel-Weng R, Birbaumer N, Plewnia C (2009) Enhancement of planning ability by transcranial direct current stimulation. J Neurosci 29:7271–7277

Dockery CA, Liebetanz D, Birbaumer N, Malinowska M, Wesierska MJ (2011) Cumulative benefits of frontal transcranial direct current stimulation on visuospatial working memory training and skill learning in rats. Neurobiol Learn Mem 96:452–460

EEA (2010) Intensified global competition for resources. European Environment Agency, Copenhagen, pp 1–6

Elliott R, Baker SC, Rogers RD, O'Leary DA, Paykel ES, Frith CD, Dolan RJ, Sahakian BJ (1997) Prefrontal dysfunction in depressed patients performing a complex planning task: a study using positron emission tomography. Psychol Med 27:931–942

Farah MJ, Illes J, Cook-Deegan R, Gardner H, Kandel E, King P, Parens E, Sahakian B, Wolpe PR (2004) Neurocognitive enhancement: what can we do and what should we do? Nat Rev Neurosci 5:421–425

Flower K, Li L, Chen CY, Baggott MJ, Galloway GP, Mendelson J (2010) Efficacy, safety, and ethics of cosmetic neurology far from settled. Clin Pharmacol Ther 88:461–463

Franke AG, Bonertz C, Christmann M, Huss M, Fellgiebel A, Hildt E, Lieb K (2011) Non-medical use of prescription stimulants and illicit use of stimulants for cognitive enhancement in pupils and students in Germany. Pharmacopsychiatry 44:60–66

Fu X, Giavalisco P, Liu X, Catchpole G, Fu N, Ning ZB, Guo S, Yan Z, Somel M, Pääbo S, Zeng R, Willmitzer L, Khaitovich P (2011) Rapid metabolic evolution in human prefrontal cortex. Proc Natl Acad Sci USA 108:6181–6186

Galert T, Bublitz JC, Heuser I, Merkel R, Repantis D, Schöne-Seifert B, Talbot D (2009) Das optimierte Gehirn. Ein Memorandum zu Chancen und Risiken des Neuroenhancements. Gehirn Geist 11:40–48

Goethals I, Audenaert K, Jacobs F, Van de Wiele C, Ham H, Pyck H, Vandierendonck A, Van Heeringen C, Dierckx R (2005) Blunted prefrontal perfusion in depressed patients performing the Tower of London task. Psychiatry Res 139:31–40

Goto Y, Grace AA (2005) Dopaminergic modulation of limbic and cortical drive of nucleus accumbens in goal-directed behavior. Nat Neurosci 8:805–812

Grafman J, Litvan I, Massaquoi S, Stewart M, Sirigu A, Hallett M (1992) Cognitive planning deficit in patients with cerebellar atrophy. Neurology 42:1493–1496

Greely H, Sahakian B, Harris J, Kessler RC, Gazzaniga M, Campbell P, Farah MJ (2008) Towards responsible use of cognitive-enhancing drugs by the healthy. Nature 456:702–705
Gurría A (2010) Presentation of the PISA 2010 results. Organisation for Economic Co-operation and Development, Paris
Jacobs R, Harvey AS, Anderson V (2007) Executive function following focal frontal lobe lesions: impact of timing of lesion on outcome. Cortex 43:792–805
Jaeggi SM, Buschkuehl M, Jonides J, Perrig WJ (2008) Improving fluid intelligence with training on working memory. Proc Natl Acad Sci USA 105:6829–6833
Jurado MB, Rosselli M (2007) The elusive nature of executive functions: a review of our current understanding. Neuropsychol Rev 17:213–233
Kimberg DY, Aguirre GK, Lease J, D'Esposito M (2001) Cortical effects of bromocriptine, a D-2 dopamine receptor agonist, in human subjects, revealed by fMRI. Hum Brain Mapp 12: 246–257
Klingberg T (2010) Training and plasticity of working memory. Trends Cogn Sci 14:317–324
Kroger JK, Sabb FW, Fales CL, Bookheimer SY, Cohen MS, Holyoak KJ (2002) Recruitment of anterior dorsolateral prefrontal cortex in human reasoning: a parametric study of relational complexity. Cereb Cortex 12:477–485
Kroner S, Krimer LS, Lewis DA, Barrionuevo G (2007) Dopamine increases inhibition in the monkey dorsolateral prefrontal cortex through cell type-specific modulation of interneurons. Cereb Cortex 17:1020–1032
Kuo MF, Paulus W, Nitsche MA (2008) Boosting focally-induced brain plasticity by dopamine. Cereb Cortex 18:648–651
La Rue A (2010) Healthy brain aging: role of cognitive reserve, cognitive stimulation, and cognitive exercises. Clin Geriatr Med 26:99–111
Lambourne K, Tomporowski P (2010) The effect of exercise-induced arousal on cognitive task performance: a meta-regression analysis. Brain Res 1341:12–24
Lang N, Siebner HR, Ernst D, Nitsche MA, Paulus W, Lemon RN, Rothwell JC (2004) Preconditioning with transcranial direct current stimulation sensitizes the motor cortex to rapid-rate transcranial magnetic stimulation and controls the direction of after-effects. Biol Psychiatry 56:634–639
Larriviere D, Williams MA (2010) Neuroenhancement: wisdom of the masses or "false phronesis"? Clin Pharmacol Ther 88:459–461
Larriviere D, Williams MA, Rizzo M, Bonnie RJ (2009) Responding to requests from adult patients for neuroenhancements: guidance of the Ethics, Law and Humanities Committee. Neurology 73:1406–1412
Lazeron RH, Rombouts SA, Scheltens P, Polman CH, Barkhof F (2004) An fMRI study of planning-related brain activity in patients with moderately advanced multiple sclerosis. Mult Scler 10:549–555
Leh SE, Petrides M, Strafella AP (2010) The neural circuitry of executive functions in healthy subjects and Parkinson's disease. Neuropsychopharmacology 35:70–85
Lemon N, Manahan-Vaughan D (2006) Dopamine D1/D5 receptors gate the acquisition of novel information through hippocampal long-term potentiation and long-term depression. J Neurosci 26:7723–7729
Liebetanz D, Fregni F, Monte-Silva KK, Oliveira MB, Amancio-dos-Santos A, Nitsche MA, Guedes RC (2006a) After-effects of transcranial direct current stimulation (tDCS) on cortical spreading depression. Neurosci Lett 398:85–90
Liebetanz D, Klinker F, Hering D, Koch R, Nitsche MA, Potschka H, Loscher W, Paulus W, Tergau F (2006b) Anticonvulsant effects of transcranial direct-current stimulation (tDCS) in the rat cortical ramp model of focal epilepsy. Epilepsia 47:1216–1224
Liebetanz D, Koch R, Mayenfels S, Konig F, Paulus W, Nitsche MA (2009) Safety limits of cathodal transcranial direct current stimulation in rats. Clin Neurophysiol 120:1161–1167
Martinez-Aran A, Penades R, Vieta E, Colom F, Reinares M, Benabarre A, Salamero M, Gasto C (2002) Executive function in patients with remitted bipolar disorder and schizophrenia and its relationship with functional outcome. Psychother Psychosom 71:39–46

McNab F, Varrone A, Farde L, Jucaite A, Bystritsky P, Forssberg H, Klingberg T (2009) Changes in cortical dopamine D1 receptor binding associated with cognitive training. Science 323: 800–802

Merkel A (1998) The role of science in sustainable development. Science 281:336–337

Merkel A (2007) Opening address at the World Economic Forum, Report. Presidency of the European Union, Davos-Klosters

Monte-Silva K, Kuo MF, Thirugnanasambandam N, Liebetanz D, Paulus W, Nitsche MA (2009) Dose-dependent inverted U-shaped effect of dopamine (D2-like) receptor activation on focal and nonfocal plasticity in humans. J Neurosci 29:6124–6131

Monte-Silva K, Liebetanz D, Grundey J, Paulus W, Nitsche MA (2010) Dosage-dependent non-linear effect of L-dopa on human motor cortex plasticity. J Physiol 588:3415–3424

Nagano-Saito A, Liu J, Doyon J, Dagher A (2009) Dopamine modulates default mode network deactivation in elderly individuals during the Tower of London task. Neurosci Lett 458:1–5

Newman SD, Carpenter PA, Varma S, Just MA (2003) Frontal and parietal participation in problem solving in the Tower of London: fMRI and computational modeling of planning and high-level perception. Neuropsychologia 41:1668–1682

Nitsche MA, Fricke K, Henschke U, Schlitterlau A, Liebetanz D, Lang N, Henning S, Tergau F, Paulus W (2003) Pharmacological modulation of cortical excitability shifts induced by transcranial direct current stimulation in humans. J Physiol 553:293–301

Nitsche MA, Doemkes S, Karakose T, Antal A, Liebetanz D, Lang N, Tergau F, Paulus W (2007) Shaping the effects of transcranial direct current stimulation of the human motor cortex. J Neurophysiol 97:3109–3117

Nitsche MA, Boggio PS, Fregni F, Pascual-Leone A (2009) Treatment of depression with transcranial direct current stimulation (tDCS): a review. Exp Neurol 219:14–19

Normann C, Boldt J, Maio G, Berger M (2010) Options, limits and ethics of pharmacological neuroenhancement. Nervenarzt 81:66–74

OECD (2010) PISA 2009 results: what students know and can do – student performance in reading, mathematics and science, vol I. OECD, Paris

Olesen PJ, Westerberg H, Klingberg T (2004) Increased prefrontal and parietal activity after training of working memory. Nat Neurosci 7:75–79

Ottowitz WE, Dougherty DD, Savage CR (2002) The neural network basis for abnormalities of attention and executive function in major depressive disorder: implications for application of the medical disease model to psychiatric disorders. Harv Rev Psychiatry 10:86–99

Ouellet MC, Beauchamp MH, Owen AM, Doyon J (2004) Acquiring a cognitive skill with a new repeating version of the Tower of London task. Can J Exp Psychol 58:272–288

Owen AM (1997) Cognitive planning in humans: neuropsychological, neuroanatomical and neuropharmacological perspectives. Prog Neurobiol 53:431–450

Owen AM, Downes JJ, Sahakian BJ, Polkey CE, Robbins TW (1990) Planning and spatial working memory following frontal lobe lesions in man. Neuropsychologia 28:1021–1034

Paffenbarger RS Jr, Hyde RT, Wing AL, Hsieh CC (1986) Physical activity, all-cause mortality, and longevity of college alumni. N Engl J Med 314:605–613

Penner MR, Roth TL, Barnes CA, Sweatt JD (2010) An epigenetic hypothesis of aging-related cognitive dysfunction. Front Aging Neurosci 2:9

Pereira AC, Huddleston DE, Brickman AM, Sosunov AA, Hen R, McKhann GM, Sloan R, Gage FH, Brown TR, Small SA (2007) An in vivo correlate of exercise-induced neurogenesis in the adult dentate gyrus. Proc Natl Acad Sci USA 104:5638–5643

Phillips AG, Ahn S, Floresco SB (2004) Magnitude of dopamine release in medial prefrontal cortex predicts accuracy of memory on a delayed response task. J Neurosci 24:547–553

Quednow BB (2010) Ethics of neuroenhancement: a phantom debate. BioSocieties 5:153–156

Rainville C, Amieva H, Lafont S, Dartigues JF, Orgogozo JM, Fabrigoule C (2002) Executive function deficits in patients with dementia of the Alzheimer's type: a study with a Tower of London task. Arch Clin Neuropsychol 17:513–530

Rasser PE, Johnston P, Lagopoulos J, Ward PB, Schall U, Thienel R, Bender S, Toga AW, Thompson PM (2005) Functional MRI BOLD response to Tower of London performance of first-episode schizophrenia patients using cortical pattern matching. Neuroimage 26:941–951
Rektorova I, Srovnalova H, Kubikova R, Prasek J (2008) Striatal dopamine transporter imaging correlates with depressive symptoms and tower of London task performance in Parkinson's disease. Mov Disord 23:1580–1587
Repantis D, Laisney O, Heuser I (2010a) Acetylcholinesterase inhibitors and memantine for neuroenhancement in healthy individuals: a systematic review. Pharmacol Res 61:473–481
Repantis D, Schlattmann P, Laisney O, Heuser I (2010b) Modafinil and methylphenidate for neuroenhancement in healthy individuals: a systematic review. Pharmacol Res 62:187–206
Ridding MC, Ziemann U (2010) Determinants of the induction of cortical plasticity by non-invasive brain stimulation in healthy subjects. J Physiol 588:2291–2304
Rinaldi A, Mandillo S, Oliverio A, Mele A (2007) D1 and D2 receptor antagonist injections in the prefrontal cortex selectively impair spatial learning in mice. Neuropsychopharmacology 32:309–319
Rissman J, Gazzaley A, D'Esposito M (2008) Dynamic adjustments in prefrontal, hippocampal, and inferior temporal interactions with increasing visual working memory load. Cereb Cortex 18:1618–1629
Salehi B, Cordero MI, Sandi C (2010) Learning under stress: the inverted-U-shape function revisited. Learn Mem 17:522–530
Schlaug G, Renga V (2008) Transcranial direct current stimulation: a noninvasive tool to facilitate stroke recovery. Expert Rev Med Devices 5:759–768
Seamans JK, Floresco SB, Phillips AG (1998) D1 receptor modulation of hippocampal-prefrontal cortical circuits integrating spatial memory with executive functions in the rat. J Neurosci 18:1613–1621
Shallice T (1982) Specific impairments of planning. Philos Trans R Soc Lond B Biol Sci 298: 199–209
Shallice T, Burgess PW (1991) Deficits in strategy application following frontal lobe damage in man. Brain 114(Pt 2):727–741
Siebner HR, Lang N, Rizzo V, Nitsche MA, Paulus W, Lemon RN, Rothwell JC (2004) Preconditioning of low-frequency repetitive transcranial magnetic stimulation with transcranial direct current stimulation: evidence for homeostatic plasticity in the human motor cortex. J Neurosci 24:3379–3385
Siebner HR, Hartwigsen G, Kassuba T, Rothwell JC (2009) How does transcranial magnetic stimulation modify neuronal activity in the brain? Implications for studies of cognition. Cortex 45:1035–1042
Stagg CJ, Nitsche MA (2011) Physiological basis of transcranial direct current stimulation. Neuroscientist 17:37–53
Stagg CJ, Best JG, Stephenson MC, O'Shea J, Wylezinska M, Kincses ZT, Morris PG, Matthews PM, Johansen-Berg H (2009) Polarity-sensitive modulation of cortical neurotransmitters by transcranial stimulation. J Neurosci 29:5202–5206
Stefan K, Wycislo M, Gentner R, Schramm A, Naumann M, Reiners K, Classen J (2006) Temporary occlusion of associative motor cortical plasticity by prior dynamic motor training. Cereb Cortex 16:376–385
Sussman S, Pentz MA, Spruijt-Metz D, Miller T (2006) Misuse of "study drugs": prevalence, consequences, and implications for policy. Subst Abuse Treat Prev Policy 1:15
Takano Y, Yokawa T, Masuda A, Niimi J, Tanaka S, Hironaka N (2011) A rat model for measuring the effectiveness of transcranial direct current stimulation using fMRI. Neurosci Lett 491: 40–43
Thirugnanasambandam N, Grundey J, Paulus W, Nitsche MA (2011) Dose-dependent nonlinear effect of L-DOPA on paired associative stimulation-induced neuroplasticity in humans. J Neurosci 31:5294–5299
Tyson PJ, Laws KR, Roberts KH, Mortimer AM (2004) Stability of set-shifting and planning abilities in patients with schizophrenia. Psychiatry Res 129:229–239

Ungvari GS, Xiang YT, Tang WK, Shum D (2008) Prospective memory and its correlates and predictors in schizophrenia: an extension of previous findings. Arch Clin Neuropsychol 23:613–622

Unterrainer JM, Owen AM (2006) Planning and problem solving: from neuropsychology to functional neuroimaging. J Physiol Paris 99:308–317

Utz KS, Dimova V, Oppenlander K, Kerkhoff G (2010) Electrified minds: transcranial direct current stimulation (tDCS) and galvanic vestibular stimulation (GVS) as methods of non-invasive brain stimulation in neuropsychology – a review of current data and future implications. Neuropsychologia 48:2789–2810

van den Heuvel OA, Veltman DJ, Groenewegen HJ, Cath DC, van Balkom AJ, van Hartskamp J, Barkhof F, van Dyck R (2005) Frontal-striatal dysfunction during planning in obsessive-compulsive disorder. Arch Gen Psychiatry 62:301–309

Vines BW, Nair DG, Schlaug G (2006) Contralateral and ipsilateral motor effects after transcranial direct current stimulation. Neuroreport 17:671–674

Wilens TE, Adler LA, Adams J, Sgambati S, Rotrosen J, Sawtelle R, Utzinger L, Fusillo S (2008) Misuse and diversion of stimulants prescribed for ADHD: a systematic review of the literature. J Am Acad Child Adolesc Psychiatry 47:21–31

Wilkinson LS, Humby T, Killcross AS, Torres EM, Everitt BJ, Robbins TW (1998) Dissociations in dopamine release in medial prefrontal cortex and ventral striatum during the acquisition and extinction of classical aversive conditioning in the rat. Eur J Neurosci 10:1019–1026

Yanagisawa H, Dan I, Tsuzuki D, Kato M, Okamoto M, Kyutoku Y, Soya H (2010) Acute moderate exercise elicits increased dorsolateral prefrontal activation and improves cognitive performance with Stroop test. Neuroimage 50:1702–1710

Yerkes RM, Dodson JD (1908) The relation of strength of stimulus to rapidity of habit-formation. J Comp Neurol Psychol 18:459–482

Ziemann U, Ilic TV, Pauli C, Meintzschel F, Ruge D (2004) Learning modifies subsequent induction of long-term potentiation-like and long-term depression-like plasticity in human motor cortex. J Neurosci 24:1666–1672

Zimerman M, Hummel FC (2010) Non-invasive brain stimulation: enhancing motor and cognitive functions in healthy old subjects. Front Aging Neurosci 2:149

Part II
Philosophical and Ethical Aspects of Cognitive Enhancement

Chapter 9
Better Brains or Bitter Brains? The Ethics of Neuroenhancement

Kirsten Brukamp

Abstract The topic of enhancement emerges as a novel, contemporary problem in medical ethics. In particular, neuroenhancement reveals itself as a challenging subject due to the advancements of neuroscience. At present, pharmacological neuroenhancement is widely debated, but only scarce empirical data exist regarding its prevalence. Arguments for and against neuroenhancement relate to various disciplines, such as medicine, anthropology, sociology, and classical ethics. Medical considerations caution against the use of pharmaceutical neuroenhancers because of medical risks, the lack of evidence-based medicine, and financial challenges to health care systems. Perspectives on neuroenhancement from the humanities involve the concepts of human nature, virtue ethics, liberty, and justice. The purposes behind neuroenhancement are disputable with regard to their social value. In conclusion, neuroenhancement appears to remain a controversial phenomenon.

Keywords Neuroenhancement • Medical ethics • Medical risks • Evidence-based medicine • Liberty • Justice

9.1 Overview

This short overview summarizes the contents of the paper and provides the section titles in parentheses: A brief outline sketches potential "Approaches in Medical Ethics" that help to address the moral problems of pharmacological neuroenhancement. A pragmatic method is chosen in order to weigh arguments from various standpoints. The introduction to "Definitions and Classifications" clarifies the terminology related to neuroenhancement.

K. Brukamp (✉)
Institute for History, Theory, and Ethics of Medicine, RWTH Aachen University, University Hospital Aachen, Aachen, Germany
e-mail: kbrukamp@ukaachen.de

E. Hildt and A.G. Franke (eds.), *Cognitive Enhancement: An Interdisciplinary Perspective*, Trends in Augmentation of Human Performance 1,
DOI 10.1007/978-94-007-6253-4_9,

The following subdivisions provide pro and con arguments from different perspectives and disciplines, such as medicine, anthropology, sociology, and classical ethics. With regard to the medical field, the section "Medical Risks" argues against the spread of pharmacological neuroenhancement because of medical complications, while the section "The Lack of Evidence-Based Medicine" highlights the lack of evidence for the claim that neuroenhancement is beneficial. The subsequent section, "Challenges to Health Care Systems," discusses the potential economic difficulties that desires for enhancement may bring about. Considering the humanities, the section "Human Nature and Virtue Ethics" presents concepts that critics of neuroenhancement frequently refer to in their arguments. The section on "Debatable Purposes" sketches the differential purposes of neuroenhancing *versus* recreational drug use, differences that are oftentimes not explicitly addressed in the ethical debate. The section "Liberty and Justice" discusses fundamental concepts in practical philosophy and points out how their application to pharmacological neuroenhancement yields ambivalent results. Finally, the "Conclusions" recapitulate the considerations, which result in an overall rather skeptical assessment of pharmacological neuroenhancement.

9.2 Approaches in Medical Ethics

Neuroenhancement increasingly proves itself as a relevant topic in society, and it carries important financial implications with it. Neuroenhancement is a problem for the discipline of medical ethics, and medical ethics is expected to fulfill the following roles, among others:

1. to *recognize* topics of discussion in medical ethics that possess high relevance for society as a whole and to *inform* the public about them,
2. to *structure and summarize arguments* in order to improve the level of deliberations and make them more efficient, and
3. to provide a weighted *conclusion* or to give practical *advice*, when they are sought after, based on the knowledge gained in the previous steps.

Approaches on how to tackle problems in medical ethics differ according to the theories, values, and principles that are applied. One possibility is to begin with an ethical metatheory, such as virtue ethics, deontology, or consequentialism (Beauchamp and Childress 2009), and to develop a consistent system out of this meta-theory to solve existing problems. However, this method has neither proven practical nor efficient in the discussion of clinical case studies. As a result, the approach of principlism has become a strong and well-known current in modern medical ethics. According to principlism, a few predominant and far-reaching principles appear to serve as excellent means for balancing the interests of patients, namely the principles of autonomy, beneficence, non-maleficence, and justice (Beauchamp and Childress 2009).

In bioethics, as elsewhere, people tend to dwell on the apparent dichotomy between conservative and liberal positions. These respective positions may be summarized succinctly and colloquially as "It is best that people stay as they were" *versus* "Let the people decide for themselves." Nevertheless, opinions in the neuroenhancement debate, as with other bioethical topics, are not necessarily divided according to the highly superficial stereotypes of these dichotomous positions (Parens 2005).

Here, the focus will be on discourse and pragmatism (Racine 2010) in order to reach actual answers to practical questions. Some specific ones arising from the topic ask: Should people use pharmacological neuroenhancement? Should neuroenhancers be over-the-counter medication? Should physicians prescribe pharmacological neuroenhancers? How should the consequences be dealt with legally, ethically, and financially? In the following sections, after an introductory clarification of terms, pro and con arguments from different disciplines and viewpoints will be presented in order to arrive at a credible conclusion.

9.3 Definitions and Classifications

Enhancement comprises strategies to improve one's bodily appearance or functioning by medical means, although the client is healthy and not considered a patient. This latter judgment occurs relative to a presumed objective standard of normalcy, which is, of course, debatable, culture-specific, and ever-changing. Medical knowledge is utilized outside the classical medical system, and enhancement transcends the typical medical purposes of therapy, prevention, rehabilitation, and palliative care. The desire for enhancement is subjective, and it aims at personal satisfaction and mood improvement. Examples include aesthetic surgery, doping with performance-enhancing drugs in sports, and anti-aging medication. Neuroenhancement is correspondingly an enhancement concerning the nervous system. Given that potential options for neuroenhancement are still in development, two preliminary ways of classification currently make sense (Brukamp and Groß 2012):

1. *Functions*: Neuroenhancement may be geared towards diverse functions of the brain. In particular, the following pursuits are feasible:

   1.1. *Cognitive enhancement*: Its goal is to improve wakefulness, attention, concentration, intelligence, memory, and executive functions. Examples for medication candidates include methylphenidate, modafinil, anti-dementia medication, and amphetamine derivatives.

   1.2. *Mood enhancement*: Its aim is to brighten emotional states. Examples are selective serotonin reuptake inhibitors (SSRIs), such as fluoxetine.

   1.3. *Moral enhancement*: It is meant to improve social behavior, and one of its presumed mechanisms of action is to increase a subject's empathy for others. Candidates include oxytocin and 3,4-methylenedioxymethamphetamine

(MDMA), the latter of which possesses an additional mood enhancement component. Nevertheless, at present, medical knowledge is insufficient to precisely utilize potential moral enhancers for this particular purpose. Therefore, moral enhancement to improve social behavior remains speculative overall.

1.4. *Various effects*: Moreover, some neurotropic medications have been noted to lead to extraordinary sensory impressions and transient feelings of pleasure. Others increase the pain threshold in healthy individuals. Many medications that act on the central nervous system may be regarded as neuroenhancers depending upon the purpose and the context in which they are used.

2. *Methods*: Neuroenhancement may involve varying techniques and devices. Only a subset of them is debated because of moral concerns. In a broad sense, the term "neuroenhancement" also refers to such innocuous methods as sufficient sleep, adequate nutrition, physical exercise, the use of mnemonic techniques, and brain training. Since these methods cause no or few moral problems, they are neglected in the ethical discussion. However, the moral issues of the following types of neuroenhancement in the narrow sense have been recognized:

   2.1. *Pharmacological enhancement*: At the present time, the use of medication remains the most frequent strategy for attaining neuroenhancement. The drugs utilized either originate from medication intended for therapeutic purposes, or they stem from illegal substances.

   2.2. *Non-invasive technology*: An enhancement potential may exist for transcranial stimulation techniques that are currently in development, such as transcranial magnetic stimulation (TMS) or transcranial direct current stimulation (tDCS).

   2.3. *Implanted technology*: Neuroenhancement might also involve invasive technology in the future, e.g. in the format of brain implants.

In the following, only pharmacological neuroenhancement will be specifically addressed because this approach currently has the highest prevalence. Still, most arguments can be extrapolated and modified in order to apply to the other ethically controversial means of neuroenhancement as well.

9.4 Medical Risks

From a medical perspective, neuroenhancement possesses indisputable risks. Consequently, medical considerations inevitably give rise to counterarguments against its use (Brukamp and Groß 2012). Potential risks of pharmacological neuroenhancement include the following:

1. *Side effects*: Every medication has a characteristic side effect profile. Although some side effects may be negligible or acutely tolerable, long-term effects cannot reliably be predicted for the vast majority of novel pharmaceutical

agents. One difficulty in estimating long-term effects stems from the fact that the mechanisms of action are oftentimes not fully understood. The effects of neurotropic medications cannot usually be switched off immediately by simply stopping the medication. Depending on the substance, the effect may cease only after several days. For most neuropharmaceuticals in medical use today, it is dangerous to not use a gradual dose reduction in order to taper or fade out the medication.

2. *Risk of addiction*: Some medications show a high potency to induce physical or psychological dependence.
3. *Loss of efficacy over time*: In some cases, medications lose their therapeutic effects during prolonged use. Higher and higher doses, or therapeutic combinations with other medications, are then required to achieve the same effect. Undesirable side effects can be augmented this way.
4. *Overestimation of one's own abilities*: Some medications do not improve objective performance, but they merely induce the subjective impression of better functioning instead. Modafinil apparently shows this effect: In a study, experimental examinees were subjected to prolonged sleep deprivation. Subjects with simultaneous modafinil use felt that their wakefulness increased although their cognitive performance was not better, and they overestimated their own abilities (Repantis et al. 2010b). This phenomenon provides a strong case against neuroenhancement strategies because it may endanger the safety and health of oneself and of others. For the same reason, alcohol is discouraged before driving, or altogether prohibited in some jurisdictions, because users fail to realize their own impairments and do not adjust their behavior accordingly in traffic.

Physicians' safety concerns limit the prescription of neuroenhancers in practice. A study found that primary care physicians in North American urban areas were surprisingly uncomfortable with prescribing enhancers (Banjo et al. 2010):

> [...] the most prominent concerns physicians expressed were issues of safety that were not offset by the benefit afforded the individual [...] It has become routine for safety to be raised and summarily dismissed as an issue in the debate over pharmacological cognitive enhancement; the observation that physicians were so skeptical [...] suggests that such a conclusion may be premature. Thus, physician attitudes suggest that greater weight be placed upon the balance between safety and benefit in consideration of pharmacological cognitive enhancement. (Banjo et al. 2010)

9.5 The Lack of Evidence-Based Medicine

Evidence-based medicine is a practice of medicine that is informed by documented evidence for the efficacy of its interventions. The available evidence is ranked according to well-defined categories of quality standards. In particular, meta-analyses are appreciated most, followed by randomized and controlled trials, studies, reviews, and expert opinions, in declining order of respect (Sackett et al. 1996).

Reviews of empirical studies examined the potentially discernible effects of methylphenidate, modafinil, and acetylcholinesterase inhibitors in healthy humans (Repantis et al. 2010a, b). The conclusions were as follows: Methylphenidate improves memory, and modafinil increases wakefulness, both under normal conditions and after sleep deprivation, but the latter also potentially induces unsubstantiated overconfidence in one's own abilities (Repantis et al. 2010b). Aside from these results, no valid inferences are possible due to a lack of data, specifically in the case of acetylcholinesterase inhibitors (Repantis et al. 2010a; see also Chap. 3 by D. Repantis, this volume).

This relative lack of evidence is disturbing in light of the quality measures for medication that is intended for medical therapies: The later is typically tested in randomized, controlled, double-blind trials, and it is compared to other medication that has already been shown to be effective. Therefore, the hurdles are higher for the new drug, and it has to be proven to be as valuable and successful in comparison to an already established treatment regimen.

Pharmacological neuroenhancers probably possess relevant placebo effects. Positive outcomes may not be caused by the supposedly active pharmaceutical substance, but rather by the overall treatment conditions, including the psychological *sequelae* that arise from the patient's or client's attributions or perceptions. Such effects have even been demonstrated for antidepressants in severely depressed patients (Kirsch et al. 2008).

Which criteria does medication have to fulfill in order to become fully accepted in medicine and justified for reimbursement by health insurance companies? To give an example, antihypertensive medication is, at first sight, assessed regarding its ability to lower blood pressure. Nevertheless, this goal is not the only one that needs to be achieved in order to count this medication among the truly deserving, effective health care measures. For the latter, additional benefits need to be shown. Antihypertensive medication should prevent the numerous complications of elevated blood pressure, e.g. atherosclerosis, coronary artery disease, myocardial infarction, cardiac failure, arrhythmia, stroke, and the ensuing medical fatalities.

Therefore, antihypertensive medication that merely lowers blood pressure is not considered particularly valuable in scientific medicine. The prevention of severe complications and fatalities are considered worthier goals. Long-term medical studies aim at showing these benefits, and they require more time and labor in order to prove these distant benefits. For the overall assessment, minor short-term effects need to be neglected in favor of significant beneficial or detrimental long-term outcomes.

An application of this concept from evidence-based medicine to pharmacological neuroenhancers provides a very strong challenge to their use (Brukamp and Groß 2012). First of all, enhancing short-term effects are far from obvious at this time. Even if these were shown, the neuroenhancers might still not possess the profound positive medium-term or long-term effects on the quality of life that their proponents envision. The likelihood is rather low that scientific studies will demonstrate that neuroenhancing medications mediate long-term enhancement across cognitive modalities, improve grades and careers overall, and lead to an entirely more desirable life. According to the principles of evidence-based medicine, all these benefits would need to be proven in comparison to or on top of other, already widely accepted and utilized strategies with lower risks.

9.6 Challenges to Health Care Systems

The desire for enhancement poses a challenge to health care systems because of its potential to deplete assets. Consequently, enhancement might be rejected for this reason. Health care systems are primarily responsible for taking care of patients with medical conditions who need help in deserving situations. When resources are limited, the majority of these resources should be channeled to needy patients. Enhancement endeavors, and neuroenhancement in particular, may divert financial resources away from the classical medical systems as such. For example, additional expenses arise from dealing with complications of pharmaceuticals:

> [...] the additional demand for healthcare services needed to mitigate the overall impact of widespread non-medically indicated uses of prescription drugs may place further pressure on current professional and healthcare resources. (Racine and Forlini 2009)

Physicians invest personal resources in the dyadic physician-patient relationship in terms of time, attention, and interest. Addressing enhancement may distract these personal resources away from physicians' priorities. Such a distraction would occur if physicians followed a recommendation like "Neurologists should respond to a request for neuroenhancement as they would respond to a chief complaint." (Larriviere et al. 2009) This recommendation is prominently called "guidance for physicians" by its inventors (Larriviere et al. 2009). The advice seems particularly at odds with the following concurrent statement: "The liability risks associated with prescribing medications for neuroenhancement are uncertain." (Larriviere et al. 2009) Such liability risks leave physicians in a weak legal position when they actually support their patients' desire to enhance.

The neuroenhancement debate calls attention to the problem that commercial interests encroach on the traditional act of prescribing medication, which is initiated by the physicians, not the patients. In light of the poor medical evidence in support of neuroenhancers, some pharmaceutical companies attempt to replace objective information with marketing. Direct-to-consumer advertising (DTCA), also called direct-to-consumer marketing, is well-established in the United States of America and in New Zealand. It entails direct advertising of pharmaceutical drugs to consumers, instead of to physicians only. The introduction of some types of direct-to-consumer advertising in Canada has been met with considerable criticism. Its economic potential is estimated as high, and its opponents argue that a significant rise in broadcast advertising results in higher spending on medication (Mintzes et al. 2009). Direct-to-consumer advertising does work in the way that it was intended:

> Our results suggest that more advertising leads to more requests for advertised medicines, and more prescriptions. If DTCA opens a conversation between patients and physicians, that conversation is highly likely to end with a prescription, often despite physician ambivalence about treatment choice. (Mintzes et al. 2003);

> Direct-to-consumer marketing of self-referred imaging services, in both print advertisements and informational brochures, fails to provide prospective consumers with comprehensive balanced information vital to informed autonomous decision making. Professional guidelines and oversight for advertising and promotion of these services are needed. (Illes et al. 2004)

9.7 Human Nature and Virtue Ethics

Critics of neuroenhancement frequently refer to the value that lies in human nature. Key principles of such arguments are the maintenance of, or search for, authenticity, truth, originality, personality, and identity as a human being (Brukamp and Groß 2012). The use of pharmacological neuroenhancers is perceived as deception, self-deception, forgery, or cheating. According to this position, neuroenhancement merely results in an imitation of traits that should be formed by natural means and that are only considered valuable if they have a natural origin. An example for such an argument reads as follows:

> As the power to transform our native powers increases, both in magnitude and refinement, so does the possibility for 'self-alienation' – for losing, confounding, or abandoning our identity. I may get better, stronger, and happier – but I know not how. I am no longer the agent of self-transformation, but a passive patient of transforming powers. Indeed, to the extent that an achievement is the result of some extraneous intervention, it is detachable from the agent whose achievement it purports to be. 'Personal achievements' impersonally achieved are not truly the achievements of persons. (The President's Council on Bioethics 2003)

In contrast, an argument in favor of neuroenhancement relies on a different anthropological assumption about human nature: Humans constantly seek self-improvement. This quest to improve oneself appears as a seemingly incontrovertible human trait, as a core value for humans. Enrichments in daily life are generally accepted, such as measures to increase comfort and bodily embellishments. Merely sticking with the biologically given facts could then equate to an is-ought fallacy in the moral realm. Therefore, the proponents of neuroenhancement argue that medications are just an extension of the methods by which the same goals are pursued. The target of self-improvement remains the same, while the means are broadened to include pharmaceutical support (Galert et al. 2009). The following citation gives an example of such a position:

> Education and training [...] may be labeled as 'conventional' means of enhancing cognition. [...] By contrast, methods of enhancing cognition through 'unconventional' means [...] are nearly all to be regarded as experimental at the present time. [...] They may eventually come to have important consequences for society and even, in the longer run, for the future of humankind. [...] From [...] a comprehensive viewpoint, the inadequacies of some aspects of the current regulatory and policy framework become apparent, as it treats different modes of enhancement differently even though, arguably, there is no good justification for doing so. (Bostrom and Sandberg 2009)

Nevertheless, this latter argument is challenged by the fact that people frequently arrive at logically separate moral judgments about ends (or goals) *versus* means. This distinction has been widely defended in general normative ethics for a long time.

According to virtue ethics, morality originates from within. It depends on a person's personality, her formed habits and persistent traits. Character is more important than individual actions. While good behavior is generally time-invariant, it can be adapted to situations. In its classical form, virtue ethics goes back to Greek and Roman times, but it has seen a successful revival in recent decades. A related contemporary concept is the pursuit of self-improvement by natural means. Bettering one's habits can strengthen one's self-esteem and bring about greater satisfaction and happiness. This process influences and gradually constitutes one's identity. The methods of forming habits can take on manifold formats: behavioral practice, cognitive training, reflective reasoning, sports, sleep, or meditation. These ways of shaping behavior are socially widely accepted and have been used across cultures and time.

Supporters of virtue ethics and natural self-improvement ascribe value to the progressive shaping of behavior, to striving for improvement and perfection. Positive changes are achieved by gradual movements in the right direction. On the contrary, pharmacological neuroenhancement is regarded as an external, artificial measure that merely serves as a means of self-deception. Consequently, it is customarily rejected.

One counterargument to the latter position asserts that enhancement would be recommendable in the format of moral enhancement, even if natural self-improvement surpassed pharmacological cognitive and emotional enhancement. Regarding moral behavior, the thresholds and obstacles for achieving behavioral corrections are usually much higher than for cognitive and emotional contexts. However, this argument is invalid because moral enhancement concerns only specific subgroups of the population that are amenable to therapies proper:

At present, the improvement of social behavior is only desirable and advisable for people with significant impairments in their moral fabric. This applies predominantly to criminal offenders who commit crimes because of an underlying psychiatric disorder associated with aggression, such as an anti-social personality disorder (American Psychiatric Association 2000; World Health Organization 2007). In these cases, an offer of medication, in addition to or instead of other retaliation measures, may be beneficial for both themselves and society as a whole.

Such an approach is not wish-fulfilling enhancement, though. Rather, it is a medically justified treatment since personality disorders constitute psychiatric diagnoses with clear-cut medical treatment options (American Psychiatric Association 2000; World Health Organization 2007). In short, therapies targeted against criminal behavior as part of psychiatric disorders need to be correctly designated as moral treatments, not moral enhancements.

9.8 Debatable Purposes

Supporters of neuroenhancement often emphasize the common desire to pursue success. The concept of success is certainly subjective, and humans search for success in a number of different ways, for example through competition. Pharmacological neuroenhancers promise apparent advantages for competition in schools, at universities, or at work. An additional motive for use is the goal to experience pleasure and fun, e.g. by refueling energy for partying and dancing for many hours in a row. In such instances, those who do not use neuroenhancers are then regarded as socially inept or potential failures.

Survey studies found an association between poor performance and neuroenhancer use (Franke et al. 2011). This phenomenon seems to suggest that underachievers feel more pressure to utilize medications than the top of the class. However, an alternative interpretation proposes that a group somewhat disadvantaged from the outset is inclined to use a variety of drugs at the same time, including legal, illegal, and prescription drugs. Drugs may then lead to further negative effects on users' performance in school or at work. Overall, empirical data on the prevalence of neuroenhancement are scarce (Hildt 2011). One informal survey reporting a rather high prevalence (Maher 2008) does not meet the quality criteria of research methods in empirical social science in order to draw meaningful conclusions.

The pursuit of success can become strained. Some users of pharmacological neuroenhancers seem to care mostly about self-interest, such as through competition, careerism, conformity, effort, work, and tangible achievements. Factors that may be underrepresented in this framework are relaxation, release of tension, play, individualism, and spiritual experience.

One important question in this context is: For which purposes do healthy people use non-medical drugs voluntarily? The answer has apparently changed over time. During the 1960s and 1970s, most people gravitated towards drug use for relaxation purposes. Some of them sought a spiritual element that would help them to identify meaning in their lives. One goal was to display disdain with the mainstream culture. Drug use seemed to open up an alternative to the perceived restrictions and limitations in society. To some followers, it promised to provide a break from conformism and from the hunt for power and money. Doubts remain whether these expectations were fulfilled, in light of the side effects of illegal drugs.

A look at the purposes for neuroenhancer use today shows a completely different picture because these purposes are, in part, contrary to those of recreational drugs. Neuroenhancers supposedly enable people to better comply with the demands of the mainstream attitudes on achievements in school, academia, and the workplace. Consumers wish to attain common, widely accepted goals. For example, self-perceived underachievers seize apparent opportunities in order to integrate more completely into society.

The discrepancy between these two groups of voluntary drug users leads to the question whether both purposes have their respective value, depending on the context. In some situations, people want to exercise their individualism by distinguishing themselves from the mainstream, whereas others seek to achieve a high level of success in order to adhere to the norms of the predominant culture. In any event, the discrepancy shows that the fashions of drug use are culture-relative and change with time. This phenomenon may instill a sense of caution against pharmacological neuroenhancement.

9.9 Liberty and Justice

One of the major arguments in favor of pharmacological neuroenhancement, which has frequently been put forth, is the argument of liberty. Personal autonomy in the setting of democratic freedom should guarantee individualism and the pursuit of distinct goals. The conclusion is that everyone may decide for herself whether or not to use neuroenhancement. Proponents then take a "why not?" position and support a so-called responsible version of drug use (Metzinger 2006; Greely et al. 2008).

Nevertheless, in democratic societies, legitimate interests of others place limitations on the unconstrained pursuit of individual goals. Which arguments may tilt the balance against the acceptance of pharmacological neuroenhancers?

One argument relies on the concept of distributive justice (Brukamp and Groß 2012). It claims that personal financial resources determine access to neuroenhancing medication. If neuroenhancers indeed conferred advantages, wealthy groups in society would gain more such advantages. In contrast, the proponents of pharmacological neuroenhancement assert that it would help to create equal opportunities because it corrects injustice sown by distinctive biological traits. In reply, the opponents propose that this view ignores social inequalities and their root causes. The following quotation, actually phrased by supporters of neuroenhancement, illustrates the difficulties with it from the perspective of distributive justice:

> Whether the cognitive enhancement is substantially unfair may depend on its availability, and on the nature of its effects. [...] If cognitive enhancements are costly, they may become the province of the rich, adding to the educational advantages they already enjoy. [...] Policy governing the use of cognitive enhancement in competitive situations should avoid exacerbating socioeconomic inequalities [...] In developing policy for this purpose, problems of enforcement must also be considered. In spite of stringent regulation, athletes continue to use, and be caught using, banned performance-enhancing drugs. (Greely et al. 2008)

A further argument against pharmacological neuroenhancement asserts that social pressure may cause it to spread from small groups to larger groups and maybe even to the majority of the population, despite its potential risks and complications. Such a coercion phenomenon particularly applies to vulnerable groups (World Medical Association 1964/2008), such as children. Some candidate pharmacological enhancers like methylphenidate are used to treat diseases with established medical diagnoses in children, such as attention-deficit hyperactivity disorder (ADHD). Nevertheless, this therapeutic use cannot readily lead to the conclusion that a use in healthy children for enhancement purposes is either safe or desirable. Neuropsychological studies reveal that the brain's vulnerability changes over time. Tampering with natural development without a proper medical reason may result in long-term effects that cannot be foreseen at this time.

As a vulnerable group (World Medical Association 1964/2008), children merit special attention and protection. They are not capable of informed consent, and they are subject to their parents' surrogate decisions. Schoolteachers may persuade well-meaning, but naive parents to comply with perceived social standards, thereby exerting social pressure on parents to have their children use neuroenhancers. The following citations illustrate ethical difficulties with neuroenhancement concerning children in particular:

> Behavior-modifying agents would allow parents, teachers, or others to intervene directly in a child's neurochemistry when that child behaves in a way that defies their standards of conduct. In some cases, the children clearly benefit; in other cases, they do not. In all cases, the use of such drugs to shape behavior raises serious questions concerning the liberty of children. (The President's Council on Bioethics 2003);

> Schools should be prevented from coercion [. . .] into adopting stimulants for enhancement purposes. (Singh and Kelleher 2010)

9.10 Conclusions

Neuroenhancement, as a topic in medical ethics, is worthy of careful consideration for two reasons: First, it is a specific example of enhancement, a concept that will probably be discussed even more intensely in the future, due to the fact that increasingly efficient medical systems provide an apparent excess of resources. These resources could either be utilized for enhancement purposes that benefit some, or they could improve general health standards for the general population. Second, results from neuroscientific research provide insight into the functioning of the brain and challenge human identity. Basic research results from cognitive psychology and neuromedicine are beginning to yield options for translation into medical practice.

Arguments, both those in support of and those opposed to neuroenhancement, relate to various disciplines. From a medical standpoint, several perspectives justify a critical stance, be it because of medical risks, deficits in evidence-based medicine, or challenges to health care systems. Considering society as a whole, skeptical arguments also stem from anthropology, sociology, and classical ethics, and they touch on the concepts of human nature, virtues, purposes, liberty, and justice. In conclusion, neuroenhancement should be discouraged at present.

Consequently, the specific answers posed at the beginning may be answered as follows: Pharmacological neuroenhancers should not become over-the-counter medication, and physicians should refrain from prescribing them as neuroenhancers at the present time. The pressure is on the proponents to demonstrate medical evidence for beneficial effects. In addition, they ought to present legal and financial plans for how to potentially employ neuroenhancers under certain circumstances.

Questions remain regarding how the issues of distributive justice and social coercion can be resolved and which agencies in health care systems will pay for neuroenhancement and deal with ensuing medical side effects and complications in society. On the other side, opponents need to draw attention to the numerous problems with neuroenhancement, which relate to multiple fields such as medicine, anthropology, sociology, and classical ethics. These problems defy an easy solution.

References

American Psychiatric Association (2000) Diagnostic and statistical manual of mental disorders DSM-IV-TR. 4th edn, Text revision. American Psychiatric Association, Arlington

Banjo OC, Nadler R, Reiner PB (2010) Physician attitudes towards pharmacological cognitive enhancement: safety concerns are paramount. PLoS One 5(12):e14322

Beauchamp TL, Childress JF (2009) Principles of biomedical ethics. Oxford University Press, Oxford/New York

Bostrom N, Sandberg A (2009) Cognitive enhancement: methods, ethics, regulatory challenges. Sci Eng Ethics 15:311–341

Brukamp K, Groß D (2012) Neuroenhancement – a controversial topic in contemporary medical ethics. In: Clark PA (ed) Contemporary issues in bioethics. InTech, Rijeka

Franke AG, Bonertz C, Christmann M, Huss M, Fellgiebel A, Hildt E, Lieb K (2011) Non-medical use of prescription stimulants and illicit use of stimulants for cognitive enhancement in pupils and students in Germany. Pharmacopsychiatry 44(2):60–66

Galert T, Bublitz C, Heuser I, Merkel R, Repantis D, Schöne-Seifert B, Talbot D (2009) Das optimierte Gehirn. Gehirn und Geist 11:1–12

Greely H, Sahakian B, Harris J, Kessler RC, Gazzaniga M, Campbell P, Farah MJ (2008) Towards responsible use of cognitive-enhancing drugs by the healthy. Nature 456:702–705

Hildt E (2011) Neuroenhancement bubble? – neuroenhancement wave! Am J Bioeth Neurosci 2(4):44–47

Illes J, Kann D, Karetsky K, Letourneau P, Raffin TA, Schraedley-Desmond P, Koenig BA, Atlas SW (2004) Advertising, patient decision making, and self-referral for computed tomographic and magnetic resonance imaging. Arch Intern Med 164:2415–2419

Kirsch I, Deacon BJ, Huedo-Medina TB, Scoboria A, Moore TJ, Johnson BT (2008) Initial severity and antidepressant benefits: a meta-analysis of data submitted to the Food and Drug Administration. PLoS Med 5(2):e45

Larriviere D, Williams MA, Rizzo M, Bonnie RJ, on behalf of the AAN Ethics, Law and Humanities Committee (2009) Responding to requests from adult patients for neuroenhancements. Guidance of the Ethics, Law and Humanities Committee. Neurology 73:1406–1412

Maher B (2008) Poll results: look who's doping. Nature 452:674–675

Metzinger T (2006) Intelligente Drogenpolitik für die Zukunft. Gehirn und Geist 1–2:32–37

Mintzes B, Barer ML, Kravitz RL, Bassett K, Lexchin J, Kazanjian A, Evans RG, Pan R, Marion SA (2003) How does direct-to-consumer advertising (DTCA) affect prescribing? A survey in primary care environments with and without legal DTCA. Can Med Assoc J 169(5):405–412

Mintzes B, Morgan S, Wright JM (2009) Twelve years' experience with direct-to-consumer advertising of prescription drugs in Canada: a cautionary tale. PLoS One 4(5):e5699

Parens E (2005) Authenticity and ambivalence: toward understanding the enhancement debate. Hastings Cent Rep 35(3):34–41

Racine E (2010) Pragmatic neuroethics: improving treatment and understanding of the mind-brain. MIT Press, Cambridge, MA

Racine E, Forlini C (2009) Expectations regarding cognitive enhancement create substantial challenges. J Med Ethics 35:469–470

Repantis D, Laisney O, Heuser I (2010a) Acetylcholinesterase inhibitors and memantine for neuroenhancement in healthy individuals: a systematic review. Pharmacol Res 61:473–481

Repantis D, Schlattmann P, Laisney O, Heuser I (2010b) Modafinil and methylphenidate for neuroenhancement in healthy individuals: a systematic review. Pharmacol Res 62(3):187–206

Sackett DL, Rosenberg WM, Gray JA, Haynes RB, Richardson WS (1996) Evidence-based medicine: what it is and what it isn't. Br Med J 312(7023):71–72

Singh I, Kelleher KJ (2010) Neuroenhancement in young people: proposal for research, policy, and clinical management. Am J Bioeth Neurosci 1(1):3–16

The President's Council on Bioethics (2003) Beyond therapy: biotechnology and the pursuit of happiness. The President's Council on Bioethics, Washington, DC

World Health Organization (2007) International statistical classification of diseases and related health problems. 10th revision. Version for 2007. apps.who.int/classifications/apps/icd/icd10online. Accessed 4 May 2011

World Medical Association (1964/2008) Declaration of Helsinki – ethical principles for medical research involving human subjects. Helsinki, 1964/2008. www.wma.net/en/30publications/10policies/b3/index.html. Accessed 4 May 2011

Chapter 10
Cognitive Enhancement – To What End?

Michael Hauskeller

Abstract If human enhancement consists in the making of better humans, then we obviously need to have some idea of what "better" humans would be like and in what respect they would be "better." However, it can easily be shown that what counts as better is in fact highly context-dependent, so that there is no universal measure for human improvement. Cognitive enhancement is usually justified as boosting performance, but whether it is desirable for people to perform better very much depends on what they are getting better at, what end the improvement serves, and who benefits from it. Even an enhancement for an alleged "common good" can be, all things considered, highly undesirable.

Keywords Cognitive enhancement • Human improvement • Happiness • Freedom • Common good

In 1998, the eminent molecular biologist and Nobel laureate James Watson challenged critics of non-therapeutic human germ line interventions by posing the rhetorical question: "If we could make better human beings by knowing how to add genes, why shouldn't we do it?"[1] Indeed, why shouldn't we? Put like this, it seems decidedly irrational to object. Almost by definition, there cannot be anything wrong with making better human beings. How can it possibly be wrong to create something that is *better*? Thus to suggest that an enhancement could be bad, seems like a contradiction in terms. If it were bad it wouldn't be an enhancement. Perhaps we don't know how to make better humans yet, but that is a purely practical problem,

[1] Engineering the Human Germline, Symposium at UCLA, 20 March 1998.

M. Hauskeller (✉)
Faculty of Sociology and Philosophy, University of Exeter, Exeter, UK
e-mail: M.Hauskeller@exeter.ac.uk

E. Hildt and A.G. Franke (eds.), *Cognitive Enhancement: An Interdisciplinary Perspective*, Trends in Augmentation of Human Performance 1,
DOI 10.1007/978-94-007-6253-4_10,

which doesn't affect the desirability of the general project. Perhaps it will turn out that "adding genes" does not have the desired effect, but again that doesn't mean that we shouldn't at least try. The question is, if we *knew* how to improve our nature and *knew* that it was safe to do so (that it had no detrimental side effects), then why should we not do it? The answer seems to be obvious: there is no such reason. In the absence of any negative side effects, human enhancement is necessarily good and hence desirable.

Let us call this position *meliorism* (Caplan 2006). Watson's question is, of course, designed to silence critics of meliorism by making them appear irrational and foolish. But in fact the question conceals at least three problems. The first problem is the assumption that the only knowledge we need in order to realize the meliorist programme is *technical* in nature. It is assumed that we already know the end, we all agree on the desirability of that end and that we only need to discover the appropriate means for achieving it. Perhaps we don't know yet how to make better humans, but we *do* know what being a better human would consist in. But the question is, do we really?

The second problem concerns the attitude which the question endorses. Even if we accept that humans can be more or less good, and that better humans are imaginable than the ones we have now, it is not obvious that striving to *make* better humans is a good thing (cf. Hauskeller 2011).

The third problem arises from the question of how we can *acquire* the technical knowledge that is needed to perform human enhancements safely and with precision. Given that we know what we want, how do we figure out how we get what we want? And can we justify what we need to do in order to find out?

10.1 What Does "Better" Mean in a Human?

While the two last problems are certainly in need of thorough reflection, in this paper I'm only going to address the first problem concerning the notion of "better" human beings. Talking about "making *better* human beings" implies that humans are not only different from each other, but that there are better and worse ways of being human. It only makes sense to speak of better humans if there are, at least theoretically, good humans and not-so-good humans. This means that there must be some standard by which to measure the quality of a human. But is there? And if there is, what might this standard be? When a car company promises to "make better cars" we have a fairly good idea of what they have in mind: cars that are safer, more comfortable, more economical, or more elegant. But even with cars, what you regard as better depends on what you regard as important in a car. Whether you think that Japanese cars are better than German cars or the other way around depends on your personal preferences. Perhaps Japanese cars are more economical and German cars are safer. It would be futile, though, to argue about which car is better as such, that is, *as a car*. That would only be possible if cars had one and only one purpose or function, for instance to get us from one place to another as fast as possible without leaving the ground. If that were the case we would all have to agree that, if car A is faster than car B, then car A is the better car. Yet even cars have more than one purpose. What we expect of a car is not always the same.

However, cars are at least purpose-built, so it is not unreasonable to ask what constitutes a good car (and hence, what constitutes a "better" car). Humans, on the other hand, are not purpose-built, at least not yet, and if we cannot even agree on what makes a good car, it seems highly unlikely that we will be able to agree about what makes a good human being. Watson himself had high hopes that through genetic intervention we may finally get rid of "stupid children" and "ugly girls."[2] So, according to Watson, making better humans would involve making humans generally more intelligent and more beautiful. Others may think that being a morally good person is more important and would not regard anything as an improvement that does not also improve our willingness to act morally. Some will argue that, on the contrary, as long as we adhere to old-fashioned moral ideals, which are nothing but prejudices, we will never be truly advanced. Some will say that it doesn't matter either way, because the only thing that *really* matters is whether or not we are happy. Better humans will be happier humans, and the best humans would be those who get the maximum amount of pleasure for as long as possible. This last suggestion is actually intuitively the most plausible of them all, since it can be argued that if there is one thing by which all so-called enhancements can be judged, it must be happiness. It is, after all, the one thing that appears to be intrinsically valuable, whereas everything else that we value is either valuable as a means to the end of happiness or not valuable at all.

Happiness or human well-being, however, depends on many factors and it is not at all obvious what we need to change in human beings to increase their happiness or at least the likelihood of their being happy. Will it help to provide smart drugs for children and students, cosmetic surgery, or performance-enhancing drugs that make us faster, stronger, and generally more skillful? The Oxford philosopher and leading transhumanist Nick Bostrom suggests that there are three areas where real enhancement is possible. The first is the extension of life in good health, which, Bostrom believes, everyone desires if they are honest (Bostrom 2008). The second is the refinement of our cognitive abilities: our intelligence, our memory, our alertness. The third is the enhancement of our emotions, "eagerly sought by many." It is obvious though that enhancement in the last two areas cannot just consist in having "more." It is quite obvious that having a vastly improved memory can easily become a burden. Who would want to remember every single detail of their lives, who always register every single detail? And rightly so, because such a memory can seriously hamper one's ability for abstract thinking and adequate functioning in a social environment (Luria 1987). And being very intelligent also is not always a blessing. Recent studies have shown that the higher the IQ of a child at the age of 8 is, the more likely it is that she will suffer from depressive symptoms during puberty from the age of 13 onwards (persisting in males even beyond the age of 17)

---

[2]DNA. British Documentary, 2003.

(Glaser et al. 2011). That is probably not what one wishes one's child to go through. And what about emotions? Should we be less aggressive, more loving? Will that always be a good thing? Will it make us happier or just shallower? Or perhaps we should transform ourselves in such a way that we generally have deeper emotions, that we feel more intensely. But that wouldn't work either, because sometimes we wish for less intensity (to feel anger, envy, or pain more intensely does not seem to be generally desirable).

A popular solution to these ambiguities is to see the real improvement in an extension of human freedom. Emotional enhancement is then understood as increasing the ability to *control* one's emotions (Bostrom 2005), and memory enhancement as increasing the ability to "remember important things when you want to" (Savulescu 2001, 420). Those abilities are considered to be "general purpose means," which are useful for any plan of life (Buchanan et al. 2000, 167; Savulescu 2001). So it seems that a better human being is one that has more control over things: what they feel, what they remember, when they die (it is argued that if immortality begins to get burdensome we can always kill ourselves). So being enhanced means basically having more *control* over one's life (by means of gaining more control over one's body and mind). Control is a good thing, the best short of the happiness that it will ensure. But is that really so? Is control always good? There are reasons to doubt that. It is quite obvious that the attempt to gain control over a thing is sometimes self-defeating. Seeking to acquire control cannot work because of the nature of what is sought to be controlled. Memories, for instance, strike me as one of those things. Our memories are an essential part of our personal identity, which cannot exist in a moment, cut off from the past and the future, but necessarily stretch back in time to incorporate the life that we have lived. We know who we are by means of the memories we have. If we lose our memory we lose the context that defines what we are. This is shown nicely in Michel Gondry's film *Eternal Sunshine of the Spotless Mind*, which is about memory erasure and how this affects the lives of those who undergo it. By voluntarily erasing their memories of each other the two protagonists destroy an important part of their identity. Likewise, in Christopher Nolan's *Memento* the main protagonist Leonard who suffers from anterograde amnesia is literally a nobody who, because of his condition, is incapable of entering into the various human relationships that make up our lives: genuine love, hate, friendship, or enmity, seem no longer possible without memories. Memories are clearly enormously important because they hold our lives together by connecting our present selves to our past selves, and this would no longer be the case once we had *control* over our memories. Once we can choose what we want to remember and what to forget, the connection has been severed, and our identity has in effect also become subject to choice (our own choice, but possibly also the choice of others). Control over them turns memories into fictions, and when our memories are fictitious our lives are too. In other words, controlled memories are not memories at all.

Like memories, emotions are by their very nature uncontrollable. An emotion that we could fully control would no longer be an emotion, because having an emotion means *being moved* by something – something that is beyond our control.

It is precisely for this reason that emotions connect us to the world; that they make things real for us. Thus having an emotion does not mean making ourselves move, or causing ourselves to be moved. But if that is correct, then we cannot possibly be in control of our emotions. If we can choose to be happy or to be sad at will, then happiness and sadness become fabrications that have lost touch with reality. Perhaps that is also true for other things, such as happiness itself. If it is, i.e., if a certain lack of control is a requirement of happiness, then making better humans in the sense of happier humans is downright impossible.

10.2 Cognitive Enhancement

Let us now have a closer look at what is normally referred to as cognitive enhancement. Cognitive enhancements are, roughly put, all interventions that, through the manipulation of the human brain, improve the human knowledge situation by facilitating or accelerating knowledge acquisition, processing, storage, application, or range. We can distinguish between pharmaceutical, neurotechnological, and genetic means of enhancement. While the latter is still largely science fiction, at least as far as humans are concerned, pharmaceutical and neurotechnological enhancement devices are already being used. Ritalin, which was initially prescribed to treat ADHD, or modafinil (designed to treat narcolepsy) are now widely used to enhance concentration and wakefulness (which appears to be useful for many cognitive tasks). Other drugs such as Dexedrine® or Adderall® can be used to similar effect. Beta-blockers, which reduce anxiety, and mood enhancers, such as the anti-depressant drug Prozac®, can also be understood as cognitive enhancers in a wider sense in so far as they make it easier for people to adhere to their tasks and to use their cognitive faculties effectively. Happier and less angst-ridden people are more likely to be alert and to perform well. In any case, the primary purpose of both pharmaceutical cognitive enhancers and mood enhancers in terms of their actual *usage* is the same, namely, to boost performance and thus to enhance productivity. This is actually the very reason why they are considered enhancements in the first place. So it is not because Ritalin® improves our ability to concentrate that we see it as an enhancement drug, but rather because by means of improving our ability to concentrate, it allows us to *perform better* in situations that involve the completion of certain cognitive tasks. Only in relation to the task which is meant to be performed can the effected change in a person's abilities be seen as an enhancement. In other words, we have become better (if we have), not as human beings, but as performers of a certain task or pursuers of a certain goal. We may have chosen this goal for ourselves, or it may have been imposed on us. Either way, whether we think that the enhancement is desirable, ultimately depends (or should depend) on whether or not we think that the task is worth performing, the goal worth pursuing. That an intervention helps someone to perform better is in itself not a good reason to support and endorse the intervention. We always need to ask what a better performance in a specific context is *good* for, and of course also, for *whom* it is good. (Sometimes a better performance may only be good for the profits of pharmaceutical companies.) Often, it will increase our chances to compete with others, which might be good for the individual, but not necessarily in the long term interests of the community in which the individual lives, and for the individual only if she happens to live in a competitive society. If the guiding principle were not competition, but, say, cooperation, the situation might be very different, so that other performance requirements would hold. Generally speaking, what counts as cognitive enhancement is highly context-dependent. It depends on what someone wants or what the goal is. This is why forgetting can be as much an enhancement as remembering, high intelligence as much as low intelligence. This is quite obvious when you look at the intervention from the perspective of the individual and with regard to their desire, not to perform well, but to be reasonably happy. Thus having an artificial hippocampus that boosts memory capacity need not be a blessing at all. It may equally well be experienced as a burden. And whether a drug that partially erases (or dampens) one's memory is an enhancement or not, depends on so many aspects of the situation that we can only decide the question by focusing on particular aspects and ignoring others. Thus, if we suffer from the haunting memory of a traumatic experience, a drug that helps us forget may be seen as an enhancer, and the successful, chemically induced forgetting as an enhancement – but only if we ignore everything else and look only at the specific suffering that has now disappeared. If, instead, we took into account the inevitable distortion of our self-image and our relation to reality, we would perhaps be more hesitant to call the effected change, all things considered, an enhancement. But perhaps we are not interested in the perspective of the individual at all but want to base our assessment on the consequences an intervention has for society as a whole. In that case, too, we may come to the conclusion that we would all be better off if we were less intelligent and remembered less.

We may also want to take into account what people are actually going to do with improved cognitive abilities. Very often we find descriptions of already-available or merely envisaged cognitive enhancement technologies, both in the popular media and in scientific journals, informed by the assumption that those technologies will eventually be used to benefit humanity. Fairly common are statements such as the following from an article that appeared 3 years ago in *Time Magazine* (Szalavitz 2009): "Indeed, it would be hard to argue against promoting the use of an intelligence enhancer if it were risk-free and available to everyone. Imagine a legion of cancer researchers on smart drugs, racing toward a cure. Or how about a better class of Wall Street executives, blessed with improved thinking and wiser judgment?" That would be nice indeed. Unfortunately, it is not a very likely scenario. Legions of extremely smart cancer researchers racing toward a cure? Far too good to be true. Wise Wall Street Executives? Sounds more like a contradiction in terms. It seems far more likely that the latter will use their improved brains to find even more effective ways to amass wealth, and the former may also feel that they have got far better things to do than spending all their time and energy on finding a cure for cancer. The assumption that improved cognitive abilities will naturally be used for the common good is hardly convincing. It ignores human selfishness, which we have no reason to believe is less common or less articulated among the more intelligent than it is among the cognitively less fortunate. That is why the philosopher Julian Savulescu, one of the most prominent promoters of human enhancement in the UK, has recently demanded that research into, and development of cognitive enhancement be complemented by an exploration of the possibility "of biomedical means of *moral* enhancement" (Persson and Savulescu 2010). According to Persson and Savulescu, "biomedical moral enhancement, were it feasible, would be the most important biomedical enhancement. Without moral enhancement, other techniques of biomedical enhancement seem likely to increase global injustice" (S. 12). And not only that: it would also make the world a far more dangerous place, with super smart terrorists being in a much better position to successfully follow through on their evil schemes. And because it is so important, Persson and Savulescu suggest that moral enhancement be made compulsory. Interestingly, this last statement has prompted John Harris, who is usually in total agreement with Savulescu, to write a surprisingly sharp response, in which he basically attacks Savulescu and his co-author for their lack of trust in the self-purifying power of cognitive enhancement and, of course, for their apparent willingness to sacrifice human freedom for more security (Harris 2011). The argument between Savulescu and Harris shows nicely how what we are willing to regard as an enhancement does not merely depend on the overall context, but also on the value system that we happen to endorse. If we value human freedom more than anything else, then we will see certain changes as enhancements that we would not regard as enhancements if we valued security more. However, even if we are pretty clear about our values, the essential contextuality of every concrete biomedical intervention, cognitive or otherwise, makes it difficult, perhaps impossible, to decide once and for all whether an intervention should, ultimately, count as an enhancement or not. If I value individual freedom, I should, it seems, welcome interventions that help me and others get more control over our lives. But the trouble is that whatever manipulation of my body, including my brain, helps me gain more control, is likely to be usable by others to gain more control over *me*. As C.S. Lewis pointed out more than 50 years ago (Lewis 1955, 68–70), every power that "we" acquire is a power that can equally well be used against us (just think of the atomic bomb). If we can construct brain-computer interfaces that, say, allow army pilots to control their machines by thought alone, then the possibility of manipulating the soldiers' minds directly by means of the same device is never far away – not to mention the danger of allowing people to set in motion deadly devices by a purely mental act, which may make killing other human beings even easier than it already is. None of this is probably in the interest of the individual. Is it in the interest of society instead? Is it, as some have argued, for the common good? And if it were, would that make it permissible or even desirable?

10.3 Enhancement for the Common Good?[3]

In a paper published last year in AJOB Neuroscience (Vedder and Klaming 2010), Anton Vedder and Laura Klaming (both work at the Tilburg Law School) argue that the neurotechnological improvement of eyewitness memory through transcranial magnetic stimulation (TMS) would be an enhancement "for the common good" and that many of the objections commonly raised against cognitive enhancement in general would cease to apply if we looked at it from the perspective of the common good rather than from that of the individual. So let's see how convincing this claim is.

Unfortunately, Vedder and Klaming say very little about what, in their view, *constitutes* the common good, except that an enhancement for the common good would be one that is "neither primarily self-regarding nor self-serving and potentially benefits society as a whole" (22). It is not immediately clear, though, what kind of enhancement should count as beneficial for society, and for what reason exactly. Nor is it clear whether and under which circumstances common good should take precedence over individual good.

Vedder and Klaming discuss the neurotechnological improvement of eyewitness memory as a paradigmatic example of an enhancement for the common good. They do not explicitly argue their case, which suggests that they believe the connection to be obvious and undeniable. However, we can reconstruct their *implicit* argument as follows:

1. Eyewitness testimony "plays an important role in the apprehension, prosecution and adjudication of criminals" because the decisions made by law enforcement officials rely heavily on it.
2. Relying on eyewitness testimony can only be justified if it can be trusted, i.e., if there is sufficient reason to believe that it is accurate.
3. The accuracy of an eyewitness's testimony depends on the accuracy of her memory, which, however, is notoriously malleable and hence unreliable.
4. Therefore, any means of improving the accuracy of memory is desirable with respect to the purpose of apprehending, prosecuting and adjudicating criminals.
5. Since it is in everybody's interest that criminals are found out and get convicted (and innocents do not), improving eyewitness memory is therefore beneficial for all of us (except perhaps for criminals), that is, for society as a whole.

The crucial step of this argument is of course the last one, where the interests of a particular societal subgroup (i.e., law enforcement officials) are identified with the interests of society as a whole. The problem with this step is that there are all sorts of possible technical and legal innovations that appear good for law enforcement

[3] A version of this last section has already been published as Hauskeller (2010).

officials (the police, prosecutors, lawyers and judges) by allowing them "to do a better job" (22), but which we would be very hesitant to regard as good for society as a whole.

Thus, it would arguably be a lot easier to apprehend and convict criminals if the doings of all citizens and visitors were permanently monitored by the police. For the sake of the common good, we should therefore strive to make state surveillance as widespread and thorough as possible. "Enhanced interrogation techniques" no doubt also help the police and prosecutors do a better job, as does the practice of detaining suspects without trial for as long as needed to ascertain their guilt or innocence. In general, individual legal rights often present an obstacle to law enforcement, which makes it appear immensely desirable to be permitted to suspend them. Although all these measures are clearly in the interest of law enforcement, and all upright citizens presumably have an interest in seeing the law enforced, many people would still disagree with the proposition that they are all "for the common good." This indicates that there must be something wrong with Vedder and Klaming's reasoning. The fault lies in the assumption that just because we all share a certain interest (e.g. that crime be prevented), we must also have a derivative interest in whatever serves that first interest. The reason why this is not so is that we have, in fact, various interests, which cannot all be fully satisfied because they are, to a certain extent, mutually exclusive. We may have a strong interest in being protected from crime, but we also have a strong interest in having our privacy and autonomy respected. (It's again the conflict between the value of security and the value of freedom that I pointed out earlier.) However, since we cannot be *fully* protected from crime without accepting a drastic infringement on our privacy and autonomy, the best we can hope for is a compromise that allows *both* interests to be satisfied to the greatest possible extent. Thus we can never infer from the fact that one of our interests is served by a particular practice that this practice is for the common good, i.e., in our own best interest, unless we know for certain that none of our other interests are violated or threatened by it. For the common good is nothing if not the individual good of all people.

Although Vedder and Klaming concede that "the protection of privacy and autonomy of individuals seems important" (22), they leave open whether eyewitness memory improvement by means of TMS should be made mandatory or remain voluntary, i.e., whether or not we should allow eye witnesses to refuse the enhancement – and this despite the fact that admittedly TMS may cause "unpredictable responses such as unwanted or even traumatic memories" (ibid.), which is surely not in the interest of the witness. The authors' willingness even to *consider* forcing people, in the name of the common good, to undergo a treatment that is not entirely without risks, shows clearly enough how dangerous it can be to adopt the perspective of a presumed 'common good' that Vedder and Klaming recommend we *always* adopt when we look at "cognitive enhancement in general" (ibid.). There is a tendency here to view the common good as something absolute that exists irrespective of what is good for the individuals concerned, that is more important than the latter, and that, therefore, occasionally requires that the merely individual good be sacrificed.

Now, if that is not regarded as entirely out of the question with respect to innocent bystanders of a crime, it would appear that we are even more justified to make sure that law enforcement is assisted as best as possible when we know or suspect that the witness was not innocent at all, but in fact actively involved in committing the crime. When Vedder and Klaming highlight the importance of accurate eyewitness testimony, they disregard the fact that eyewitness reports are often unreliable not because the witness forgot what really happened, but rather because they, for one reason or another, choose to *lie* about it. In that case, memory enhancement would do nothing to secure greater accuracy. Instead, we would need to apply some kind of truth serum. (That truth serums have proved unreliable in the past is of no account because, first, we may develop better ones in future, and second, because TMS is, according to Vedder and Klaming, not entirely reliable either.) After the Mumbai massacre in November 2008, in which more than 170 people died, Indian officials were planning to use such a truth serum on the sole surviving gunman, Azam Amir Kasab, to make him disclose the details of the attack.[4] I don't know whether they really did, and if they did, whether they succeeded, but the point is that if we accept the argument proposed by Vedder and Klaming, then we must conclude that such a 'veracity enhancement' would clearly be for the common good and hence desirable. It also appears reasonable to say that, given the circumstances, we would be more justified to administer the drug by force. For that is what the advancement of the 'common good' seems to require. And why stop here? It is no doubt in the public interest that perpetrators are found out and punished, but surely it would even be better if they didn't commit any crimes in the first place. If that is the case, then developing and distributing a pill that prevents people from committing crimes (in accordance with Savulescu's suggestions for moral enhancement) would clearly be an enhancement for the common good in Vedder and Klaming's sense.

Now I'm not saying that improving eyewitness memory in the way proposed by Vedder and Klaming is such a bad thing. Perhaps it is, and perhaps it isn't. What I'm worried about is the *argument* that the authors use to support their claim that the use of TMS would be an "enhancement for the common good," and the *suggestion* that we generally look at enhancement proposals from a "common good perspective." Any notion of common good that is worth its salt must be informed by what is good for the individual, and the connection must always be born in mind and properly considered. The concession that even common good enhancements must always be voluntary is not sufficient. Once memory enhancement (TMS), veracity enhancement (truth serums), or morality enhancement (no-crime pill) are available, it's going to be very hard to refuse them. For why would anyone refuse to assist the law, unless they got something to hide?

---

[4] *Scientific American*, 4 December 2008.

References

Bostrom N (2005) In defense of posthuman dignity. Bioethics 19(3):202–214
Bostrom N (2008) Why I want to be a posthuman when I grow up. In: Gordijn B, Chadwick R (eds) Medical enhancement and posthumanity. Springer, Dordrecht, pp 107–136
Buchanan A, Brock DW, Daniels N, Wikler D (2000) From chance to choice. Cambridge University Press, Cambridge
Caplan A (2006) Is it wrong to try to improve human nature? In: Miller P, Wilsdon J (eds) Better humans? The politics for human enhancement and life extension. Demos, London, pp 31–39
Glaser B et al (2011) Age- and puberty-dependent association between IQ score in early childhood and depressive symptoms in adolescence. Psychol Med 41:333–343
Harris J (2011) Moral enhancement and freedom. Bioethics 25(2):102–111
Hauskeller M (2010) Human enhancement and the common good. AJOB Neurosci 1(3):37–39
Hauskeller M (2011) Human enhancement and the giftedness of life. Philos Pap 40(1):55–79
Lewis CS (1955) The abolition of man. Macmillan, New York
Luria AR (1987) The mind of a mnemonist. Harvard University Press, Cambridge, MA
Persson I, Savulescu J (2010) Moral transhumanism. J Med Philos 35(6):656–669
Savulescu J (2001) Procreative beneficence: why we should select the best children. Bioethics 15(5–6):413–426
Szalavitz M (2009) Popping smart pills: the case for cognitive enhancement. TIME Magazine, 6.1.09
Vedder A, Klaming L (2010) Human enhancement for the common good: using neurotechnologies to improve eyewitness memories. AJOB Neurosci 1(3):22–33

Chapter 11
Nano-bionic Devices for the Purpose of Cognitive Enhancement: Toward a Preliminary Ethical Framework

Frédéric Gilbert

Abstract This chapter examines the emerging ethical challenges raised by implementation of nanotechnology in brain devices for enhancement purposes in subjects with healthy brains. This chapter will proceed in five steps. The first section introduces brain implants and discusses how their status may be changed by nanotechnologies for enhancement purposes. The second section explores whether the ethics of nano-bionic devices for cognitive enhancement purposes in healthy, informed subjects might be helped by referring to the treatment-enhancement distinction. Such a distinction could serve to illuminate guidelines and policies. The third section examines whether the designs for nano-bionic devices for cognitive enhancement raise a number of intrinsically new ethical problems if applied to healthy subjects. The fourth section looks at whether nano-bionic devices used for the purpose of enhancement introduce novel ethical difficulties to the informed consent of healthy and free subjects. The fifth section sketches the preliminary ethics that have to be established before a healthy individual could undergo an informed consent process for invasive nano-bionics brain intervention.

Keywords Cognitive enhancement • Nanotechnology • Brain implants • Ethics • Informed consent

11.1 Introduction

In recent years, treatments using brain devices have attracted great interest in the medical community, not only for their effectiveness in terms of treatment, but also for their potential future applications. In the near future, developments in medical

---

F. Gilbert (✉)
School of Philosophy, Faculty of Arts, ARC Centre of Excellence for Electromaterials Science (ACES), University of Tasmania, Hobart, TAS, Australia
e-mail: frederic.gilbert@utas.edu.au

E. Hildt and A.G. Franke (eds.), *Cognitive Enhancement: An Interdisciplinary Perspective*, Trends in Augmentation of Human Performance 1,
DOI 10.1007/978-94-007-6253-4_11, © Springer Science+Business Media Dordrecht 2013

nanotechnologies have the potential to make these brain device technologies less invasive and more localized, less costly and more reliable. These possibilities will play major roles in the performance of the implants while expanding the safety and the range of applications (Andrews 2009). With these great hopes for nano-bionic devices, designed for the brain, come various ethical challenges (Berger et al. 2008).

Human brain implants can be described as surgically invasive devices. They are commonly used for the following treatments: to remediate various symptoms of innate or acquired loss of sensory functions (e.g. cochlear implants, retinal implants) (Andrews 2009); to restore movement in cases of paralysis (e.g. motor neuroprosthetics) (Kennedy and Bakay 1998; Hochberg et al. 2006); and to treat neuronal diseases (e.g. Deep Brain Stimulation implants). According to MEDTRONIC, the largest manufacturer of Deep Brain Stimulation devices, since 1995 over 80,000 people worldwide have been treated with their brain implants (Medtronic 2011). Additionally, as of 2009, approximately 188,000 people worldwide have received cochlear implants according to the U.S. Food and Drug Administration (Davis 2009).

Deep Brain Stimulation (DBS) implants are versatile and can be applied to many types of neurological dysfunction. The current applications of DBS treatments have targeted Parkinson's symptoms, such as tremor, rigidity, stiffness, slowed movement, or even walking problems. Others DBS applications include treatment for diseases, such as dystonia, cluster headaches, minimally conscious state, and phantom limb pain (Clausen 2010). The technique consists of implanting a battery-operated neuromedical device that delivers an electrical stimulation to a specific area of the patient's brain. DBS technologies have also been used to treat psychiatric conditions (most of them are still in the experimental phase), such as severe depression, obsessive-compulsive disorder, Tourette's syndrome, obesity, epilepsy, Alzheimer's disease, impulsive and aggressive behavior and addiction (Franzini et al. 2005; Kuhn et al. 2007; Heinze et al. 2009; Clausen 2010).

In terms of enhancements, Hamani et al. (2008) and Adrian et al. (2010) have reported serendipitous correlations between DBS and cognition (e.g. memory), while Berney et al. (2002) and Houeto et al. (2002) have observed positive mood modification (e.g. decrease in major depressive tendencies). In addition, Tsai et al. (2010) and Haq et al. (2010) reported functionality enhancements (i.e. increased libido). These discoveries have sparked great interest and speculation, especially with regard to future nanomedicine implementations.

Researchers are currently developing a set of new generation brain implants that use nanomaterials. A nanometer is one-billionth of a meter. The rapid expansion of nanomaterial for use in brain implants holds great promise. For example, biosensors, which are used for targeted drug delivery and as platforms for neural cell growth (Lee and Parpura 2009). Researchers at the Australian Center of Excellence in Electromaterial Science (ACES) have already engineered biodegradable polymers such as cortex drug delivery to treat epilepsy in rodents (Wilz et al. 2008; Halliday and Cook 2009). Moreover, members of ACES are also involved in the world's first-in-human trial using an "intelligent brain implant" with a seizure warning for epileptic patients. The brain implant can predict when an epileptic seizure will occur (Snyder et al. 2008; NeuroVista 2011) and issue an alert. Current research at ACES involves the development of brain implants that can detect and deliver drugs for use in humans. These current applications allow us to consider seriously the expansion of applications of these devices to other neural diseases. It is quite possible to imagine, in the near future, the use of biodegradable polymers for cortex drug delivery associated with an intelligent brain device to treat various neuronal diseases or an intelligent brain implant that can predict and send warnings for other neuronal dysfunctions, such as the imminent onset of an aggressive phase, manic phase, addictive rush, etc. These future applications are still speculative; nonetheless, it is important to develop an ethical framework around such future development given that such treatment could directly affect how healthy individuals improve their well-being (Gilbert et al. 2012).

According to the constitution of the World Health Organization (1946), "health is a state of complete physical, mental and social well-being and not merely the absence of disease or infirmity." The constitution adds that "the enjoyment of the highest attainable standard of health is one of the fundamental rights of every human." What if the use of nano-bionic devices could achieve the highest attainable physical, mental and social well-being? In others terms, what if one day, versatile brain implants using nanomaterials were not strictly reserved for treating patients with neurodegenerative diseases and psychiatric conditions? Given this possible extension of the uses of nano-bionic devices, it behooves us to ask: Is there something intrinsically unethical about using these devices for cognitive enhancement purposes in healthy, fully informed, free, consenting subjects? Why would nanotechnologically-based, human enhancement be morally impermissible if it respects the individual's right to choose for her own physical integrity?

11.2 Much Ado About Nothing?

What is enhancement? For its proponents, enhancement is merely a form of treatment which could help people be *better* rather than *well*. From a scientific point of view, it is not clear if any significant enhancement effects exist in healthy subjects (Repantis et al. 2008). To this unknown, there is an emerging literature that highlights the "exaggeration" regarding the use of cognitive enhancement (Partridge et al. 2011), exaggeration that has encouraged speculative assumptions that reduce ethics to approximate debate (Gilbert 2011). That being said, this section examines whether the ethics of nano-bionic devices for the purpose of cognitive enhancement in healthy, informed subjects might be helped by referring to the treatment-enhancement distinction, which could then serve to establish guidelines and policies for the legitimate use of such devices.

There are difficulties in precisely defining a distinction between enhancement and treatment with respect to nano-bionic devices. On one hand, the use of the concept of treatment is universally associated with ethical values: treatments are intended to restore health or to prevent sickness. On the other hand, enhancement is a polysemic concept that has numerous uses in academic literature that are not necessarily associated with ethics. When the notion of enhancement is used in philosophy, ethical values feature prominently. For instance, transhumanism, a group of futurist philosophical theories, regards enhancement as "one of the most important and challenging issues of the new century" (Bostrom and Savulescu 2009); other schools of thought point out that, "the challenges of human enhancement" make it important to think about how research is conducted (Juengst et al. 2003). However, when enhancement is used in science, the concept refers to any amelioration of a state without reference to ethical value. This use can be illustrated by random academic examples such as "adenovirus enhancement" (Curiel et al. 1991); "enhancement of transfection efficiency of naked plasmid DNA in skeletal muscle" (Taniyama et al. 2002); "enhancement of secreted phosphoprotein" (Noda et al. 1990), etc. Under these scientific descriptions, the notion of enhancement does not involve any ethical value.

The use of the concept of enhancement in the above scientific examples is restricted to biomedical mechanisms. However, it has to be remembered that even in these restricted senses, ethical questions arise. For instance, if science has focused on enhancing one's genes that contribute to memory in particular, then, at the same time, science has also enhanced one's memory overall. The former use of "gene enhancement" may seem to have little ethical value, but the latter "memory enhancement" does raise questions of ethical value.

This analysis helps us to see that the treatment-enhancement distinction might be difficult to uphold (Harris 2007). However, one could still defend a traditional way of preserving the distinction by arguing that medicine should be limited to treatment and should not engage in enhancement. How strong is this position?

It is not a controversial claim to affirm that every treatment aims to enhance a certain state. It is the case that all treatments are legitimated only by the assumption that they are likely to improve a patient's quality of life. So, every treatment constitutes a form of enhancement somehow (Elliott 2003). To be sure, if a convincing line is drawn between the treatment (medical treatment to restore or preserve a patient's quality of life) and the enhancement of otherwise normal traits (clearly to improve normal functioning), this line would not coincide with what medicine is morally obliged to do or not do (Daniels 2000). Since treatment involves a form of enhancement, what would be ethically permissible versus impermissible may not be allowed by what exactly constitutes cognitive enhancement or treatment in terms of nano-bionic devices. To speak about the distinction between treatment versus enhancement is neither necessary nor helpful in order to find the appropriate ethical framework for the use of nano-bionic devices in healthy, informed and consenting subjects.

What really matters in ethics is not the question of whether every treatment is a form of enhancement, but rather, whether every enhancement is a form of treatment. A closer look at the universal argument, "every treatment presents a form of enhancement" (Synofzik 2009), reveals a major logical conflation. In fact, the argument is not sufficient to conclude anything about enhancement versus treatment since it does not necessarily imply that "all enhancement presents a form of treatment." It is a logical conflation to derive the second proposition from the first. In fact, if "every treatment presents a form of enhancement," an argument such as, "some enhancements are not a form of treatment," is sufficient to demonstrate the logical fallacy of "every treatment presents a form of enhancement." For instance, somatropin (human growth hormone), used in sports, is not a form of treatment; it is, rather, only a form of enhancement.[1] From this proposition, it is possible to conclude that if not every enhancement is a form of treatment, then not every treatment is a form of enhancement. In other words, some nano-bionic device treatments could be presented as a form of enhancement but not all. The same conclusion is applicable for nano-bionic devices for cognitive enhancements purposes: some of them may be presented as forms of treatment but not all. This criticism of the argument does not modify our previous point concerning enhancement of healthy, informed subjects. The distinction between treatment-enhancement is not necessary. This distinction does not help to develop an ethical framework for nano-bionic devices, which could serve to steer guidelines and policies for their legitimate use in healthy individuals. The next section will examine whether nano-bionic devices raise new ethical issues.

11.3 Novel Ethical Issue?

Long before Drexler's 1986 book, *Engines of Creation: The Coming Era of Nanotechnology* (Drexler 1986), which anticipates the use of precise molecular assembly for a wide variety of applications in order to increase the quality of human life, meliorating the human condition through advances in technology had been debated. The debate can, in part, be traced back to the Enlightenment. Already, in 1795, Condorcet, in his "Outlines of an historical view of the progress of the human mind," asked whether the "human race should be meliorated by new discoveries in the sciences" within "the improvement of the instruments which increase the power and direct exercise of [moral, intellectual and physical] faculties."[2] Condorcet's question followed his assumption:

> No bounds have been fixed to the improvement of the human faculties; that the perfectibility of man is absolutely indefinite; that the progress of this perfectibility, henceforth above the control of every power that would impede it, has no other limit than the duration of the globe upon which nature has placed us. (de Condorcet 1795)

---

[1] I assume that enhancers have an enhancement effect in any body (i.e. a healthy body and a sick body). For instance, growth hormones affect any body, not only a body with disease. However, not all treatment will have an enhancement effect (i.e. a healthy body and a sick body). For example, could a treatment with radiation or a treatment for influenza enhance a healthy body?

[2] "L'espèce humaine doit-elle s'améliorer, soit par de nouvelles découvertes dans les sciences [...]; soit enfin par le perfectionnement réel des facultés intellectuelles, morales et physiques, qui peut être également la suite, ou de celui des instruments qui augmentent l'intensité et dirigent l'emploi de ces facultés" (Condorcet 2004: 430).

This exerpt from Condorcet illustrates that modern ideas of using advances in technology for the purposes of enhancement, whether it be to enhance physical and cognitive capacities, to enhance moods or to extend human life spans, can be seen as a continuation of the enlightenment project (Hughes 2010) and not as a new proposal of the futuristic transhumanist agenda (Bostrom 1998). Enhancement is not unique to our epoch (Gilbert and Baertschi 2011), and given this observation, one could ask whether the ethics of nanotechnology in general, (Lenk and Biller-Andorno 2007; Alpert 2008; Ferrari and Nordmann 2010) and of nano-bionic device technology in particular, could bring something intrinsically novel to the debate?

What might be perceived as novel is that as the enhancement capacity increases with nano-bionic implants, neuroscience might become less interested in what brains *are*, and more interested in what neuroscience can make brains *become*. This distinction between a pre-enhanced brain (the "given" of what brains are) and what a brain is "made to be" through nano-enhancement implies that the brain can be changed or modified. As nanotechnology becomes more integrated with the brain, the idea arises that a nano implant might change the brain's intrinsic functionality. Supposedly, the kinds of changes available through nano-bionic implants will not be concerned with what brains are being *changed from*, but rather with what brains are being *changed to* – perhaps because of their paradoxical magnitude and subtlety all at once. A potential ethical novelty could result in practicing science in the interests of change rather than preserving brain functionalities (Glover 1984). However, this speculative view has it limits. Indeed, what makes nano-bionic devices any different from other brain enhancers (i.e. pharmaceutical, genetic, electrical, etc.) for making neuroscience more future-oriented than it has been to date? Such an orientation is not exclusive to nanomedicine.

Questioning the intrinsic novelty of nano-bionic devices forces us to ask whether they bring something unique to the relation between a *preoperative* person and a *postoperative* person. In others words, do nano-bionic devices shed any new light on the reasons why a person who is seeking nano-bionic enhancement must be concerned with the postoperative person who she could become? On a first approximation, it appears that at the moment of consent, the person who is seeking nano-bionic cognitive enhancement must care about the post-operative person's life as much as with any other invasive brain intervention. These ethical issues that arise from nano-bionic devices are not categorically distinct from those arising from earlier generations of brain implants. For instance, a person receiving DBS has to pass through a process of consent; this process raises similar ethical issues regarding personality alteration (Gilbert 2012; 2013). In general terms, nano-bionic devices are not unique in that matter. Nano-bionic devices do not seem to intrinsically impose new disproportionate burdens on the postoperative person for the sake of the person consenting to enhance herself. Like other cases of invasive intervention, nano-bionic devices do not exclusively concern the preoperative challenges regarding cognitive enhancement, but rather the potential postoperative issues, such as personality changes, which could occur after the intervention. Although not novel, it needs to be mentioned that this risk of damaging the physical and psychological integrity of the postoperative person might be morally wrong to ignore; at least the risk ought not to be ignored because of some *laissez faire* policies. Seeking nano-bionic cognitive enhancement requires any patient to be fully informed of the risks involved before, during and after the intervention, but such intervention does not raise any fundamentally new ethical question.

Also, if there is a genuine ethical problem associated with the use of nano-bionic devices for enhancement capacities in healthy subjects, it is not on account of the risks of transformation of personal identity. Indeed, it would be a philosophical blunder to believe that the risks of upsetting personal identity through nano-bionic devices are categorically different from those achieved through pharmacological enhancements (or other forms of enhancers). Concerns that nano-bionic technology may threaten personal integrity are sometimes overstated. Comparatively speaking, a glance at how religion or cultural identity can permanently affect personality tells us that nano-bionic devices may not be more damaging to the integrity of one's person than, say, unconscious social pressure in a healthy individual's life.

To help ensure that the debate concerning the use of nano-brain devices for enhancement is not slowed down by numbers of publications that try in vain to find unique or authentic ethical issues in terms of enhancement, I adopt Nordmann's view that ethical literature should be based on its application rather than on any speculative scenario (Nordmann 2007). I believe that using hypothetical cases to ascertain ethical attitudes toward the risks and benefits of nano-neuro implants for enhancement may lead to social misunderstanding, not to mention social panic (e.g. the impact of somatic nuclear transfer for reproductive purpose[3]). An attempt to involve concrete ethical cases will shape the direction of ethical research toward a rational nanotechnology enhancement debate.

In terms of novelties, it seems that nano-brain devices are not intrinsically unique in terms of enhancement, but I believe they raise a number of issues that need to be considered in future ethical guidelines and clinical trials. Let us look more closely at these issues in the following sections.

11.4 Challenge to Informed Consent

As a necessary precondition for informed consent, a healthy individual seeking cognitive enhancement must be (1) a volunteer, chosen without manipulation, undue influence or coercion (2) competent to appreciate the risks and benefits, (3) able to understand alternative treatment options, and (4) free of psychiatric conditions or psychiatric comorbidity. Admitting that the efficacy and safety of nano-bionic

[3]The debate on somatic nuclear transfer for reproductive purpose (reproductive cloning) has been monopolized and depicted without accuracy by the media. This phenomenon has dragged ethical questions on speculative ground and shifted the focus away from issues of safety and efficacy.

devices could be reasonably possible for cognitive enhancement, if a subject lacks one of these four minimum ethical requirements, then the subject should be excluded from invasive neurosurgery.

As an essential element of informed consent – the appreciation of the risks of harm (that subjects must assume) may not be obvious at the time of the decision. Potential patients may be legally competent and free to choose concerning their own physical and psychological integrity, but they may not be able to make meaningfully autonomous decisions regarding consent to nano-bionic cognitive enhancement. When in consultation for cognitive enhancement intervention, a healthy patient may feel invulnerable to safety issues given the possibility of quick beneficial effects. Indeed, healthy populations might underestimate the long-term risk effects of nanomaterials in the pursuit of quick results. A look at the phenomenal increase of the use of botulinum toxin type A or sildenafil, the former for beauty enhancement, the latter for sexual performance, shows that many healthy individuals, with freedom of choice, are not hindered by the potential for long-term risks if there are immediate, tangible benefits (Turner and Sahakian 2006). An incorrect estimation of the risk may result in a notable increase in claims that allege failure to obtain proper informed consent before intervention; this is an issue that plastic surgery is facing (Tebbetts and Tebbetts 2002).

If nano-bionic device clinics for enhancement purposes of healthy subjects become a reality, practitioners have to ensure that a patient's comprehension of the risks and uncertainties is evaluated at the time of obtaining consent. Official national and international guidelines should be developed by multidisciplinary experts. Guidelines could provide information to assist potential patients' decision-making.

Nano-bionic devices for cognitive enhancement purposes could be ethically approved for healthy subjects if subjects make informed decisions and are not coerced. Although not entirely sufficient, this approach to conditions for access to nano-bionic devices will ensure ethical approval for the enhancement field – safety being the core ethical prerequisite. The nano-bionic device enhancement field should avoid being inspired by pure mercantile profit and aggressive marketing; it should avoid the less-than-noble model of the field of plastic surgery, where recently, pathology labeled 'plastic surgery addiction' has been observed (Wright 1986; Phillips et al. 2001). Risk of addiction in relation to enhancement should not be ignored. In terms of nano-bionic implants, the need to continually enhance oneself could lead to risky addiction. Given that a certain level of electric stimulation might result in a certain level of enhancement, who should decide and adjust the implant stimulation level? Famous studies of rat brains using positive reinforcement produced by electrical stimulation have demonstrated addictive, often lethal behavior outcomes (Olds and Milner 1954; Wright 1987). Similar compulsive self-stimulation in human has been already reported in the literature (Portenoy et al. 1986). Postoperative follow-up issues raise many complex questions.

A study by Macoubrie (2005) of public perceptions of nanotechnology found that the public anticipated major benefits from nanotechnology. How do such perceptions arise in public opinion? Can such opinion influence an individual's informed consent process? Healthy individuals use the media as a primary source of information about new technologies and their possible applications. Although the media have an important role in disseminating information, they also play an influential role in disseminating *mis*information. Can positive media depictions of the potential of nanotechnology for enhancement purposes influence a patient's informed consent? A review of DBS media studies might be able to answer this question. Bell et al. (2010) have reported an insightful study on the impact of enthusiastic media portrayals of DBS on patients' hopes and expectations. They concluded that healthcare providers view media portrayals of DBS as "playing a key role in establishing expectations for DBS patients and for public in general" (Bell et al. 2010). From the point of view of healthy patients who go through an informed consent process because they are willing to enhance their memory, the use of an easily available optimistic depiction (Racine et al. 2007) – both in the medical literature and in the popular media – might be a challenge to truly informed consent. Ford (2009) suggests that the overtly optimistic reports about new neurosurgical innovations generate an "educational vulnerability" for patients. Ford affirms that very often when patients approach neurosurgical techniques, they have already been preconditioned by overly optimistic portrayals of novel brain interventions, and this compromises truly informed consent. Informed consent is an important mechanism for respecting patient autonomy, but in order to reach this ambitious goal, the effect of exposure to unbalanced media reports must be considered.

Based on these observations, enhancement providers would have the responsibility of designing a process for obtaining fully informed consent, while avoiding the exploitation of unrealistic hopes built by optimistic portrayals that do not engage in any ethical examination. Healthy patients seeking enhancement must be educated about the realistic potential of nano-brain devices because they may harbor misconceptions about the potential for enhancement efficacy and underestimate the safety risks.

11.5 Recourse to Reasonable Medical Alternatives

The invasive nature of nano-bionic device procedures exacerbates bona fide concerns of safety, as compared to non-invasive pharmacological interventions, such as selective serotonin reuptake inhibitor drugs or beta-receptor blockers. Since pharmacological intake implies less severe risks and adverse complications in both the short and long term, it stands to reason that a non-invasive method is undeniably preferable to surgical interventions for cognitive enhancement.

Although the application of nano-bionic devices does not seem to raise novel ethical problems, significant ethical concerns are nonetheless at stake. Besides the technical difficulties involved in making risk assessments, which try to project the likelihood and extent of harm that could result for healthy patients seeking nano-bionics enhancement, there are also ethical difficulties to face. These include questions such as:

1. Under what circumstances should healthy individuals have the right to receive nano-bionic devices for cognitive enhancement purposes?
2. What level of prospective benefit could justify such invasive nano-bionic interventions in healthy patients?
3. How should nano-bionic devices be regulated to assure the primary concern for patient safety, without stopping research in this field altogether?

These questions must be thought through and given serious ethical attention before applications for nano-bionic devices can be approved for use in healthy human subjects.

Nano-bionic devices should be seen as a last solution, the appropriateness of which depends on what outcomes a patient wants to achieve through this invasive intervention. Before any invasive intervention could be regarded as an option, it must be proven that it enhances mood as well as functional and cognitive abilities in healthy subjects in general; moreover, it must be proven – with a high level of significance – that it could enhance the very individual asking for the intervention. Any benefit that can result from an individual's enhancement in particular does not necessarily entail that the whole population would be identically enhanced given the same intervention. Not only should it be established that an intervention could enhance the particular individual, but it must also be demonstrated that it is the only alternative left in order to reach that individual's aim. Nano-bionic enhancers must offer an individual significant advantages over non-enhanced individuals; indeed, the level of improvement is often well below the level of normal function (Repantis et al. 2008). Additionally, nano-bionic devices must be proven to be more effective than pharmacological measures available to the individual.

Severe short and long-term side effects have to be eradicated on both a physical and psychological level before nano-bionic devices can be considered a viable option for healthy subjects. Yet, to address the ethics of these requirements, nano-bionic devices must prove themselves, and part of doing so means they must pass scientific muster. At this point, assurances of the safety and efficacy of nano-bionic devices are not borne out in the literature. Current reporting on DBS is particularly vulnerable to researcher/investigator bias because of an excessive reliance on single-patient case reports (Schlaepfer and Fins 2010; Gilbert and Ovadia 2011). Although the risk that selective publishing poses is by no means unique to DBS, it is essential that safety and efficacy be proven, and for this, higher powered studies will be needed. Ethical patient selection for such studies will of course be a challenge. Alternatively, without higher powered studies, investigator bias may drive development of this research.

Rather than imposing a total ban on further nano-bionic implant developments or endorsing total freedom, ethical regulations (informed and uncoerced choice, restriction to cases where there are no other alternatives available, competence to appreciate the risk, no psychiatric conditions or comorbidity) should be evaluated with priority.

11.6 Conclusion

There is a wide spectrum of positions in the human enhancement debate. For the advocates of human enhancement, the debate should be solved with respect to a fundamental right: the freedom of choice regarding psychological and physical integrity. What could be wrong with a healthy person enhancing her own brain functions with a nano-device if it harms no one other than perhaps the individual? Following this preliminary investigative chapter, I conclude that regulating enhancement technology for enhancement purposes is not directly infringing on the fundamental right of freedom of choice with regard to physical and psychological integrity. Although there might not be a right to prevent one from deciding to engage in risky interventions that could result in damaging physical and psychological integrity, there is a duty to prevent one from deciding to engage in an unethically invasive intervention that could damage one's physical and psychological integrity. Although nano-bionics for cognitive enhancement is not unethical *per se*, there are sufficient reasons for restricting the kind of interventions surgeons are requested to do. It is a corollary from the necessity to regulate the informed consent that each person's freedom to decide must be limited so as not to directly inflict unnecessary bodily injury on one's own physical and psychological integrity (i.e. the person willing to undergo enhancement), nor on any other (e.g. a surgeon). As seen above, regarding the deliberative informed consent process, it is not clear where the realm of autonomy ends and where the realm of heteronomy begins. Allowing nano-bionic enhancement without the four minimum requirements for a proper informed consent framework is simply unethical. Individual autonomy could not survive without health and safety concerns. If a surgical brain procedure is proven to be safe and effective, the enhancement debate becomes more of a matter for healthcare policymakers rather than an issue based exclusively on the right to freedom of choice.

No one can predict for what purposes bionic devices built with nanomaterials will be used in a few decades or even what nano-bionic devices will become in the history of cognitive enhancement in terms of ethics. Given their current stage of development, nano-bionic technologies for enhancement purposes are more speculative than actual. It would be premature to advocate invasive nano-neurosurgical interventions strictly on the basis of freedom of choice. If one day nano-bionic devices could be used for enhancement purposes, policymakers must ensure that their use will not heighten behavioral consumption patterns and dependency on technology, which are unfortunately related to individuals having freedom of choice.

Acknowledgement This research was funded by the Australian Centre of Excellence for Electromaterials Science (ACES). Thanks to Susan Dodds, Eliza Goddard and Timothy Krahn.

References

Adrian W, Laxton Tang-Wai DF, McAndrews MP et al (2010) A phase I trial of deep brain stimulation of memory circuits in Alzheimer's disease. Ann Neurol 68(4):521–534

Alpert S (2008) Neuroethics and nanoethics: do we risk ethical myopia? Neuroethics 1:55–68

Andrews RJ (2009) Neuromodulation: deep brain stimulation, sensory neuroprostheses, and the neural-electrical interface (In: Sharma HS (ed) Nanoneuroscience and nanoneuropharmacology). Prog Brain Res 180:127–139

Bell E, Maxwell B, McAndrews MP, Sadikot A, Racine E (2010) Hope and patients' expectations in deep brain stimulation: healthcare providers' perspectives and approaches. J Clin Ethics 21:112–124

Berger F, Gevers G, Siep L, Weltring KM (2008) Ethical, legal and social aspects of brain-implants using nano-scale materials and techniques. Nanoethics 2:241–249

Berney A, Vingerhoets F, Perrin A, Guex P, Villemure J-G, Burkhard PR, Benkelfat C, Ghika J (2002) Effect on mood of subthalamic DBS for Parkinson's disease: a consecutive series of 24 patients. Neurology 59:1427–1429

Bostrom N (1998) What is transhumanism? http://www.nickbostrom.com/old/transhumanism.html. Accessed 18 Apr 2012

Bostrom N, Savulescu J (2009) Human enhancement ethics: the state of the debate. In: Savulescu J, Bostrom N (eds) Human enhancement. Oxford University Press, Oxford

Clausen J (2010) Ethical brain stimulation – neuroethics of deep brain stimulation in research and clinical practice. Eur J Neurosci 32:1152–1162

Condorcet N (2004) Tableau historique des progrès de l'esprit humain. Projets, esquisee, fragments et notes (1772–1974). Schandeler J.-P, et Crépel P (eds) Paries: Institut National d'Etudes Démographique

Curiel DT, Agarwal S, Wagner E, Cotton M (1991) Adenovirus enhancement of transferrin-polylysine-mediated gene delivery. Proc Natl Acad Sci USA 88:8850–8854

Daniels N (2000) Normal functioning and the treatment-enhancement distinction. Camb Q Healthc Ethics 9:309–322

Davis J (2009) Peoria's first cochlear implant surgery has grandfather rediscovering life. Peoria Journal Star, 2009-10-29. http://www.pjstar.com/features/x876590686/Peoria-s-first-cochlear-implant-surgery-has-grandfather-rediscovering-life. Accessed 7 Mar 2011

de Condorcet N (1795) Outlines of an historical view of the progress of the human mind [1793]. J. Johnson, London

Drexler EK (1986) Engines of creation: the coming era of nanotechnology. Anchor Books, New York, http://e-drexler.com/d/06/00/EOC/EOC_Table_of_Contents.html

Elliott C (2003) Better than well: American medicine meets the American dream. Norton, New York

Ferrari A, Nordmann A (2010) Beyond conversation: some lessons for nanoethics. Nanoethics 4:171–181

Ford PJ (2009) Vulnerable brains: research ethics and neurosurgical patients. J Law Med Ethics 37:73–83

Franzini A, Marras C, Ferroli P, Bugiani O, Broggi G (2005) Stimulation of the posterior hypothalamus for medically intractable impulsive and violent behavior. Stereotact Funct Neurosurg 83:63–66

Gilbert F (2011) Working while under the influence of performance enhancing drugs: is one "more responsible"? Am J Bioeth Neurosci 2:57–59

Gilbert F (2012) The burden of normality: from 'chronically ill' to 'symptom free'. New ethical challenges for deep brain stimulation postoperative treatment. J Med Ethics. doi:10.1136/medethics-2011-100044

Gilbert F (2013) Deep brain stimulation for treatment resistant depression: postoperative feeling of self-estrangement, suicide attempt and impulsive-aggressive behaviours, Neuroethics. doi: 10.1007/s12152-013-9178-8

Gilbert F, Baertschi B (2011) Neuroenhancement: much ado about nothing? Am J Bioeth Neurosci 2:45–47

Gilbert F, Ovadia D (2011) Deep brain stimulation in the media: over-optimistic media portrayals calls for a new strategy involving journalists and scientifics in the ethical debate. J Integr Neurosci 5:16. doi:10.3389/fnint.2011.00016

Gilbert F, Harris AR, Kapsa RMI (2012) Efficacy testing as primary purpose of phase 1 clinical trials: is it applicable to first-in-human bionics and optogenetics trials? Am J Bioeth Neurosci 3:20–22

Glover J (1984) What sort of people should there be? Penguin, New York

Halliday AJ, Cook MJ (2009) Polymer-based drug delivery devices for neurological disorders. CNS Neurol Disord Drug Targets 8:205–221

Hamani C, McAndrews MP, Cohn M, Oh M, Zumsteg D, Shapiro CM, Wennberg RA, Lozano AM (2008) Memory enhancement induced by hypothalamic/fornix deep brain stimulation. Ann Neurol 63:119–123

Haq I, Foote K, Goodman W, Ricciuti N, Ward H, Sudhyadhom A, Jacobson C, Siddiqui M, Okun M (2010) A case of mania following deep brain stimulation for obsessive compulsive disorder. Stereotact Funct Neurosurg 88:322–328

Harris J (2007) Enhancing evolution. The ethical case for making better people. Princeton University Press, Princeton

Heinze HJ, Heldmann M, Voges J, Hinrichs H, Marco-Pallares J, Hopf JM, Müller UJ, Galazky I, Sturm V, Bogerts B, Münte TF (2009) Counteracting incentive sensitization in severe alcohol dependence using deep brain stimulation of the nucleus accumbens: clinical and basic science aspects. Front Hum Neurosci 3:22

Hochberg L, Serruya M, Friehs G, Mukand J, Saleh M, Caplan A, Branner A, Chen D, Penn R, Donoghue J (2006) Neuronal ensemble control of prosthetic devices by a human with tetraplegia. Nature 442:164–171

Houeto JL, Mesnage V, Mallet L, Pillon B, Gargiulo M, du Moncel ST et al (2002) Behavioural disorders, Parkinson's disease and subthalamic stimulation. J Neurol Neurosurg Psychiatry 72:701–707

Hughes J (2010) Technoprogressive biopolitics and human enhancement. In: Jonathan M, Sam B (eds) Progess in bioethics. MIT Press, Hong Kong, pp 163–188

Juengst ET, Binstock RH, Mehlman M, Post SG, Whitehouse P (2003) Biogerontology, "anti-aging medicine," and the challenges of human enhancement. Hastings Cent Rep 33:21–30

Kennedy PR, Bakay RA (1998) Restoration of neural output from a paralyzed patient by a direct brain connection. Neuroreport 9:1707–1711

Kuhn J, Lenartz D, Huff W, Lee S, Koulousakis A, Klosterkoetter J, Sturm V (2007) Remission of alcohol dependency following deep brain stimulation of the nucleus accumbens: valuable therapeutic implications? J Neurol Neurosurg Psychiatry 78:1152–1153

Lee W, Parpura V (2009) Carbon nanotubes as substrates/scaffolds for neural cell growth (In: Sharma HS (ed) Nanoneuroscience and nanoneuropharmacology). Prog Brain Res 180: 111–125

Lenk C, Biller-Andorno N (2007) Nanomedicine-emerging or re-emerging ethical issues? A discussion of four ethical themes. Med Health Care Philos 10:173–184

Macoubrie J (2005) Informed public perceptions of nanotechnology and trust in Government Woodrow Wilson International Center for Scholars, the Pew charitable Trusts. http://www.wilsoncenter.org/sites/default/files/macoubriereport1.pdf. Accessed 18 Apr 2012

Medtronic (2011) Deep brain stimulation for movement disorders. http://professional.medtronic.com/therapies/deep-brain-stimulation/index.htm. Accessed 7 Mar 2011

NeuroVista (2011) NeuroVista receives approval for expanding Australian clinical study. http://www.neurovista.com/docs/2011-01-30_PressRelease_ApprovalforExpandedTrial.pdf. Accessed 17 Mar 2011

Noda M, Vogel RL, Craig AM, Prahl J, DeLuca HF, Denhardt DT (1990) Identification of a DNA sequence responsible for binding of the 1,25-dihydroxyvitamin D3 receptor and 1,25-dihydroxyvitamin D3 enhancement of mouse secreted phosphoprotein 1 (Spp-1 or osteopontin) gene expression. Proc Natl Acad Sci USA 87:9995–9999

Nordmann A (2007) If and then: a critique of speculative nanoethics. Nanoethics 3:31–46

Olds J, Milner P (1954) Positive reinforcement produced by electrical stimulation of septal area and other regions of rat brain. J Comp Psychol 47:419–427

Partridge BJ, Bell SK, Lucke JC, Yeates S, Hall WD (2011) Smart drugs "as common as coffee": media hype about neuroenhancement. PLoS One 6(11):10.1371/journal.pone.0028416

Phillips KA, Grant J, Siniscalchi J, Albertini RS (2001) Surgical and nonpsychiatric medical treatment of patients with body dysmorphic disorder. Psychosomatics 42:504–510

Portenoy KR, Jarden JO, Sidtis JJ, Lipton RB, Foley KM, Rottenberg DA (1986) Compulsive thalamic self-stimulation: a case with metabolic, electrophysiologic and behavioral correlates. Pain 27:277–290

Racine E, Waldman S, Palour N, Risse D, Illes J (2007) "Currents of hope": neurostimulation techniques in U.S. and U.K. print media. Camb Q Healthc Ethics 16:312–316

Repantis B, Schlattmann P, Laisney O, Heuser I (2008) Antidepressants for neuroenhancement in healthy individuals: a systematic review. Poiesis and Praxis 6:139–174

Schlaepfer TE, Fins J (2010) Deep brain stimulation and the neuroethics of responsible publishing, when one is not enough. JAMA 303:775–776

Snyder DE, Echauz J, Grimes DB, Litt B (2008) The statistics of a practical seizure warning system. J Neural Eng 5:392–401

Synofzik M (2009) Ethically justified, clinically applicable criteria for physician decision-making in psychopharmacological enhancement. Neuroethics 2:89–102

Taniyama Y, Tachibana K, Hiraoka K, Aoki M, Yamamoto S et al (2002) Development of safe and efficient novel nonviral gene transfer using ultrasound: enhancement of transfection efficiency of naked plasmid DNA in skeletal muscle. Gene Ther 6:372–380

Tebbetts JB, Tebbetts TB (2002) An approach that integrates patient education and informed consent in breast augmentation. Plast Reconstr Surg 110:971–978

Tsai HC, Chen SY, Tsai ST, Hung HY, Chang CH (2010) Hypomania following bilateral ventral capsule stimulation in a patient with refractory obsessive-compulsive disorder. Biol Psychiatry 68:7–8

Turner DC, Sahakian BJ (2006) Neuroethics of cognitive enhancement. Biosocieties 1:113–123

Wilz A, Pritchard EM, Li T, Lan JQ, Kaplan DL, Boison D (2008) Silk polymer-based adenosine release: therapeutic potential for epilepsy. Biomaterials 29:3609–3616

World Health Organization (1946) Preamble to the constitution of the World Health Organization as adopted by the International Health Conference. http://www.who.int/about/definition/en/print.html. Accessed 18 Apr 2012

Wright MR (1986) Surgical addiction: a complication of modern surgery? Arch Otolaryngol 112:870–872

Wright HN (1987) Characterization of olfactory dysfunction. Arch Otolaryngol 112:870–872

Chapter 12
Cognitive-Enhancing Drugs, Behavioral Training and the Mechanism of Cognitive Enhancement

Emma Peng Chien

Abstract In this chapter, I propose the mechanism of cognitive enhancement based on studies of cognitive-enhancing drugs and behavioral training. I argue that there are mechanistic differences between cognitive-enhancing drugs and behavioral training due to their different enhancing effects. I also suggest possible mechanisms for cognitive-enhancing drugs and behavioral training and for the synergistic effects of their simultaneous application.

Keywords Cognitive enhancement • Mechanism • Cognitive-enhancing drugs • Behavioral training • Synergistic effects

12.1 Finding the Mechanism of Cognitive Enhancement

Studies on cognitive enhancement aim to find effective ways to improve the cognitive functions of healthy subjects as well as unhealthy subjects. There are two main approaches to exploring effective cognitive-enhancing methods. The first one seeks to try different possible enhancers with different combinations of strengths and durations in order to find more effective ways from a variety of trials. The second approach is to find the underlying mechanisms of how enhancers modulate cognitive functions so that enhancers can be used more effectively. These two ways of finding effective cognitive enhancers are interdependent. On the one hand, data on how different cognitive enhancers work help to build and examine different proposals of the mechanism of cognitive enhancement. On the other hand, some ideas about the underlying mechanism of cognitive enhancement help to decide the types and the quantities of enhancers to experiment on without trying all variations

E.P. Chien (✉)
Department of Philosophy, University of Alberta, Edmonton, AB, Canada
e-mail: chien1@ualberta.ca

E. Hildt and A.G. Franke (eds.), *Cognitive Enhancement: An Interdisciplinary Perspective*, Trends in Augmentation of Human Performance 1,
DOI 10.1007/978-94-007-6253-4_12,

of them. Both approaches of finding effective enhancing methods are necessary and worth pursuing. This chapter will contribute to the inquiry of finding effective enhancing methods by taking up the second direction.

In order to locate the mechanism of cognitive enhancement, I will start by reflecting on the synergistic effects of the simultaneous application of cognitive enhancing drugs and behavioral training, whose mechanism may play a crucial role in finding the general mechanism of cognitive enhancement, as suggested by Husain and Mehta in a review article (Husain and Mehta 2011). Several studies show that the resultant enhancing effects from the concurrent application of drugs and behavioral trainings are greater than the enhancing effects of applying either drugs alone or behavioral trainings alone (Knecht et al. 2004; Berthier et al. 2009). I suggest that the synergistic effects indicate that drugs and behavioral training modulate cognitive functions in different ways. I will justify this suggestion by discussing the difference of the enhancing effects and the mechanisms of cognitive-enhancing drugs and behavioral training. My argument regarding the difference between drugs and behavioral training is as follows: If cognitive-enhancing drugs and behavioral training enhance cognitive functions through the same mechanism, cognitive-enhancing drugs and behavioral training have the same enhancing effects. Cognitive-enhancing drugs and behavioral training do not have the same enhancing effects. Thus, cognitive-enhancing drugs and behavioral training enhance cognitive functions through different mechanisms. After arguing for the differences between the mechanism of cognitive-enhancing drugs and behavioral training, I will propose the mechanism for cognitive-enhancing drugs and for behavioral training and an explanation for the synergistic enhancing effects based on the conclusion of my argument. I suggest that this proposal could be a basis for the general mechanism of cognitive enhancement.

12.2 The First Premise: Thesis on the Mechanisms and the Effects of Cognitive Enhancement

My first premise is: if cognitive-enhancing drugs and behavioral training enhance cognitive functions through the same mechanism, cognitive-enhancing drugs and behavioral training have the same enhancing effects. Cognitive functions correlate with the excitation or inhibition of their corresponding neural networks. When drugs and/or behavioral training enhance a cognitive function, the corresponding neural network of this cognitive function is being modulated in a way that the cognitive function performs better. For instance, Ritalin enhances one's attention through its modulation of the corresponding neural network of attention (Husain and Mehta 2011). In addition, different dosages of enhancers may result in different strengths of enhancing effects; different types of enhancers may have different effects, such as the length of retention of the enhancing effects and the influence on brain activities after the end of enhancing treatment (Berthier et al. 2009;

Klingberg et al. 2005; Knecht et al. 2004; Maguire et al. 2003; Olesen et al. 2004). Because the effects of a cognitive enhancer are dependent on the way the enhancer modulates the corresponding neural network of the target cognitive function, we can then say that if two cognitive enhancers modulate the target neural network in the same way, (that is to say, if these two enhancers have same mechanisms,) then these two enhancers would have the same enhancing effects, such as achieving the same strength of enhancing effects, possessing the same duration of retention of the enhancing effects, and influencing brain activation patterns after the end of enhancing treatment in the same way.

12.3 The Second Premise: Different Enhancing Effects of Cognitive-Enhancing Drugs and Behavioral Training

The second premise of my argument states: cognitive-enhancing drugs and behavioral training do not have the same enhancing effects. There are four main differences between the enhancing effects of drugs and that of behavioral training.

First, the effects of behavioral training last longer than those of drugs after the end of enhancer usage. Berthier et al. (2009) show that even both memantine and constraint-induced aphasia therapy (CIAT) can achieve the same degree of improvement in patients with chronic poststroke aphasia after the same length of treatment; however, CIAT has retention effects of up to 3 weeks after the end of treatment, whereas memantine does not. In other studies on working memory, even though the retention of the enhancing effects of levodopa on working memory can last up to a month after the treatment (Knecht et al. 2004), behavioral training can last longer (up to 3 months) after the treatment (Klingberg et al. 2005).

Second, behavioral training changes brain activation patterns and neuronal growth. Studies on superior memorizers show that the strategy used by superior memorizers during their training change their brain activation patterns when performing tasks on memory (Maguire et al. 2003). Other studies on working memory training show that such training increases the density of dopamine receptors in the subjects' brains (McNab et al. 2009) and increased the activation of working memory-related brain activities (Olesen et al. 2004).

Third, drugs and behavioral training have different specificity on enhancing targets. For instance, levodopa treatment enhances subjects' ability to encode stimulus salience, while the repetition of behavioral training, which is required for the subjects to acquire the target cognitive skill, decreases salience (Knecht et al. 2004).

Fourth, studies show that only users with low performance are helped by the use of cognitive-enhancing drugs, while the same drugs may reduce the cognitive function of subjects who are already high-performing (Husain and Mehta 2011). On the other hand, even though behavioral training may not benefit the subjects with high performance, it does not seem to damage the subjects' cognitive performance as some cognitive-enhancing drugs do.

These differences between cognitive-enhancing drugs and behavioral training suggest that these two enhancing methods have different enhancing effects. Among the differences are: different retention rates, different influences on brain activation patterns and neuronal growth, different specificities, and different influences on subjects with varying cognitive abilities.

12.4 Conclusion: Different Mechanisms of Cognitive-Enhancing Drugs and Behavioral Training

I conclude that cognitive-enhancing drugs and behavioral training enhance cognitive functions through different mechanisms. Furthermore, I propose that drugs enhance cognitive functions by modulating only parts of the neural network that correspond to the cognitive functions, while behavioral training enhances cognitive functions by modulating a larger neural network within which the target neural network is only a part. The neural networks of cognitive functions usually involve more than one kind of neurochemical pathway. For instance, the neural network of working memory involves the neurochemical pathways of dopamine, noradrenaline acetylcholine, and serotonin, and the neural network of affective processes involves the neurochemical pathways of dopamine and serotonin (Cools et al. 2008; Harmer 2008; Luciana et al. 2001; Robbins and Arnsten 2009). When we use drugs to enhance cognitive functions, we usually use only one kind of drug. For instance, levodopa, which is the precursor of dopamine, is used to enhance working memory (Knecht et al. 2004). In this way, levodopa enhances working memory through modulating part of the neural network of working memory. On the other hand, behavioral training usually involves a neural network larger than the target neural network. For instance, working memory training involves not only working memory but also vision, attention, and, possibly, executive functions to complete the training.

The proposed mechanistic difference between cognitive-enhancing drugs and behavioral training provides an explanation for why there are synergistic effects when applying drugs and behavioral training at the same time. There are synergistic effects because drugs and behavioral training modulate different parts of neural networks independently. Thus, when drugs and behavioral training are employed at the same time, drugs and behavioral training can enhance the part that the other enhancer fails to enhance. Such a combination results in better enhancing results.

In addition, the proposed mechanistic difference between cognitive-enhancing drugs and behavioral training also suggests some explanations for the differences of the enhancing effects of these two enhancing methods. First, the enhancing effects of behavioral training last longer than those of drugs because behavioral training modulates not only the target neural network but also the interaction between the target neural network and other neural networks. I suggest that the enhancement of the interaction between neural networks helps to maintain the enhancing effects on the target neural network. Second, it is also possible that the interaction between different neural networks that result from behavioral training makes it easier to change brain activation patterns and neuronal growth. Third, the fact that drugs and behavioral training have different specificity may be due to the range of neural networks that they influence. For instance, it is easier for levodopa to enhance the subjects' ability of encoding stimulus salience than behavioral training because levodopa acts on a more specific range of neural network, which may be what is required for the encoding of stimulus salience. Fourth, the negative effects of cognitive-enhancing drugs on subjects with high performance may result from the trade-off between different cognitive functions, as Husain and Mehta (2011) suggest. The reason why behavioral training does not have the same negative effects may be that the interaction between different neural networks resulting from behavioral training has already reduced the trade-off between different neural networks or different cognitive functions.

To conclude, given the interdependency between theories and empirical studies, the proposed mechanistic difference of cognitive-enhancing drugs and behavioral training proposed in this chapter provides a basis for finding more effective enhancing methods. This difference also needs to be examined by empirical studies. If empirical studies do agree with the proposed mechanistic difference, this proposal could then be a good start for developing the general mechanism of cognitive enhancement.

Acknowledgement In preparing the present version of this chapter I benefited from the written comments from Elisabeth Hildt, Sheila Madary, and Andreas G. Franke, and comments and suggestions from the participants in the conference on cognitive enhancement in Mainz, Germany, February 2011. I am grateful to the German Federal Ministry of Education and Research for supporting the trip to the conference and the preparation of this chapter.

References

Berthier ML, Green C, Lara JP, Higueras C, Barbancho MA, Dávila G, Pulvermüller F (2009) Memantine and constraint-induced aphasia therapy in chronic poststroke aphasia. Ann Neurol 65:577–585

Cools R, Roberts AC, Robbins TW (2008) Serotoninergic regulation of emotional and behavioural control processes. Trends Cogn Sci 12:31–40

Harmer CJ (2008) Serotonin and emotional processing: does it help explain antidepressant drug action? Neuropharmacology 55:1023–1028

Husain M, Mehta MA (2011) Cognitive enhancement by drugs in health and disease. Trends Cogn Sci 15:28–36

Klingberg T, Fernell E, Olesen PJ, Johnson M, Gustafsson P, Dahlstrom K, Gillberg CG, Forssberg H, Westerberg H (2005) Computerized training of working memory in children with ADHD – a randomized, controlled trial. J Am Acad Child Adolesc Psychiatry 44:177–186

Knecht S, Breitenstein C, Bushuven S, Wailke S, Kamping S, Flöel A, Zwitserlood P, Ringelstein EB (2004) Levodopa: faster and better word learning in normal humans. Ann Neurol 56:20–26

Luciana M, Burgund ED, Berman M, Hanson KL (2001) Effects of tryptophan loading on verbal, spatial and affective working memory functions in healthy adults. J Psychopharmacol 15:219–230

Maguire EA, Valentine ER, Wilding JM, Kapur N (2003) Routes to remembering: the brains behind superior memory. Nat Neurosci 6:90–95

McNab F, Varrone A, Farde L, Jucaite A, Bystritsky P, Forssberg H, Klingberg T (2009) Changes in cortical dopamine D1 receptor binding associated with cognitive training. Science 323:800–802

Olesen PJ, Westerberg H, Klingberg T (2004) Increased prefrontal and parietal activity after training of working memory. Nat Neurosci 7:75–79

Robbins TW, Arnsten AFT (2009) The neuropsychopharmacology of fronto-executive function: monoaminergic modulation. Annu Rev Neurosci 32:267–287

Chapter 13
What Is Cognitive Enhancement and Is It Justified to Point Out This Kind of Enhancement Within the Ethical Discussion?

Roland Kipke

Abstract The term "cognitive enhancement" distinguishes one area of mental capacities: the cognitive capacities in contrast to non-cognitive capacities. The widespread use of this term within the ethical debate about pharmacological enhancement suggests that the enhancement of cognitive properties raises particular ethical questions that are different from the ethical questions raised by other kinds of neuro-enhancement. Or, the suggestion is at least that cognitive enhancement raises the ethical questions with a higher intensity. This article examines whether these suggestions are correct. The purpose of this article is therefore not to examine the ethical questions raised by cognitive enhancement but to examine the question as to whether the ethical problems of cognitive and non-cognitive enhancement are significantly different and, therefore, whether the concentration on cognitive enhancement within the bioethical debate is justified. The result of this examination is: the suggestion is not correct, the ethical questions raised by the different forms of pharmacological enhancement are in most respects equal or similar, and the concentration on cognitive enhancement is largely not justified.

Keywords Cognitive enhancement • Non-cognitive enhancement • Cognition • Ethics • Terminology

13.1 The Two Meanings of "Cognitive Enhancement"

In the ethical discussion on the use of medical means for the improvement of mental properties, different terms serve as the central term. They do not have exactly the same meaning. Some authors speak of "neuro-enhancement", others

R. Kipke (✉)
Internationales Zentrum für Ethik in den Wissenschaften (IZEW),
University of Tübingen, Tübingen, Germany
e-mail: kipke@izew.uni-tuebingen.de

E. Hildt and A.G. Franke (eds.), *Cognitive Enhancement: An Interdisciplinary Perspective*, Trends in Augmentation of Human Performance 1,
DOI 10.1007/978-94-007-6253-4_13, © Springer Science+Business Media Dordrecht 2013

of "brain doping" or "mind doping." And many people use the term "cognitive enhancement." I am concerned with what "cognitive enhancement" means, and above all, whether it is justified to single out this kind of enhancement that the term "cognitive enhancement" refers to: Are we correct to act on the assumption that cognitive enhancement poses particular ethical questions that are different from the ethical questions posed by non-cognitive types of neuro-enhancement?

The term "cognitive enhancement" obviously has different meanings. Some open questions are related to the term "enhancement": What does "enhancement" actually mean and is enhancement clearly distinguishable from therapy? These questions have been debated for many years; however, that is not the topic of this article. I use the term "enhancement" in the sense of altering healthy human properties by medical means, alterations that are subjectively evaluated as improvements, at least at the moment when the person conducts the enhancement measure. Moreover, I assume we can reasonably differentiate between enhancement and therapy, although there is some grey area (Lenk 2002).

However, I am concerned with the first part of the term: "cognitive." What does it mean? The term "cognitive enhancement" is used in two different ways:

A. For one, the term covers the entire field of interventions in a healthy brain by medical (so far only pharmacological) means in order to enhance mental traits or abilities. These mental traits are of cognitive, emotional and motivational nature (Synofzik 2006; Persson and Savulescu 2008; Förstl 2009; Hyman 2011). This is the problem with using the term "cognitive enhancement": "cognitive" traits are only traits and processes that have to do with thinking, perception, knowledge and memory:

   > Cognition can be defined as the processes an organism uses to organize information. This includes acquiring information (perception), selecting (attention), representing (understanding) and retaining (memory) information, and using it to guide behavior (reasoning and coordination of motor outputs). (Bostrom and Sandberg 2009)

   Cognition is not identical with the realm of mental processes and properties, but mental traits and processes are more than cognitive. In science and philosophy as well as in everyday speech, it is common to distinguish between cognitive and non-cognitive functions and abilities. Cognitive abilities make up only one part of the human psyche; the enhancement of mental traits is not only related to cognitive abilities. On the contrary, the enhancement of emotional traits, of mood and emotional balance by using anti-depressants is an important part of the enhancement of mental properties. The debate concerning this kind of mental enhancement began with the use of Prozac, an anti-depressant, and Peter Kramer's well-known book, *Listening to Prozac* (1997). Therefore, the first use of "cognitive enhancement" is simply incorrect. We should not use it in this way.
B. Secondly, the term "cognitive enhancement" refers only to the particular enhancement that aims to improve cognitive abilities. As we have seen, this use of the term is more adequate. We can find the term used in this way in many publications, and, in this paper, I also use it in this way.

As an overall term, I use "neuro-enhancement," which is also widely used in the ethical debate. While the term "cognitive enhancement" designates only one part of the medical enhancement of mental properties, "neuro-enhancement" encompasses the enhancement of cognitive and non-cognitive properties.[1]

And what is the term for the opposite of cognitive enhancement, the neuro-enhancement that aims to improve non-cognitive properties? Sometimes we can find the term "mood enhancement". However, this term is too narrow because people who use anti-depressants not only try to improve their mood but also their social skills, their self-confidence and so on. Therefore, I call the opposite of cognitive enhancement simply "non-cognitive enhancement."

13.2 The Question Regarding the Particular Ethical Role of Cognitive Enhancement

Although we can differentiate various kinds of neuro-enhancement, we can ask whether this differentiation is ethically relevant. Although the second use of "cognitive enhancement" is adequate, we have to ask whether this widespread use is justified. We have to ask whether it is legitimate to single out this kind of neuro-enhancement in the ethical discussion. It would be justified if cognitive enhancement raised more or other ethical problems than non-cognitive enhancement does. Publications that only speak of cognitive enhancement seem to assume this particular feature of cognitive enhancement and the necessity or at least adequacy of its own ethical discussion (Schleim and Walter 2007; Bostrom and Sandberg 2009; Cakic 2009; Tännsjö 2009; Biedermann 2010; Goodman 2010).

If cognitive enhancers do not lead to particular problems, we do not have any reason to point them out within the ethical discussion. In fact, I think that we have hardly any good reason for this particular consideration of cognitive enhancement. To justify this thesis, I will go very briefly into major ethical issues that are discussed in the context of neuro-enhancement and examine whether these are raised in a particular way – either less or more – by cognitive enhancers.

The main ethical problems that are discussed in relation to neuro-enhancement can be classified in several ways. But, it is not implausible to differentiate these nine fields of ethical problems:

1. Efficacy and side effects
2. Personal authenticity
3. Quality of happiness

[1] To be more precise we should speak of "psycho-enhancement" or "mental enhancement" because the aim of these measures is not to improve neurons but mental or psychological properties. Moreover, the nervous system is the basis for more than our mental properties. However, "psycho-enhancement" and "mental enhancement" are not widely used terms in this discussion.

4. Development of the character and self-awareness
5. Accountability of achievement
6. Fairness in competition
7. Social pressure
8. Social justice
9. Changes of values and the idea of mankind

In order to prevent misunderstandings, I would like to emphasize that my concern is not to prove or analyze these ethical problems. I only say: If these problems exist, then almost all of them exist for cognitive enhancers as well as for non-cognitive enhancers.

13.2.1 Efficacy and Side Effects

First, in contrast to the expectations and fears in the ethical discussion on neuro-enhancement, the currently used and discussed neuro-enhancers do not have a great effect on mental properties. This is true for means that are considered as cognitive enhancers as well as for those considered to be non-cognitive enhancers. Antidepressants do not have reliable effects on emotional traits; anti-dementia drugs do not show a positive effect on the memory; and stimulants like Ritalin hardly improve the ability of attention (Repantis et al. 2009; Lieb 2010). At most, Modafinil has a verifiable effect on attention (Lieb 2010).

Second, the pills that are presently used with the aim of enhancing mental capacities not only have no positive effect on the targeted mental properties, but most of the means can also have serious side effects. That also holds true for both cognitive and non-cognitive enhancers. Only the risk of addiction may be different because cognitive enhancers have an impact on the dopaminergic system, which is involved in all of the known substances with addiction potential (Heinz et al. 2012). However, the addictive potential of non-cognitive enhancers has not been well investigated to date.

Altogether we can detect some differences, but there is no significant difference between cognitive and non-cognitive enhancement that would justify their different ethical consideration in this respect. Neither of them can meet the high expectations with regard to efficacy and both have the potential to harm the people who use them. However, this situation could possibly change with the progress of pharmacology or even with the development of technical devices.

If the efficacy and safety of neuro-enhancers are not given to a certain extent, the other ethical problems remain of secondary importance because not very many people would use these enhancers. In the following discussion of other ethical questions, the problems of lacking efficacy and safety are presumed to be solved.

13.2.2 Personal Authenticity

Many authors recognize that one risk of neuro-enhancement is its potential to lead to a reduced authenticity of personality. In most cases, this fear is articulated with regard to non-cognitive enhancers (Krämer 2009; Schmidt-Felzmann 2009). The drug-induced change of feelings or emotional dispositions is expected to impair authenticity. Accordingly, cognitive enhancers are not under that suspicion and are, therefore, less problematic in this respect. Is this assumption correct?

To find an answer we first have to clarify what authenticity is. "Authenticity" is a difficult concept. I see three important aspects that are all addressed in our language when we talk about somebody being authentic: (a) The constancy of traits of the person, (b) the identification of the person with his traits, (c) the consistency between the traits (Kipke 2011).

(a) The constancy of traits of the person: Being authentic in the first sense means that the personality remains the same and his traits remain the same. If this is what is meant by "authenticity," and the change in the traits by neuro-enhancers is the problem, no ethically relevant difference between cognitive and non-cognitive enhancers is recognizable. The possible problem does not lie in the kind of mental traits but in their temporal extension. Both kinds of enhancers change the personality.
(b) The identification of the person with his traits: in this sense, one is authentic if he identifies himself with his characteristics. It is difficult to recognize any problem here because the identification with the desired traits is just the motive for producing or altering them by way of drugs. In any case, if there ever were a problem, then it would apply to both the cognitive enhancers and non-cognitive enhancers. The relevant problem here is the relationship between the person and his traits but not the kind of traits.
(c) The consistency between the traits: here, the authenticity of the person lies in the coherence of the characteristics of a person – both synchronous and diachronic. Since the traits or skills produced by neuro-enhancers do not necessarily emerge from biographical learning and development processes but are rather produced abruptly and possibly without reference to previous ways of life and personal development, a risk of inauthenticity in terms of incoherence is possible (Kipke 2011). If this is the case, then the problem relates to cognitive and non-cognitive enhancers. Therefore, the way of developing the traits would be at stake here, but not the kind of traits.

In short, with regard to the aspect of authenticity, we cannot find a relevant difference between the different groups of neuro-enhancers.

13.2.3 Quality of Happiness

Some authors fear that the happiness people seek through neuro-enhancement is shallow and superficial.[2] Because enhancers that target emotional qualities are mainly associated with the creation of happiness, they seem to be affected primarily by this concern. Cognitive enhancers often seem to be excluded from this criticism because they are used to improve cognitive performance rather than feelings of happiness. Cognitive enhancers thus appear to be less problematic in this regard.

But it is not that simple. Of course, cognitive enhancers also pursue the goal of happiness. Although it is not easy to make binding statements about happiness in general, it should be clear that happiness not only persists in positive feelings but also in a particular view of one's life and in the fulfillment of important desires. Cognitive enhancers have, of course, the goal of increasing the happiness that resides in the fulfillment of important desires: first, desires for personal qualities that are important to the person, and second, desires for success in competition.

If the quality of happiness resulting from the use of non-cognitive enhancers is questionable, then the quality of happiness resulting from the use of cognitive enhancers is as well. The kind of happiness at stake may be different, but in both cases the question is: Is the alleged happiness really happiness? For example, authenticity in terms of biographical coherence should be an important factor in the experience of happiness. There is no principal difference in the ethical questions raised by cognitive enhancers and other neuro-enhancers.

Furthermore, according to psychological research, an important factor in achieving happiness is the experience of self-efficacy and the resulting self-confidence (Bandura 1994). Compared to traditional methods of mental self-forming, neuro-enhancement has the fundamental problem that it does not allow higher levels of self-efficacy experience, that is, the experience of changing oneself by one's own forces. Moreover, in this respect, we cannot find any principal difference between the different groups of enhancers (Kipke 2011).

Finally, according to some authors, happiness also lies in the acceptance of how we are. In this view, striving to alter the conditions of our lives and personalities is a threat to happiness (President's Council 2003; Sandel 2007). In this respect, too, the kind of improved traits does not make a difference.

[2] President's Council (2003, 305): "What is to be particularly feared about the increasingly common and casual use of mind-altering drugs, then, is not that they will induce us to dwell on happiness at the expense of other human goods, but that they will seduce us into resting content with a shallow and factitious happiness." Cf. the considerations of Christoph Rehmann-Sutter on "authentic happiness": Rehmann-Sutter (2008). Cf. also Krämer (2009, 213).

13.2.4 Development of Character and Self-Awareness

In part, neuro-enhancement is criticized because it prevents the development of certain properties, such as self-discipline, perseverance, self-control, etc. (President's Council 2003), or at least it does not promote them. The idea is that only those people who develop their personal qualities through laborious and time-consuming mental work can develop these qualities.

Here, too, is apparently no difference between the different groups of neuro-enhancers. The criticism is not related to the types of enhanced traits but to the way they are achieved. If this claim is true, then it applies to properties such as concentration and memory as well as properties of self-confidence, emotional balance, and the like.

In the same way, neuro-enhancement can be criticized because it does not provide the opportunity to increase experiences and properties like self-awareness and self-efficacy. For example self-awareness: Unlike self-improvement through mental work, neuro-enhancement does not function without relatively strong self-attention, a higher attention to one's own traits and personality (Kipke 2011). Therefore, self-improvement through mental work has a tendency to increase self-awareness, and self-awareness is something widely appreciated. It is possible that the insights yielded by the improvement of non-cognitive traits are more interesting because people often have more of a lack of knowledge and understanding of their emotions than of their cognitive abilities. But, the principal problem that neuro-enhancement does not provide this opportunity of increasing self-awareness arises from not only non-cognitive neuro-enhancers but cognitive enhancers as well.

13.2.5 Accountability of Achievement

Because the properties in question are not just existent or are not developed by mental work, for some authors the question is whether the resultant benefits are attributable to the person at all, whether they are his achievement and whether the person deserves appreciation for this achievement (Stier 2009; Lieb 2010).

Is there a difference between the different neuro-enhancers in this respect? At first glance, there seems to be. The reason for this impression is that if we speak of "achievement" in this context, we think of cognitive achievement. But, even if we often use the term "achievement" in this narrow sense, the related ethical question is the same for non-cognitive enhancement. If someone improves his social and communicative skills by way of neuro-enhancers and performs well, this is also an achievement (e.g. a successful negotiation or a successful presentation at a party). And if the question of attributability of this achievement is posed by cognitive enhancers, it also arises in the cases in which the achievement is caused by non-cognitive enhancers. Therefore, here too, cognitive enhancers do not play a special role.

13.2.6 Fairness in Competition

The question of fairness in competition is closely linked to the previous question concerning accountability of achievement. But, while the question of accountability was situated only on an individual level, the question of fairness in competition brings the social dimension into play. Success in competition depends on specific mental abilities. Consequently, the question arises whether the competition for success in university and career, for recognition and for access to money and power, will be jeopardized. This is because it does not take place under fair conditions (if some people improve their abilities and thus their chances by using neuro-enhancers).

In the ethical discussion, the question of jeopardized competition is often only related to cognitive enhancement. However, this restriction is hard to understand since we approach competition with more than just cognitive abilities. For success in professional life, we need social, communicative and emotional skills, as well as cognitive abilities. Those people who cannot communicate appropriately, who present themselves with too little self-confidence or show too little empathy usually do not have much success (Brehm 2001; Boyatzis et al. 2002; Tangney et al. 2004). This is particularly true for many fields of highly-qualified work. These fields are those in which neuro-enhancement is currently the most wide-spread (DAK 2009).

Only in the case of exams in school and university settings, in which merely propositional knowledge is required and, therefore, only cognitive abilities such as understanding, attention and memory are necessary, does cognitive enhancement play a more important role than non-cognitive enhancers do. Cognitive enhancement raises the particular issue concerning the integrity of exam results. However, outside of those particular situations, we compete with more than our mental capacities. And if social and emotional skills can be improved by neuro-enhancers, the question of the distortion of competition is the same.

13.2.7 Social Pressure

A further concern is related to social competition. The point at issue here is not the corruption of competition but its increase. The idea is, first, that social competition produces a pressure to perform, which can be a strain, and, second, that the use of cognitive enhancers is only an expression of the high level of this pressure to perform and that it increases this pressure even more. Even if a certain level of pressure to perform is widely considered as fruitful for efficient work and an innovative society, many authors are worried that an increasing pressure to perform is too stressful and destructive for many people.[3]

---

[3]Even authors who are not generally critical of neuro-enhancement consider this as a big problem; cf. e.g. Schöne-Seifert (2006, 281).

Particularly problematic can be the pressure on those who actually do not want to use such enhancers. They would be forced to resort to these means against their conviction in order to maintain their career chances. Doping in high-performance sports gives credence to this assumption (Bette and Schimank 2006). The mere suspicion that competitors are taking enhancers produces the pressure to do the same.

Also, these questions are sometimes discussed particularly in terms of cognitive enhancement. Again, the assumption is made that only or at least primarily cognitive skills are decisive for career success. As we already know, this assumption is wrong. Therefore, the concentration on cognitive enhancement within the ethical debate is wrong. If social pressure is a problem with cognitive enhancers, then it is also a problem for non-cognitive enhancers.

13.2.8 Social Justice

The question of social justice is one of the key issues discussed in terms of neuro-enhancement. The concern of some authors is that neuro-enhancement increases social injustice and creates a two-class society or widens the gap between the existing classes even more. Only wealthier people can afford the expensive neuro-enhancers and, thus, increase their already-existing advantages through improved mental abilities, thereby gaining better chances in life. The people who were previously economically or educationally disadvantaged, thus, fall further behind (President's Council 2003).

In addition, this question is often limited to cognitive enhancers. In this view, social opportunities are primarily determined by cognitive abilities. The pharmacologically augmented IQ is used as the main example for a reason for growing injustice, but this is a one-sided view. As was already shown above, we have social opportunities not only because of our cognitive abilities but also as a result of our non-cognitive skills. The probability is very small that someone will have success in the long run only with high cognitive abilities but low emotional competences such as empathy. So, if social justice is threatened by cognitive enhancers, then it is also threatened by non-cognitive enhancers. Again, cognitive enhancement does not demand a special ethical assessment.

13.2.9 Changes of Values and the Idea of Mankind

Negative consequences for society could also consist in changes of our values and the ideas of ourselves as human beings. Since neuro-enhancement is a technical shortcut, a widely-cited concern is that it undermines the value of hard work. It could create or strengthen the fallacious idea that we can have success without effort. In reality, we cannot gain something without pain (Juengst 1998). A related idea is that neuro-enhancement could lead to an overestimation of the value of efficiency, which is regarded as erring and harmful for human well-being (Boldt and Maio 2009). A third concern refers to the idea of mankind that can be negatively affected by neuro-enhancement. Neuro-enhancement can support the idea that we are only living mechanisms and that human action is merely the outcome of neuro-chemical processes. By using neuro-enhancers we exercise a mechanistic anthropology, so to speak. Thereby, the important concepts of freedom and responsibility could perish (Freedman 1998; Gesang 2007).

Now, it is not any more surprising that we cannot detect any ethically relevant difference between the types of neuro-enhancers in this respect either. If the presented concerns are justified, then they do so due to the general nature of neuro-enhancers, their quick result without mental effort and their nature as neuro-chemical or technical means, but not due to the particular kind of mental properties that are changed by these means.

13.3 Two General Arguments

In addition, I have two general, non-ethical arguments against a separate discussion on cognitive enhancement. One argument deals with the possibility of a differentiation of the various mental properties; the other is related to the factual impact of existing supposed cognitive enhancers.

13.3.1 The Possibility of Differentiation

The separated discussion of cognitive enhancement assumes that the different neuro-enhancers have effects either on cognitive abilities or on non-cognitive properties. The basic assumption is that it is possible to separate emotion and cognition clearly so that we can selectively improve cognitive properties. In contrast, neuroscientific and psychological research show that the two areas are closely linked. Emotion affects cognition, and cognition affects emotion (Dolan 2002). This general link between the two areas is especially important for the skills of concentration and memory, which are two of the main focuses of cognitive enhancement. What appeals to us emotionally fits into the semantic and episodic memory more easily. Our concentration depends on our emotional state (Hänze 2009). Social sensitivity (empathy) correlates with a relatively high IQ (Asendorpf 2009). If these mental domains are that closely linked and deeply interdependent, it is highly implausible that we could improve one domain without affecting the other.

13.3.2 The Impact of Existing Supposed Cognitive Enhancers

In addition, the empirically detectable side effects of current enhancers considered as cognitive enhancers go beyond the domain of cognitive abilities. For example, Ritalin (Methyphenidat), the drug often prescribed to ADHD patients (attention deficit hyperactivity disorder, ADHD), often (1:10–1:100) results in aggression, excitement, emotional lability, anxiety, depression, irritability – mental states that obviously belong to the domain of emotion (Fachinformation des Arzneimittel-Kompendiums der Schweiz 2009).

Thus, the empirically observable side effects of drugs that are supposed to be cognitive enhancers, such as Ritalin, are not limited to the area of cognitive abilities. If these side effects occur in ill patients, it seems likely that they will also occur in healthy persons. An isolated enhancement of cognitive functions without a change in the emotional dimension is therefore unrealistic. Even if one makes the counterfactual presumption of effective enhancers without severe side effects, one should not ignore the basic realities of the human psyche.

13.4 Conclusion

The results of this brief examination of the different fields of ethical problems of neuro-enhancement are: In most respects, the problems and questions raised by cognitive enhancement on the one hand and by non-cognitive enhancement on the other are the same or very similar. Neither the ethical problems relating to the individual person nor the problems relating to the social consequences make a significant ethical difference. Only in very few aspects, the ethical questions or problems seem to be different or differently large, namely, the kind of happiness that is at stake, the problem of unfairness in exams in school and university settings and, possibly, the varying importance of self-awareness, which can be prevented by the use of neuro-enhancers. These few ethical differences hardly justify a broad, separate consideration of the topic cognitive enhancement. In addition, it is generally doubtful that separating the domain of cognitive traits from the domain of non-cognitive traits is justified. If we talk about neuro-enhancement, we can and we should spotlight the whole field of neuro-enhancement.

References

Asendorpf JB (2009) Persönlichkeitspsychologie – für Bachelor. Springer, Berlin/Heidelberg
Bandura A (1994) Self-efficacy. The exercise of control. Palgrave Macmillan, New York
Bette K-H, Schimank U (2006) Doping im Hochleistungssport. Anpassung durch Abweichung, 2nd edn. Suhrkamp, Frankfurt am Main

Biedermann F (2010) Argumente für und wider das Cognitive Enhancement. Eine kritische Kurzübersicht. Ethik Med 22:317–319
Boldt J, Maio G (2009) Neuroenhancement. Vom technizistischen Missverständnis geistiger Leistungsfähigkeit. In: Müller O, Clausen J, Maio G (eds) Das technisierte Gehirn. Neurotechnologien als Herausforderung für Ethik und Anthropologie. Mentis, Paderborn, pp 383–397
Bostrom N, Sandberg A (2009) Cognitive enhancement: methods, ethics, regulatory challenges. Sci Eng Ethics 15:311–341
Brehm M (2001) Emotionen in der Arbeitswelt. Theoretische Hintergründe und praktische Einflussnahme. Arbeit – Zeitschrift für Arbeitsforschung, Arbeitsgestaltung und Arbeitspolitik 10:205–218
Cakic V (2009) Smart drugs for cognitive enhancement. Ethical and pragmatic considerations in the era of cosmetic neurology. J Med Ethics 35:611–615
DAK (2009) DAK-Gesundheitsreport 2009. Analyse der Arbeitsunfähigkeitsdaten. Syllabication: Schwertpunktthema Doping am Arbeitsplatz. www.dak.de/content/filesopen/Gesundheitsreport_2009.pdf. Last accessed 26 June 2012
Dolan RJ (2002) Emotion, cognition, and behavior. Science 298:1191–1194
Fachinformation des Arzneimittel-Kompendium der Schweiz (2009) Ritalin®/-SR/-LA Last information: May 2011. www.kompendium.ch/MonographieTxt.aspx?lang=de&MonType=fi;
Förstl H (2009) Neuro-Enhancement. Nervenarzt 80:840–846
Freedman C (1998) Aspirin for the mind? Some ethical worries about psychopharmacology. In: Parens E (ed) Enhancing human traits. Ethical and social implications. Georgetown University Press, Washington, DC, pp 135–150
Gesang B (2007) Die Perfektionierung des Menschen. deGruyter, Berlin
Goleman D, Boyatzis R, McKee A (2002) Primal leadership. Learning to lead with emotional intelligence. Harvard Business School Press, Boston
Goodman R (2010) Cognitive enhancement, cheating, and accomplishment. Kennedy Inst Ethics J 20:145–160
Hänze M (2009) Denken und Gefühl. Wechselwirkung zwischen Emotion und Kognition im Unterricht. Beltz, Weinheim/Basel
Heinz A, Kipke R, Heimann H, Wiesing U (2012) Cognitive neuro-enhancement – false assumptions in the ethical debate. J Med Ethics 38:372–375
Hyman SE (2011) Cognitive enhancement: promises and perils. Neuron 69:595–598
Juengst ET (1998) What does enhancement mean? In: Parens E (ed) Enhancing human traits. Ethical and social implications. Georgetown University Press, Washington, DC
Kipke R (2011) Besser werden. Eine ethische Untersuchung zu Selbstformung und Neuro-Enhancement. Mentis, Paderborn
Kramer PD (1997) Listening to Prozac. Penguin, New York
Krämer F (2009) Neuro-Enhancement von Emotionen. Zum Begriff emotionaler Authentizität. In: Schöne-Seifert B, Talbot D, Opolka U, Ach JS (eds) Neuro-Enhancement. Ethik vor neuen Herausforderungen. Mentis, Paderborn, 189–217
Lenk C (2002) Therapie und Enhancement. Ziele und Grenzen der modernen Medizin. Lit Verlag, Münster
Lieb K (2010) Hirndoping. Warum wir nicht alles schlucken sollten. Artemis & Winkler, Mannheim
Persson I, Savulescu J (2008) The perils of cognitive enhancement and the urgent imperative to enhance the moral character of humanity. J Appl Philos 25:162–177
President's Council on Bioethics (2003) Beyond therapy. Biotechnology and the pursuit of happiness. A report of the president's council on bioethics. University of Chicago Press, New York/Washington, DC
Rehmann-Sutter C (2008) Authentisches Glück? Ethische Überlegungen zu Neuro-Enhancement. In: Maio G, Clausen J, Müller O (eds) Mensch ohne Maß? Reichweite und Grenzen syllabication: anthropologischer Argumente in der biomedizinischen Ethik. Karl Alber, Freiburg/München, pp 242–259

Repantis D, Schlattmann P, Laisney O, Heuser I (2009) Antidepressants for neuro-enhancement in healthy individuals. A systematic review. Poiesis Prax 6:139–174

Sandel M (2007) The case against perfection: ethics in the age of genetic engineering. Harvard University Press, Cambridge, MA/London

Schleim S, Walter H (2007) Cognitive enhancement. Fakten und Mythen. Nervenheilkunde 26:83–87

Schmidt-Felzmann H (2009) Prozac und das wahre Selbst. Authentizität bei psychopharmakologischer enhancement. In: Schöne-Seifert B, Talbot D, Opolka U, Ach JS (eds) Neuro-Enhancement. Ethik vor neuen Herausforderungen. Mentis, Paderborn, pp 143–158

Schöne-Seifert B (2006) Pillen-Glück statt Psycho-Arbeit. Was wäre dagegen einzuwenden? In: Ach JS, Pollmann A (eds) No body is perfect. Baumaßnahmen am menschlichen Körper. Bioethische und ästhetische Aufrisse. Transcript, Bielefeld, pp 279–291

Stier M (2009) Neuro-Enhancement und das Problem der Verantwortung. In: Schöne-Seifert B, Talbot D, Opolka U, Ach JS (eds) Neuro-Enhancement. Ethik vor neuen Herausforderungen. Mentis, Paderborn, pp 277–293

Synofzik M (2006) Kognition à la carte? Der Wunsch nach kognitionsverbessernden Psychopharmaka in der Medizin. Ethik Med 18:37–50

Tangney JP, Baumeister RF, Boone AL (2004) High self-control predicts good adjustment, less pathology, better grades, and interpersonal success. J Pers 72:271–324

Tännsjö T (2009) Ought we to enhance our cognitive capacities? Bioethics 23:421–432

Chapter 14
No Pain, No Gain? Objections to the Use of Cognitive Enhancement on the Basis of Its Potential Effects on the Value of Achievement

Lisa Forsberg

Abstract In recent years, the prospects and perils of cognitive enhancement (CE) have been a matter of great debate. This chapter examines three objections to the use of CE. All three objections are based on potential effects that the use of CE is thought to have on the value of achievement. According to the first objection, the reason that we should oppose the use of CE is that it would cause us to no longer be responsible for our achievements. The second objection, in contrast, holds that we should oppose the use of CE because it would make us too responsible for our achievements, or the lack thereof. According to the third objection, we should oppose the use of CE because it would have character-eroding effects. This chapter examines a number of counter-arguments against each of these three objections. The chapter concludes that none of the three objections succeed as in principle objections to the use of CE.

Keywords Cognitive enhancement • Achievement • Responsibility • Authenticity • Ethics

In recent years, the prospects and perils of cognitive enhancement (CE) have been a matter of great debate. Some of the concerns expressed in the literature in relation to cognitive enhancement center on the idea that the use of cognitive enhancement would somehow undermine the value of achievement (Cole-Turner 1998; Farah 2002; President's Council on Bioethics 2003; Chatterjee 2006; Radoilska 2010). These concerns rely on the claim that cognitive enhancement technologies would bring about "gain without pain", or would allow us to take "easy shortcuts", and the intuition that using such means to achieve these effects would be wrong.

L. Forsberg (✉)
Centre of Medical Law and Ethics, King's College London, London, UK
e-mail: lisa.forsberg@kcl.ac.uk

E. Hildt and A.G. Franke (eds.), *Cognitive Enhancement: An Interdisciplinary Perspective*, Trends in Augmentation of Human Performance 1,
DOI 10.1007/978-94-007-6253-4_14,

The aim of this chapter is to examine three objections to the use of CE. All three objections oppose the use of CE on the basis of its potential effects on the value of achievement. The first objection holds that the reason that we should oppose the use of CE is that we would no longer be responsible for our achievements (Sect. 14.2). The second objection, instead, holds that we should oppose the use of CE because it would make us *too responsible* for our achievements, or the lack thereof (Sect. 14.3). The third objection holds that we should oppose the use of CE because it would have character-eroding effects (Sect. 14.4). The chapter will examine a number of counter-arguments against each of these three objections. Finally, it will be concluded that none of the three objections succeed as *in principle* objections to the use of CE.

14.1 In What Sense Might CE Pose a Threat to the Value of Achievement?

Before I examine these objections in more detail, a few notes on terminology are required. First, there has been significant debate about the question of whether a distinction between "treatment" and "enhancement" can be drawn at all and whether, even if this could be done, such a distinction would hold any moral significance (see e.g. Harris 1992; Bostrom and Savulescu 2009). Thus, any definition of CE is likely to be controversial. In addition, many authors are using the term "enhancement" to refer to slightly different things. For the purposes of this chapter, however, choosing a definition of CE is not important. I will simply stipulate that there could exist effective forms of CE interventions, and that these could be distinguished from other types of medical interventions (treatment), as well as from other forms of common practices that may have an enhancing effect on cognition (for example, coffee drinking). Second, CE interventions, for the purposes of this chapter, could involve any technique (pharmacological, surgical, deep brain stimulation, etc.) that could be efficacious in achieving an enhancing effect with regard to the cognitive capacities of healthy individuals.

All three objections to the use of CE examined in this chapter focus on the relation between "ability" and "achievement." To clarify the relation between the value of an achievement and the origin or source of the ability that is assumed to be responsible for this achievement, we could distinguish between four different categories of "ability-sources":

I. *Native/natural* abilities, that is, abilities with which we are born, where no interference *in utero* has occurred;
II. Abilities resulting from *enhancement interventions in utero*;
III. Abilities resulting from *self-enhancement*, that is, enhancement interventions undergone in adulthood where the decision to undergo the enhancement intervention is taken by the individual concerned, and;
IV. Abilities acquired through *exerted effort.*

The present chapter will be limited to individuals who themselves have the capacity to make medical treatment decisions. Therefore, category II will not feature in the analysis.

14.2 "CE Would Cause Us to No Longer Be Responsible for Our Achievements"

The first objection to the use of CE that will be examined here holds that we should oppose the use of CE it would cause us to no longer be responsible for our achievements. According to this objection, "the attainment of [excellence] by means of drugs [is] 'cheating' or 'cheap'" (President's Council on Bioethics 2003, 289). The question of whether the use of cognitive enhancement would constitute *cheating* falls outside the scope of this chapter. It has been advanced, however, that:

> [E]ven if users of [cognitive enhancement] were not cheating competitors, and even if the meaningfulness of the activity were not diminished, it might still be argued that their accomplishments do not, in some sense, fully belong to them. Users of [cognitive enhancement], much like cheaters, might take credit for achievements that are not theirs. (Goodman 2010, 155)

The concern is that the use of biomedical enhancement will cause a "shift *away* from achievements humans *undertake* through self-initiated striving, and toward those they experience by biochemical interventions, which *act on us* as passive subjects" (Fox 2005, 1150). However, a number of counter-arguments could be proposed against this objection to the use of CE.

14.2.1 We Are Even Less Responsible for What We Achieve Through Our Native/Natural Abilities

The assumption underlying the objection according to which the use of CE would cause us to no longer be responsible for our achievements appears to be that we would be less deserving of praise or blame with regard to achievements that were a result of CE than we would be with regard to achievements that were not a result of CE use. We could test this assumption through formulating a Responsibility Condition (RC), according to which

> (RC) P deserves achievement *z* by virtue of ability *y* if and only if P is responsible for ability *y*.

We could then apply RC to each of the different categories of ability-sources described above. An application of RC would yield that, in the case of self-enhanced abilities, P *would* be responsible for her abilities, by virtue of having enhanced them. In the case of self-enhanced abilities, RC would therefore be satisfied. An application of RC to native/natural abilities, however, would yield that P was *not* responsible for her abilities. In the case of natural abilities, RC would therefore *not* be satisfied. The application of RC to the different categories of ability-sources illustrates that the objection according to which the use of CE would cause us to no longer be responsible for our achievements would apply to native/natural abilities to an even greater degree than it would apply to abilities that were a result of self-enhancement. We could concede that we may be somewhat *less* responsible for that which we achieve through self-enhanced abilities than we would be for abilities that we develop through, for example, hard work. This concession would not, however, justify an *in principle* opposition to the use of CE on the basis of its effects for the value of achievement since we would be *even less* responsible for native/natural abilities and for what achievements flow from them.[1]

14.2.2 We Are Fully Responsible for Very Few (if Any) of Our Achievements

A second counter-argument against the objection to the use of CE on grounds that it would cause us to no longer be responsible for our achievements is that we are very rarely, if ever, *fully* responsible for our achievements. Therefore, it would seem implausible to assume that the relatively slight decrease in the degree of responsibility caused by the use of CE would be sufficient to prevent us from being held responsible from a moral desert point of view. As Lubomira Radoilska has observed, "[. . .] cognition is routinely considered as a human activity susceptible to kinds of appraisal, to which mere physiological processes, such as digestion, are not" (Radoilska 2010). It could be questioned, however, whether these intuitions are justified, especially in light of emerging neuroscientific evidence pointing to similarities between cognitive processes and other kinds of physiological processes and evidence of the importance of environmental influences in the development of ability. The scope of the present chapter does not allow for a detailed exploration of issues relating to free will and responsibility. However, for our purposes, it suffices to say that it would seem unreasonable to assume that we are *fully* responsible for very many of our achievements.

[1] It could be argued that the "natural" distribution of abilities ought to be valued (and preserved) because it is undesirable to interfere too much with nature, but this is an altogether different objection to the use of enhancement.

14.2.3 We Would Not Be Able to Distinguish Between Abilities Originating from Different Ability-Sources

The objection according to which the use of CE should be opposed because it would cause us to no longer be responsible for our achievements presupposes that we could distinguish between abilities originating from different ability-sources. The President's Council on Bioethics, for example, presupposes that we could distinguish achievements that were "true," "natural" products of a person from achievements that were a result of "unnatural" external help. The underlying assumption, thus, is that we could distinguish "natural" and "unnatural" achievements from (i) one another and (ii) all other forms of environmental influences. In order to make such distinctions, we would need to be able to isolate "ability variables" as well as achievements, so that "true" achievements would be a *direct function* of "ability variables." The assumption that this could be done is empirically implausible. A more appropriate approach would be to acknowledge the kinds of external influences that have aided us in each of our achievements, and to see "excellent work as something other than personal property and a sign of personal superiority" (Goodman 2010, 157).

14.3 "CE Would Make Us *Too Responsible* for Our Achievements, or the Lack Thereof"

As we have seen, the first objection to the use of CE on the basis of its potential effects for the value of achievement held that we should oppose the use of CE because it would diminish the degree to which we were responsible for our achievements. The second objection to the use of CE on the basis of its potential effects for the value of achievement holds the opposite. The underlying idea is that, "As human beings become more able to control their lives and themselves, they also become more responsible for the results" (Bostrom and Sandberg 2009, 327). The worry, thus, is that if the option to enhance is available, and this option is likely to make it easier for us to achieve certain things, we could be held responsible for our failure to enhance and for our resulting lack of achievement. Michael Sandel has, for example, proposed:

> Why, after all, do the successful owe anything to the least-advantaged members of society? The best answer to this leans on the idea of giftedness. [. . .] A lively sense [. . .] that none of us is wholly responsible for her success makes us willing to share the fruits of our talents with the less successful. (Sandel 2004, 87)

The argument that Sandel proposes may be reconstructed like this:

1. Decision-makers are able to engineer desirable traits in themselves or others;
2. The ability to engineer desirable traits in themselves or others entails responsibility for the presence or absence of desirable traits;

3. If decision-makers are responsible for the presence or absence of desirable traits in themselves or others, human solidarity will be undermined;
4. Human solidarity will be undermined.

There are, however, a number of possible counter-arguments against this objection.

14.3.1 Negative Responsibility Would Apply to Effort Too

We could (contrary to what was concluded in the previous section) stipulate that we *could* isolate 'ability variables' and achievements, to a degree that would allow us to ascertain that particular achievements were a direct function of particular abilities. If this could be done, however, the kind of negative responsibility envisaged by Sandel would apply to achievements that were a result of effort, too. Thus, if people could be held responsible for their lack of achievement when this was a result of their failure to enhance themselves, people could also be held responsible for a lack of achievement that was a result of a failure to exert an effort. Therefore, the second objection fails to provide the sought-after distinction between the kinds of "natural" achievements that achievements resulting from effort are thought to represent and achievements resulting from self-enhancement.

14.3.2 We Are Not Able to Distinguish Between Abilities of Different Ability-Sources

For the second objection, according to which the use of CE would make us too responsible for our achievements, or the lack thereof, to hold some force, we would need to be able to tell whether the absence of a particular achievement were the result of a failure to self-enhance or whether some other factor had prevented the achievement from occurring. As we saw in Sect. 14.2.3, however, it is empirically implausible to assume that we could distinguish between abilities of different ability-sources (and other forms of external influences). Unless the knowledge of which failures to achieve something were attributable to a failure on behalf of a particular individual to self-enhance were available to us, we would not have a basis on which to hold individuals responsible for their failures to self-enhance. The second objection, therefore, fails to demonstrate that CE should be opposed.

14.3.3 *Being Responsible for Our Choices Is Not the Same as Being Responsible for Bearing the Costs of These Choices*

A further reason to reject the objection that the use of CE should be opposed because it would make us too responsible for our achievements is its failure to distinguish between *being held responsible* for a choice and *being held responsible for bearing the cost* of that choice. It is submitted that even if we *could*, in certain cases, conclude that certain individuals *were* responsible for particular outcomes, this would not necessarily entail that we also conclude that they should be responsible for bearing the cost of that outcome. As Frances Kamm has noted, "We may hold someone responsible for an outcome in the sense of blaming him for that outcome, without thereby thinking that it is also his responsibility to bear the costs of his choice, for, as T.M. Scanlon has pointed out, these are two conceptually separate issues" (Kamm 2009, 122).

The second objection may indeed point to interesting questions regarding when it is appropriate for us, as a society, to hold someone responsible for their failures to avoid a particular outcome. We could, for example, consider whether it would be appropriate to (sometimes) hold individuals responsible for their failure to prevent the development of a medical condition when the medical condition is (largely) self-inflicted. These kinds of questions may be relevant in the context of, for example, resource allocation. However, there is no direct link (that would imply necessity) between being held responsible for an outcome and being held responsible for bearing the cost of that outcome in the way that the second objection presupposes. The second objection, therefore, fails to demonstrate that we should oppose the use of CE.

14.4 "No Pain, No Gain": The Use of CE Would Lead to Character Erosion

The 'no pain, no gain' objection holds that the use of enhancement technologies would, among other things, "weaken the spirit of human agency and cheapen the dignity of human activity" (Cole-Turner, cited in Fox 2005, 1150). Thus, the President's Council for Bioethics has, for example, argued that "[i]n most of our ordinary efforts at self-improvement, whether by practice, training, or study, we sense the relation between our doings and the resulting improvement, between the means used and the ends sought" (President's Council on Bioethics 2003, 292). Enhancement technologies, however,

> separate the ends sought from our understanding of the meaning those ends imply. Because they act directly on the human body and mind, biotechnological enhancements tempt us to shirk individual striving and struggle (Fox 2005, 1150).

According to proponents of this objection, the use of enhancement technologies therefore deprives individuals of the value that is associated with the use of conventional means, such as practice and effort (Fox 1150).

The idea underlying the third objection to the use of CE relies on the notion of *character erosion*. As Anjan Chatterjee has noted:

> [t]he erosion of character concern is wrapped around a 'no pain, no gain' belief. Struggling with pain builds character, and eliminating that pain undermines good character. (Chatterjee 2004, 971)

Again, a number of counter-arguments could be proposed against the third objection.

14.4.1 Unfair Discrimination Against Natural Ability

The objection against the use of CE on the basis of its potential character-eroding effects seems to entail that "the more struggle, the better." This assumption would, however, seem to unfairly discriminate against individuals endowed with significant natural abilities. M.J. Mehlman has noted that, "[a] child prodigy who produces a charming sonata with little formal training produces just as treasured a piece as does someone who [. . .] labored over the score for half a lifetime" (in Goodman 2010, 154). To advance that an individual would not be deserving of praise due to their (relative) lack of effort in producing a certain result would amount to punishing individuals for something beyond their control, namely an ability with which they were endowed more or less as a matter of luck (or the "natural lottery") at birth. In addition, as J.M. Olsen has observed, "There is no reason to assume prima facie that effort is valuable in and of itself" (Olsen 2006, 3). Indeed, as Schermer has noted, "[W]hile it is true that we admire or praise those who earn their success by hard work for their effort and endurance, on top of the admiration and praise they get for the achievement per se, this does not make all effort valuable in itself" (Schermer 2008, 359). As Mehlman has observed, "[Y]ou deserve to win a Nobel Prize if you discover the cure for cancer, whether or not you do so with the aid of cognitive enhancement drugs" (Mehlman 2004, 493).

14.4.2 Conflating Practical and in Principle Objections

As Schermer has noted, "[A]s an empirical claim, the 'no pain, no gain' objection seems to be rebutted by the mere existence of enhancement technologies [– their existence shows that it] *is* possible to earn some gains [. . .] without pain" (Schermer 2008, 358; emphasis added). However, the fact that CE could help individuals to earn some gains, such as an exam result that is identical or better than the result they would have been able to achieve through employing a conventional exam preparation strategy, such as effort/hard work without pain, does not entail that the

Fig. 14.1

Strategy		Outcome
CE	⇒	Exam result X
Conventional	⇒	Exam result X **Knowledge K**

Fig. 14.2

Strategy		Outcome
CE	⇒	Exam result X
Conventional	⇒	Exam result X **Experience E**

CE strategy would have been as effective as the conventional learning strategy, *all things considered.* As Greely et al. have observed, whether we would have reason to oppose the use of CE or not would largely depend on the "nature of its effects, that is, whether, for example, enhancement aimed at improving cognitive capacities would improve *learning*, or simply aid those who underwent it in getting a better exam result than they would otherwise have been able to get" (Greely et al. 2008, 703). The idea that CE may not be effective in achieving a result that was identical to that which could be achieved by using conventional learning strategies is illustrated in Figs. 14.1 and 14.2 below.

We could think of the use of CE and the use of conventional means, such as effort, as two different strategies that could be used to achieve a particular thing (for example, an exam result). The objection to the use of CE on the basis of its potential character-eroding effects could be understood as a claim that an individual who adopted CE as a strategy may not be able to achieve a result that was *identical* to that which they would have achieved had they adopted a conventional strategy, all things considered. CE use would therefore be an undesirable behavior. We could, for example, imagine the scenario that Greely et al. alluded to above, where individuals who adopted the conventional strategy would, in addition to the particular exam result, also acquire particular knowledge, whereas individuals who adopted the CE strategy would *not* acquire this knowledge. Thus, the results of each of the two strategies would not be identical.

The concern that the use of CE could deprive us of valuable experiences could be illustrated in a similar manner. As Goodman has observed:

> The true concern with enhancement is not that it makes activities less difficult, but that it makes them less meaningful [and although cognitive enhancements] might help one achieve an end more efficiently, [. . .] they could also deprive individuals of valuable *experiences* in the process. (Goodman 2010, 152)

We could again imagine a scenario where individuals could choose to adopt either a CE strategy or a conventional (learning/effort) strategy. The claim that individuals who adopt a CE strategy would be deprived of a valuable experience that individuals who adopted a conventional strategy would acquire is illustrated in Fig. 14.2 below.

In this scenario, individuals who adopt the conventional strategy would, in addition to the particular exam result, also acquire a particular experience (E), whereas individuals who adopt the CE strategy would *not* acquire this experience. Thus, the results of each of the two strategies would not be identical.

The concerns that individuals who adopt a CE strategy rather than a conventional (learning/effort) strategy could lose out on valuable knowledge or experiences might be justified. However, even if they *were* justified, this would merely amount to a practical objection, that is, a claim that CE is not an *effective* strategy. If this were the case, there may be practical and societal reasons for restricting CE use. One such reason might be if students used CE simply to achieve a better exam result than they would otherwise have been able to get since, as Greely et al. have noted, this would "prevent a valid measure of the competency of the examinee" (Greely et al. 2008, 704). The claim that CE may be less effective than conventional means does not, however, constitute an *in principle* objection to the use of CE. It would not entail that it would be undesirable to seek to develop more efficacious forms of CE. If we *could* develop CE that were just as efficacious in helping us achieve an outcome, identical to the outcome we would have achieved through the use of conventional (learning/effort) means, its use would not be undesirable.

14.4.3 Failing to Acknowledge That Effort Could Be Directed Towards Achieving Something More Worthwhile

If outcomes of employing CE and conventional strategies respectively *were* identical, how could we justify *not* adopting a (potentially more) effective strategy? As Goodman has observed:

> Virtues developed through hard work are not mainly valuable in themselves, but because they prepare us to accomplish more hard work. Finding a reliable, easier method takes these virtues out of context, disconnecting them from a purposeful activity. It would be foolish to forgo a reliable shortcut on account of a set of virtues that the shortcut itself renders unnecessary. (Goodman 2010, 152)

Schermer has noted that the idea that effort or suffering may sometimes enable us to learn, or to cultivate virtues that will be useful to us, does not entail that effort or suffering are *necessary* for learning or prerequisites for developing such virtues. Conversely, not all instances of effort or suffering will promote learning or virtue acquisition (Schermer 2008, 359). "Short-cuts" that minimize certain kinds of efforts or suffering (that could have a character-building potential) are accepted – and indeed widely appreciated – in many areas of contemporary life. This is because it is generally accepted that these short-cuts allow us to devote time to other activities that are more worthwhile. Thus, as Bostrom and Sandberg have observed:

> [S]ociety does not denounce athletes for wearing protective (and performance enhancing) shoes, since they enable the athletes to concentrate on interesting talents rather than on developing thick soles. In many elementary schools, calculators are disallowed in mathematics lessons, where the goal is to understand basic arithmetic, but they are allowed and increasingly necessary in the higher grades. The basics have by then been mastered, and the goal becomes to understand more advanced topics. These examples illustrate that cognitive enhancement aimed at extending and completing a person's talents may [...] [offload] irrelevant, repetitive, or boring tasks and [enable] a person to concentrate on more complex challenges that relate in more interesting ways to his or her goals and interests. (Bostrom and Sandberg 2009, 326)

It could be argued that the "no pain, no gain" objection fails to draw the distinction, proposed by Goodman, between "activities that are predominately characterized by [...] 'process goods,' [that is,] excellence in the performance of an activity, [and] 'outcome goods,' [that is,] the benefits an activity creates" (Goodman 2010, 146). Although some activities, such as chess games, largely – or even solely – consist of process goods, many activities consist both of process and outcome goods, and some consist solely of outcome goods. In these cases, the use of ineffective means to achieve a particular outcome good would be inappropriate. Instead, we should encourage the use of the most effective means available.

We can and we do make distinctions between tasks where short-cuts are acceptable and tasks where they are not acceptable. This fact suggests that we would be able to do so also with regard to different uses of effective forms of CE, if such interventions were to be developed. Moreover, as Sandberg and Bostrom have observed, "[A]n enhancement that enables an individual to solve some of society's problems would produce a positive externality: in addition to benefits for the enhanced individual, there would be spillover benefits for other members of society" (Bostrom and Sandberg 2009, 328).

Rather than demonstrating that the use of CE would have character-eroding effects, the "no pain, no gain" objection points to the importance of contemplating, "[i]n each specific case, [...] what we consider to be the most important aspect of the activity under consideration: its aim (speaking French, losing weight, strong muscles) or its effects on character" (Schermer 2008, 357). This might lead us to conclude that the use of CE would be undesirable in the context of certain activities. It does not, however, demonstrate that CE is undesirable *per se*.

14.5 Conclusion

In this chapter, I have examined three objections to the use of CE. All three objections are based on the allegedly detrimental effects that CE is thought to have for the value of achievement. The first objection holds that the reason that we should oppose the use of CE is that it would cause us to no longer be responsible for our achievements. The second objection instead holds that we should oppose the use of CE because it would make us *too responsible* for our achievements, or the lack thereof. The third objection holds that we should oppose the use of CE because it would have character eroding-effects. I have also considered a number of counter-objections against each of these three objections.

Based on this examination, it could be concluded that the three objections against the use of CE may point to potential trade-offs between certain important values (hard work, discipline, dedication) and other important values (using the most effective means to achieve a desired end, cost-effectiveness). They may also point to potential losses in situations where the use of CE may cause us to forego process goods, or valuable experiences. This shows that it will sometimes be necessary to weigh the value of, for example, individuals' experiences of performing particular activities against the value of the goods that these activities will produce, both with regard to the individual concerned and to other individuals or society. It is concluded, however, that none of the three objections against the use of CE considered in this chapter succeed in demonstrating that the use of CE is objectionable *in principle*.

References

President's Council on Bioethics (2003) Beyond therapy: biotechnology and the pursuit of happiness. PCB, Washington, DC. http://bioethics.georgetown.edu/pcbe/reports/beyondtherapy/index.html. Retrieved 18 Mar 2008

Bostrom N, Sandberg A (2009) Cognitive enhancement: methods, ethics, regulatory challenges. Sci Eng Ethics 15:311–341

Bostrom N, Savulescu J (eds) (2009) Human enhancement. Oxford University Press, Oxford

Chatterjee A (2004) Cosmetic neurology. Neurology 63:968–974

Chatterjee A (2006) The promise and predicament of cosmetic neurology. J Med Ethics 32: 110–113

Cole-Turner R (1998) Do means matter? In: Parens E (ed) Enhancing human traits: ethical and social implications. Georgetown University Press, Washington, DC

Farah MJ (2002) Emerging ethical issues in neuroscience. Nat Neurosci 5:1123–1129

Fox D (2005) Safety, efficacy, and authenticity: the gap between ethics and law in FDA decision-making. Mich State Law Rev, 1135

Goodman R (2010) Cognitive enhancement, cheating, and accomplishment. Kennedy Inst Ethics J 20(2):145–160

Greely H, Sahakian B, Harris J, Kessler RC, Gazzaniga M, Campbell P, Farah MF (2008) Commentary: towards responsible use of cognitive-enhancing drugs by the healthy. Nature 456:702–705

Harris J (1992) Wonderwoman and superman: ethics of human biotechnology. Oxford University Press, Oxford

Kamm F (2009) What is and is not wrong with enhancement? In: Bostrom N, Savulescu J (eds) Human enhancement. Oxford University Press, Oxford

Mehlman MJ (2004) Cognition-enhancing drugs. Milbank Q 82(3):483–506

Olsen JM (2006) Depression, SSRIs, and the supposed obligation to suffer mentally. Kennedy Inst Ethics J 16(3):283–303

Radoilska L (2010) An Aristotelian approach to cognitive enhancement. J Value Inq 44(3):365–375

Sandel M (2004) The case against perfection. Ethics in the age of genetic engineering. Harvard University Press, Cambridge, MA

Schermer M (2008) Enhancements, easy shortcuts, and the richness of human activities. Bioethics 22:355–363

Chapter 15
Does the Cognitive Enhancement Debate Call for a Renewal of the Deliberative Role of Bioethics?

Cynthia Forlini and Eric Racine

Abstract Discourse on the issues related to the cognitive enhancement (CE) of healthy individuals using prescription medication has been firmly grounded in academia. Results from an emerging body of research examining the perspectives of non-academic stakeholders through surveys and focus groups have added experiential facets to the ethics debate around CE that, when synthesized, can yield important messages that further complement and diverge from academic discussions. We examined published literature on stakeholder perspectives and found conclusions that diverged from current academic literature. Three points of contention were found: (1) the discussion of safety and efficacy as a rate-limiting step in the progression of CE both practically and ethically; (2) the perception of both high and low prevalence in university students and consequences of the perceived prevalence; and (3) the ambivalence and ambiguity on the acceptability of CE and the progression of the ethics debate around it. These three points of contention suggest that the ethics debate around CE in academic and in stakeholder perspectives may be running on separate parallel tracks. We propose that the discipline of bioethics needs to reaffirm its role as a meeting place for the traditional academic ethics debate on CE and the more experientially-based approach of stakeholders to enrich future deliberation.

Keywords Cognitive enhancement • Public discourse • Stakeholders • Bioethics • Policy-making

C. Forlini • E. Racine (✉)
Neuroethics Research Unit, Institut de Recherches Cliniques de Montréal (IRCM), McGill University of Montreal, Montreal, QC, Canada
e-mail: cynthia.forlini@ircm.qc.ca; eric.racine@ircm.qc.ca

E. Hildt and A.G. Franke (eds.), *Cognitive Enhancement: An Interdisciplinary Perspective*, Trends in Augmentation of Human Performance 1,
DOI 10.1007/978-94-007-6253-4_15,

15.1 Introduction

Debate on the issues related to the cognitive enhancement (CE) of healthy individuals using prescription medication has been firmly grounded in academia. The debate has yielded highly polarized viewpoints for and against CE (Caplan and Elliott 2004; Caplan and McHugh 2007; Parens 2006). These opposing points of view have been associated with the broader framework of the "culture wars", which underlies many polarized American bioethics debates like stem cell research and end-of-life care (Racine 2010). Within this debate, several points of contention have surfaced including the issue of authenticity of the individual and human achievement; autonomy and coercion; and the need for investigation of safety and efficacy of prescription drugs used for enhancement purposes (President's Council on Bioethics 2003; Farah et al. 2004; Greely et al. 2008). However, it is unclear how the bioethics debate reflects broader perspectives outside the academic literature and to what extent the culture wars captures current tensions between opposing points of view. Yet it is interesting to bear in mind some of the seminal goals of bioethics within the current CE controversy, especially the deliberative role of bioethics.

In an early definition of the field of bioethics, Potter advanced that the goal of bioethics should be to "generate wisdom, the knowledge of how to use knowledge for social good from a realistic knowledge of man's biological nature and of the biological world" (Potter 1971). Potter envisioned bioethics to be inherently deliberative and to serve as a meeting place for the biological sciences and the humanities. Other early scholars in bioethics also stressed this role of bioethics viewed as an open-ended field (Callahan 1976). Historically, bioethics has opened up ethical debates to different perspectives by serving as a meeting place. There are several examples where bioethics has bridged ethics, medicine, science, and public perspectives in times when consensus was needed and points of view diverged on "biological knowledge and human values" (Potter 1971). Even before the official establishment of bioethics, the Nuremberg Code and Declaration of Helsinki called for broader social and ethical perspectives on medicine and the biological sciences. More recent examples like the Belmont Report, the British Nuffield Council on Bioethics and the American Presidential Commission for the Study of Bioethical Issues illustrate how the field has materialized in more interdisciplinary forms. Historical and newer centers dedicated to teaching and scholarship in bioethics along with professional societies have contributed to establishing bioethics as an interdisciplinary field and a meeting place, although this goal not been fully attained (Pellegrino 2006). The key to both the historical and more recent examples is that bioethics reached out to connect different perspectives and generate new knowledge on the ethical issues in scientific research and medical practice albeit critiques that have sometimes argued, "there is a paucity of reliable and valid empirical data regarding the considered moral opinions of the public or other key stakeholders who are not empowered to sit at the table" (Kim et al. 2009; Gutmann and Thompson 1997; Racine 2003; Macpherson 2004).

To date, the ethics debate around the ethics of CE has been largely based on traditional bioethics scholarship with weak connections to broader public concerns. In this sense the case of CE is not unique, but because of the public relevance and high profile of this topic, it represents an excellent opportunity to investigate if and how the deliberative goal of bioethics has been fulfilled. This question can only be indirectly and partly addressed in this paper, but an emerging body of research examining the perspectives of stakeholders (e.g., university students, healthcare professionals and other members of the general public) through focus groups and surveys has begun to flesh out public perspectives and experiential dimensions to the ethics debate around CE.

The overlap and potential clash between academic bioethics and broader stakeholder perspectives might create gaps in the way different groups experience and understand CE. This varying experience could in turn greatly impact ethical perspectives and judgments. Failing to recognize and comprehend such potential contrasts in experience and ethical thinking could have substantial consequences on predictions and policies. In reaching out to broader stakeholder perspectives, bioethics would potentially enrich its reflection and ensure that academic concerns, biases or interests do not prevail unilaterally or inadvertently (Racine and Forlini 2009). Precedent bioethics debates raise relevant questions in this respect.

Several examples suggest that a better understanding of differences between academic ethics and broader public perspectives would potentially have had consequences for ethics reflection. For example, the direct-to-consumer marketing of genetic tests elicited many concerns in the academic bioethics debate for the proper use of genetic information from these tests to avoid stigma and anxiety on the part of consumers (Gollust et al. 2003; McCabe and McCabe 2004; Chadwick 2008). However, a recent study shows that these predictions may have been unfounded since the majority of participants having purchased a genetic test did not report any anxiety and few lifestyle changes after receiving results (Bloss et al. 2011). Understanding the significance of contrasts between academic ethics and broader public perspectives can help the ethics debate to move forward constructively as it becomes increasingly important to address and possibly take action on important issues related to CE.

In this paper, we examine and discuss three major sources of tension between academic ethics and available data on stakeholder and public perspectives on CE. We examined published literature on stakeholder perspectives, relying on survey and focus group methodology to conduct a qualitative literature review. We found five studies that corresponded to these criteria (Table 15.1). From these studies, we identified and further examined perspectives and conclusions that stood apart from current academic literature. The three points of contention we found were: (1) the discussion of safety and efficacy as a rate-limiting step in the progression of CE both practically and ethically; (2) the perception of prevalence in university students and consequences of the perceived prevalence; and (3) the ambivalence and ambiguity regarding the acceptability of CE and the progression of the ethics debate surrounding it.

Table 15.1 Studies examining stakeholder perspectives on enhancement uses of prescription medication with respective samples and methodologies

Study	Sample	Methodology
Hotze et al. *AJOB*. 2011	633 physician members of the American Medical Association	Survey
Banjo et al. *PLoS One*. 2010	212 physicians from USA and Canada	Survey
Forlini and Racine. *Neuroethics*. 2009a Forlini and Racine. *Public Understanding of Science*. 2012	65 individuals (university students, parents, healthcare professionals) in Montreal	Media prompt material and focus groups
Bergstrom and Lynoe. *Scandinavian Journal of Public Health*. 2008	Random sample of 1,000 individuals in Stockholm and 300 General practitioners (GP)	Vignettes and questionnaire
Sabini and Monterosso. *Ethics & Behavior*. 2005	185 undergraduate students from University of Pennsylvania	Vignettes and questionnaire

15.2 Is Research on Safety and Efficacy a Rate-Limiting Step?

The academic ethics debate has cited the scarcity of evidence of safety and efficacy as a major challenge of CE (Chatterjee 2004; Farah and Wolpe 2004; Hall 2004; Mehlman 2004; Greely et al. 2008). Some studies have reported positive results on the effects of stimulant medication, specifically, on healthy individuals (Elliott et al. 1997; Mehta et al. 2000; Barch and Carter 2005; Farah et al. 2008). However, with small samples and varying methodologies, these results are not a resounding call for general promotion of CE. A series of meta-analyses published by Repantis et al., show that even large scale clinical trials on prescription medications used as cognitive enhancers (e.g., anti-depressants, acetylcholinesterase inhibitors, stimulants) have not provided consistent evidence of safety and efficacy (Repantis et al. 2008, 2010a, b). Indeed, if cognitive enhancers were not safe or inefficacious, their ethical justification would be in peril, and the debate could shift toward prevention of a dangerous or ineffective practice. However, the ethical debate currently progresses in the midst of disparate evidence and encourages calls for a "programme of research into the use and impacts of cognitive-enhancing drugs by healthy individuals" (Greely et al. 2008) and "growing need for guidance on how to conduct enhancement research in an ethical manner" (Mehlman et al. 2011). At the same time, stakeholders feel a certain sense of security in the non-medical use of prescription medications for enhancement purposes even though they acknowledge that data on long-term effects are scarce (Forlini and Racine 2012).

Calling for more research in the absence of concrete evidence of safety and efficacy is an obvious and generally sound recommendation. However, the trend in the academic ethics literature has also been to consider other ethical issues in the light of an a priori role for concerns about safety and efficacy. For example, Greely et al. discuss the possibility of revisiting policies to oblige certain professions to use enhancers "[i]f particular enhancements are shown to be sufficiently safe and effective" (Greely et al. 2008). This type of reasoning positions the ethical discourse on other issues such as autonomy and coercion, fairness, as well as social and distributive justice as being contingent on having information either proving or disproving safety and efficacy. The same has been suggested with regard to policy that, "before [pharmacological cognitive enhancing] drugs are prescribed to healthy people, their long-term safety, side-effects and their effectiveness must be tested to provide important facts necessary for further decision making about their regulation" (Mohamed and Sahakian 2011). Greely et al. argue for a more decisive regulatory approval pending safety and efficacy data, "regulatory agencies should allow pharmaceutical companies to market cognitive enhancing drugs to healthy adults provided they have supplied the necessary regulatory data for safety and efficacy" (Greely et al. 2008). In this respect, the academic ethics debate implicitly and explicitly proposes that the ethical reflection on CE cannot progress any faster than research on safety and efficacy, the rate-limiting step. (Others have tended to boil down all ethical concerns of CE to matters of safety and efficacy (e.g., Caplan) although it is beyond the scope of this paper to discuss this perspective in depth.)

The perspectives and opinions of stakeholders on safety and efficacy are divided. In some respects, stakeholders share the point of view that research on safety and efficacy can be a rate-limiting step for the CE ethics debate. For instance, the design of two stakeholder studies (Banjo et al. 2010; Hotze et al. 2011) included surveys based on *hypothetical* cognitive enhancers that were previously deemed safe and efficacious. This methodological provision allowed researchers to encourage participants to think past concerns for safety and efficacy, which could have been an immediate barrier to considering other issues. Regardless of this provision, physicians in these two studies expressed concerns for safety. Banjo et al. (2010) reported that, "physicians mistrust safety claims regarding pharmaceuticals" such that it remains a chief concern despite the evidence presented. Similarly, Hotze et al. (2011) reported that safe hypothetical cognitive enhancers, "least often approved by physicians were interventions that could easily be argued as dangerous to the user," whether due to the side effects caused or the behavior produced (e.g., making a soldier more aggressive). From these responses, it can be argued that stakeholders, too, consider research on safety and efficacy a rate-limiting step, which needs to be addressed before moving along too far in the discussion.

Other pieces of data show that the concerns of stakeholders loom much larger than safety and efficacy. Results from Banjo et al. (2010) also show that safety concerns of physicians, "were not offset by the benefit afforded the individual," putting emphasis on "the balance between safety and benefit." In this fashion, participants of this study did not give a *carte blanche* for widespread use of cognitive enhancers, which would be assumed to be safe and effective. Other factors were taken into consideration indicating that, "physicians are keenly aware of the ethical landscape involved in prescribing cognitive enhancers" (Banjo et al. 2010). Indeed, other features of the ethical landscape in prescribing prescription medications as cognitive enhancers emerged in stakeholder studies. Hotze et al. (2011) described concerns of physicians, "regarding perceived threats to social justice and the meaning of human achievement." Similar concerns for social justice, but also for the responsibility of individuals, were voiced by physicians and members of the general public in a Scandinavian study by Bergstrom and Lynoe (2008). Finally, in a study centered upon the issue of fairness in using cognitive enhancers, Sabini and Monterosso (2005) found that university students used this issue to establish ethical norms about CE. Participants in this study were not questioned directly on safety but determined that cognitive enhancers, in this case methylphenidate, were only ethically acceptable as *normalizers* for individuals who performed in the bottom 10 % as opposed to *enhancers* for already high performing individuals. The results of these stakeholder studies show that though safety and efficacy remain prominent issues, the ethical concerns about social justice and the social context of CE on both individual and collective choices are also important considerations.

Research on stakeholder and public perspectives on CE clearly suggest that more information on safety and efficacy is a demand shared with academic perspectives. The difference, however, lies in the way the issue of safety and efficacy are situated within the debate on the ethics of CE. The issue of safety and efficacy does not seem to be the stumbling block that is often described in academic ethics debates. The perspectives of stakeholders may suggest a different path to organizing priorities in the ethics debate on CE. For example, safety and efficacy may be first investigated and only then, depending on the data, followed by deliberation on other issues like fairness, distribution and autonomy of the individual. An alternative model would be to first determine whether using prescription medication for performance enhancement could be fair and under what conditions, how enhancers could be justly distributed and whether anyone could be obliged to take them before seeking to fully investigate safety and efficacy. This alternative model would neither presume that safety and efficacy are stumbling blocks in addressing other matters nor presume that once safety and efficacy concerns are addressed, meaningful ethical discussion would have been exhausted. Reciprocity between academic and stakeholder perspectives could enrich the current debate on CE.

15.3 Is the Non-medical Use of Prescription Medication for Enhancement Prevalent?

The importance of debating the ethics of CE is often justified by widespread use of prescription medication for enhancement purposes, notably in reference to the use of stimulants by university students to improve academic performance. Table 15.2, adapted from a previous review (Racine and Forlini 2010) to include recent data (Franke et al. 2011), shows that between 1.3 and 11 % of university students in

Table 15.2 Prevalence rates for lifetime non-medical prescription stimulant (PS) use and PS use specifically for cognitive enhancement (CE)

Study	Sample population	PS use %	PS use for CE %
Franke et al. *Pharmacopsychiatry*. 2011	1,035 pupils in small and big cities and 512 university students of 3 departments	2.6	1.3 (cognitive enhancement)
Teter et al. *Pharmacotherapy*. 2006	4,580 college students in a large Midwestern university	8.3	5.4 (enhance concentration) 5.0 (enhance studying) 4.0 (enhance alertness)
Prudhomme White et al. *Journal of American College Health*. 2006	1,025 students at the University of New Hampshire	16.2	11.0 (enhance concentration) 8.7 (enhance studying) 3.2 (enhance grades)
Teter et al. *Journal of American College Health*. 2005	9,161 undergraduate students at the University of Michigan	8.1	4.3 (enhance concentration) 3.2 (enhance alertness)
Hall et al. *Journal of American College Health*. 2005	381 college students from the University of Wisconsin-Eau Claire	13.7	3.7 (enhance studying)
Graff Low and Gendaszek. *Psychology, Health & Medicine*. 2002	150 undergraduate students at a small, competitive college in the US	35.3	8.2 (enhance intellectual performance) 7.8 (enhance studying)

those samples used prescription stimulant medication specifically for CE. However, what is considered widespread is itself open to interpretation. Estimates of general prevalence are situated at approximately 20 % in two surveys, one conducted by a German health insurer and the other sponsored by the science journal *Nature* (Maher 2008; DAK 2009). These surveys differ significantly in methodology from the targeted public health studies that report significantly lower prevalence rates. Both the surveys and public health prevalence studies are often cited in the academic ethics debate as indicating both widespread and low prevalence.

There is a movement in the academic ethics debate that considers current evidence of prevalence too weak to support the claim of widespread CE. Lucke et al. have categorized the *Nature* poll as "anecdotal evidence" due to its convenience sampling (Lucke et al. 2010). The wide-ranging prevalence rates reported in the empirical data "indicate problems of consistency of methods and representativeness of samples" (Hall and Lucke 2010). In addition, some have indicated that these rates include other recreational uses of prescription medication and not exclusively CE uses (Lucke et al. 2011). Finally, prevalence rates, some have found, have been inappropriately associated with increasing sales of prescription medications associated with CE (Lucke et al. 2010; Outram 2010). It has been proposed that prevalence rates may even decrease with increased awareness of legal risks and potential side effects (Outram 2010). Furthermore, there is empirical evidence to support criticism of the academic ethics debate because students may actually overestimate prevalence of non-medical use of prescription stimulants among their peers (McCabe 2008). Though the prevalence of CE may be overestimated, it does not straightforwardly justify abandoning the debate on the ethics of CE, especially in light of research on stakeholder and public perspectives.

Studies on stakeholder perspectives from medical and academic contexts show that participants in these studies estimate that CE is quite prevalent. Hotze et al. reported that 61.7 % of physicians that responded to their survey received requests for CE from their patients (Hotze et al. 2011). The physicians in Banjo et al.'s study were also at ease when speaking about CE in contrast to those in the Scandinavian study who reported feeling less concerned that CE affected their practice (Bergstrom and Lynoe 2008; Banjo et al. 2010). It is possible that this difference may have cultural and temporal origins. Student perspectives show prevalence is felt to be high among student peers (Forlini and Racine 2009a). The perception of high prevalence points toward two possibilities. First, CE may be especially prevalent in medical and academic environments. Second, regardless of what the prevalence of CE actually is, the perception of stakeholders shows that issues related to CE are problematic and are a source of ethical dilemmas for the groups represented in the samples of these studies.

While debates go on with regard to the actual prevalence of CE, some do not hesitate to describe CE (e.g., the non-medical use of prescription stimulants) as a "contemporary public health problem" (Arria and DuPont 2010). These types of descriptions fuel efforts to "reduce the large and growing problem of the nonmedical use of prescription stimulants among college students" (Arria and DuPont 2010). Concurrently, policy perspectives on CE are beginning to emerge as exemplified in a discussion paper from the British Medical Association, a report from Quebec's Comission de l'éthique de la science et de la technologie (CEST), and guidance for physicians from the American Academy of Neurology (British Medical Association 2007; Commission de l'éthique de la science et la technologie 2009; Larriviere et al. 2009). All three of these reports have unanimously and clearly cited increasing prevalence of CE as one of the main motivators for policy output by their respective organizations (Outram and Racine 2011). This justification is problematic if prevalence is eventually shown to be lower than the reported rates as policies will have been generated for a "phantom practice" (Lucke et al. 2010). The perception of stakeholders provides an argument for the ethical significance of CE regardless of what the magnitude and indicates that the perception of widespread prevalence may continue to shape the ethical landscape of CE. Though the academic ethics discourse entertains healthy skepticism about prevalence figures, stakeholder perspectives indicate that research and deliberation should not be halted because of this. However, discourse could benefit from (1) being cautious and not exaggerating prevalence to justify ethical and medical attention and, at the same time, (2) not equating low prevalence with low importance for ethics debate. While an ethics issue likely becomes more important if it is also a public health concern, an ethics question cannot be boiled down to a public health concern.

15.4 What Can Ambivalence in the Ethical Perspectives of Stakeholders Mean?

The previous section described how the expectation of widespread CE in the public is informing professional associations and government reports and policies on CE. Farah et al. stated early on that, "[t]he question is therefore not whether we need policies to govern neurocognitive enhancement, but rather what kind of policies we need" (Farah et al. 2004). This attitude toward policy reflects what other authors have considered the inevitability of widespread CE and, to some extent, its social acceptance by many (Baylis and Robert 2004; Chatterjee 2004). However, this proactive attitude toward regulation has garnered opposing reactions in the academic ethics debate. Some are concerned that existing policies on CE may be premature (Lucke et al. 2011) and that ultimately, "neuroenhancing drugs should be assessed on their merits, and regulated according to the risks that they pose and the feasibility of regulating or restricting their use" (Lucke et al. 2011). Academic proponents of policy could be implicitly promoting some permutations of CE to normalize it while critics are staunchly holding their position against all forms of CE for fear of affronting human nature. Determining which approach to policy on CE is most appropriate becomes difficult with polarized academic perspectives that leave little room for the experience of stakeholders.

Stakeholder perspectives are not as clear-cut as those of authors in the academic ethics debate and have revealed different forms of ambivalence and discomfort. Stakeholders have acknowledged the presence of CE in their environments, but this did not lead them to think that it should be encouraged. A majority of physicians answering questions about hypothetical cognitive enhancers "demonstrated some ambivalence, agreeing that the medicine should be allowed but not encouraged" (Hotze et al. 2011). Another sample of physicians seemed to be uncomfortable prescribing even existing medications (methyphenidate, modafinil) for CE purposes (Banjo et al. 2010). Students felt much the same way in Sabini and Monterosso's study where the participants discussed how cognitive enhancers might be fair but "under no circumstances was a drug given a ringing endorsement" (Sabini and Monterosso 2005). A mixed sample, of students, parents and healthcare professionals displayed marked ambivalence in their initial reactions to CE as well as in the descriptive and normative frameworks discussed (Forlini and Racine 2012). For example, when asked to compare the use of methylphenidate for CE to the use of steroids in sports for performance enhancement or even caffeine consumption, a soft type of CE, participants did not draw a definitive conclusion about which was more similar. In light of these data, the term ambivalence signifies the coexistence of conflicting perspectives about the nature of CE and its acceptability in medical and academic environments.

Interpreting the ambivalence is difficult. It may be a reflection of indifference, lack of information or misunderstanding on the part of stakeholders. This type of ambivalence could be called a form of *informational ambivalence*, i.e., ambivalence caused by rather superficial misunderstandings. Alternatively, this apparent indecisiveness may be indicative of complex and evolving ethical thinking in stakeholder groups and reflect what we could call *substantial ambivalence* (Forlini and Racine 2012). Racine and Parens have both suggested that the ethics of CE may not be captured by one sole moral framework or moral test (Parens 2005; Racine 2010). Accordingly, the ambivalence of stakeholders may be a sign that the broad ethics debate around CE needs enrichment to capture the experience and reactions of different groups outside academia. In other words, ambivalence could be meaningful from a moral standpoint. However, short of simply refraining from issuing policies or making recommendations, it would remain challenging to see how substantial moral ambivalence might inform public policy (or even just the steps upstream from policy) as compared to more defined opinions we find in the academic ethics literature. Methodologically, substantial moral ambivalence is a difficult reaction to capture and interpret when considering that the term *enhancement* can be interpreted in many different ways by stakeholders, in addition to having a built-in connotation of efficacy (Forlini and Racine 2011). Given this existence of ambivalence in stakeholder perspectives and the staunch divide in the academic ethics debate on how to best address the ethical and social issues around CE, we must consider whether policy-making is a timely and appropriate step.

15.5 The Renewal of the Deliberative Role of Bioethics

The contrasts between what were designated in this article as, on the one hand, the academic ethics debate and, on the other hand, public and stakeholder perspectives were meant to bring to light differences in the way experience can inform ethical perspectives. It was not meant to illustrate any kind of misunderstanding on the part of stakeholders with regard to the published literature. Clearly, the categories of *expert* and *non-expert* or *lay citizen* fails to capture the fact that experts are not fully objective and that non-experts have valuable knowledge and experience to share. One might reframe this distinction as not between the "know" and "know not" but rather between the "know something" and "know something else." The three points of contention we identified suggest that the ethics debate around CE in academic and in public and stakeholder perspectives may be running on separate parallel tracks evolving almost independently of each other. This separate evolution could have consequences and lead to differences in how academic ethics and other stakeholders: (1) prioritize and give importance to different ethical issues; (2) justify individual and collective decisions as well as policy options on different perceptions of prevalence and (3) incorporate complex and evolving ethical thinking represented in the form of substantial moral ambivalence. In addition to the academic culture wars already at work in CE, these points of contention may be evidence of a broader and more general disconnection, this time, between academic ethics and broader stakeholder perspectives. Not acknowledging this possible gap could lead to several pitfalls in policy-making and practice-oriented scholarship.

Bioethics is already an interdisciplinary field, but its deliberative role could be reaffirmed and better established more generally (Gutmann and Thompson 1997; Racine 2003) and in the CE debate (Forlini and Racine 2009b; Racine 2010). Bioethics could reinvigorate its role as a meeting place for the traditional academic ethics debate on CE and the more experientially based approach of stakeholders. In order to play this role, Callahan has urged that, "if bioethics is to retain its vitality and be taken seriously, it will have to find a way to extricate itself from the culture wars" (Callahan 2005). Indeed, the type of wisdom about using knowledge for social good that Potter advocated could be generated *within* a bioethics that, as Callahan proposed, "expect[s] and welcome[s] struggles between opposing viewpoints" (Callahan 2005). The meeting place fostered by bioethics could help build a two-way bridge between the academic ethics and stakeholder perspectives, which would enrich the broader CE ethics debate by representing the value added of each perspective. This two-way bridge could help incorporate the human values reflected in the experience of stakeholders to "recognize the importance of the manifold factors that affect human welfare" (de Melo-Martin 2010). In the other direction, academic ethics can provide a framework to understand the principles that stakeholders are using and valuing in their experiences with CE. Bioethics need not bridge academic and stakeholder perspectives alone. Popular media are already a meeting place for public discourse on ethics and technology in a readily accessible format. Stakeholders have expressed appreciation for the information the media has provided about CE (Forlini and Racine 2012). However, there is concern that current media coverage participates in the enthusiasm and high expectations for CE in addition to potentially promoting the practice (Forlini and Racine 2009b). This observation raises the question of whether the media is a suitable locus for the bioethics debate or if it actually impedes it (Simonson 2002). Other options include fostering unmediated discussion between academics and the public (Racine et al. 2005).

Illes et al. have proposed a model to improve the communication of neuroscience research and promote self-reflection in neuroscientists on ethical issues (Illes et al. 2010). This model fills a clear gap since, in spite of the fundamental deliberative role of bioethics, general research ethics guidelines give little direction on how researchers should disseminate their research beyond peer-review journals or how and when they should consider the outcomes of their research (Zimmerman and Racine 2012). Illes et al. propose a collaborative approach involving individual researchers, academic institutions, professional organizations and research sponsors. Given the inclusiveness of such collaboration, this "Neurotalk" model may inspire ways in which bioethics can integrate academic ethics and stakeholder perspectives on CE.

15.6 Conclusion

Significant differences exist between academic bioethics and broad stakeholder perspectives on the ethics of CE, particularly on issues like safety and efficacy, perception of prevalence and moral ambivalence. These differences show that CE is a complex bioethics issue involving multiple perspectives. The issues of CE are multifaceted just as the experiences and perspectives of the many actors who are presently involved and those who may become involved in the future directions of the debate. Consequently, a monolithic ethics reflection upon CE would be detrimental to such a dynamic debate, and this calls for self-reflection both on the part of academics and others. By assuming a deliberative role and resisting polarization, bioethics could serve as a forum where perspectives are enriched and contention points are closely examined.

References

Arria AM, DuPont RL (2010) Nonmedical prescription stimulant use among college students: why we need to do something and what we need to do. J Addict Dis 29:417–426

Banjo OC, Nadler R, Reiner PB (2010) Physician attitudes towards pharmacological cognitive enhancement: safety concerns are paramount. PLoS One 5:e14322

Barch DM, Carter CS (2005) Amphetamine improves cognitive function in medicated individuals with schizophrenia and in healthy volunteers. Schizophr Res 77:43–58

Baylis F, Robert JS (2004) The inevitability of genetic enhancement technologies. Bioethics 18:1–26

Bergstrom LS, Lynoe N (2008) Enhancing concentration, mood and memory in healthy individuals: an empirical study of attitudes among general practitioners and the general population. Scand J Public Health 36:532–537

Bloss CS, Schork NJ, Topol EJ (2011) Effect of direct-to-consumer genomewide profiling to assess disease risk. N Engl J Med 364:524–534

British Medical Association (2007) Boosting your brainpower: ethical aspects of cognitive enhancement. British Medical Association, London

Callahan D (1976) Bioethics as a discipline. In: Humber J, Almeder RF (eds) Biomedical ethics and the law. Plenum Press, New York, pp 1–11

Callahan D (2005) Bioethics and the culture wars. Camb Q Healthc Ethics 14:424–431

Caplan A, Elliott C (2004) Is it ethical to use enhancement technologies to make us better than well? PLoS Med 1:e52

Caplan AL, McHugh PR (2007) Shall we enhance? A debate. In: Glannon W (ed) Defining right and wrong in brain science: essential readings in neuroethics. The Dana Press, Washington, DC, pp 271–288

Chadwick R (2008) Genetic testing and screening. In: Singer PA, Viens AM (eds) The Cambridge textbook of bioethics. Cambridge University Press, Cambridge

Chatterjee A (2004) Cosmetic neurology: the controversy over enhancing movement, mentation, and mood. Neurology 63:968–974

Commission de l'éthique de la science et la technologie (2009) Position statement on psychotropic drugs and expanded uses: an ethical perspective. Commission de l'éthique de la science et la technologie, Québec

DAK – Unternehmen Leben (2009) Gesundheitsreport 2009, http://www.dnbgf.de/fileadmin/texte/Downloads/uploads/dokumente/2009/DAK_Gesundheitsreport_2009.pdf

de Melo-Martin I (2010) The two cultures: an introduction and assessment. Technol Soc 32:5–9

Elliott R, Sahakian BJ, Matthews K, Bannerjea A, Rimmer J, Robbins TW (1997) Effects of methylphenidate on spatial working memory and planning in healthy young adults. Psychopharmacology 131:196–206

Farah MJ, Wolpe PR (2004) Monitoring and manipulating brain function: new neuroscience technologies and their ethical implications. Hastings Cent Rep 34:35–45

Farah MJ, Illes J, Cook-Deegan R, Gardner H, Kandel E, King P et al (2004) Neurocognitive enhancement: what can we do and what should we do? Nat Rev Neurosci 5:421–425

Farah MJ, Haimm C, Sankoorikal G, Chatterjee A (2008) When we enhance cognition with Adderall, do we sacrifice creativity? A preliminary study. Psychopharmacology 202:541–547

Forlini C, Racine E (2009a) Autonomy and coercion in academic "cognitive enhancement" using methylphenidate: perspectives of a pragmatic study of key stakeholders. Neuroethics 2:163–177

Forlini C, Racine E (2009b) Disagreements with implications: diverging discourses on the ethics of non-medical use of methylphenidate for performance enhancement. BMC Med Ethics 10:9

Forlini C, Racine E (2011) Considering the causes and implications of ambivalence in using medicine for enhancement. Am J Bioeth 11:15–17

Forlini C, Racine E (2012) Stakeholder perspectives and reactions to "academic" cognitive enhancement: unsuspected meaning of ambivalence and analogies. Public Underst Sci 21:606–625

Franke AG, Bonertz C, Christmann M, Huss M, Fellgiebel A, Hildt E et al (2011) Non-medical use of prescription stimulants and illicit use of stimulants for cognitive enhancement in pupils and students in Germany. Pharmacopsychiatry 44:60–66

Gollust SE, Wilfond BS, Hull SC (2003) Direct-to-consumer sales of genetic services on the Internet. Genet Med 5:332–337

Graff Low K, Gendaszek AE (2002) Illicit use of psychostimulants among college students: a preliminary study. Psychol Health Med 7:283–287

Greely H, Sahakian B, Harris J, Kessler RC, Gazzaniga M, Campbell P et al (2008) Towards responsible use of cognitive-enhancing drugs by the healthy. Nature 456:702–705

Gutmann A, Thompson D (1997) Deliberating about bioethics. Hastings Cent Rep 27:38–41

Hall W (2004) Feeling 'better than well'. EMBO Rep 5:1105–1109

Hall KM, Irwin MM, Bowman KA, Frankenberger W, Jewett DC (2005) Illicit use of prescribed stimulant medication among college students. J Am Coll Health 53(4):167–174

Hall WD, Lucke JC (2010) The enhancement use of neuropharmaceuticals: more specticism and caution needed. Addiction 105:2041–2043

Hotze T, Shaw K, Anderson E, Wynia M (2011) "Doctor, would you prescribe a pill to help me...?" A national survey of physicians on using medicine for human enhancement. Am J Bioeth 11:3–13

Illes J, Moser MA, McCormick JB, Racine E, Blakeslee S, Caplan A et al (2010) Neurotalk: improving the communication of neuroscience research. Nat Rev Neurosci 11:61–69

Kim SY, Wall IF, Stanczyk A, De Vries R (2009) Assessing the public's views in research ethics controversies: deliberative democracy and bioethics as natural allies. J Empir Res Hum Res Ethics 4:3–16

Larriviere D, Williams MA, Rizzo M, Bonnie RJ (2009) Responding to requests from adult patients for neuroenhancements. Guidance of the ethics, law and humanities committee. Neurology 73:1406–1412

Lucke JC, Bell S, Partridge B, Hall WD (2010) Weak evidence for large claims contribute to the phantom debate. BioSocieties 5:482–483

Lucke JC, Bell SK, Partridge BJ, Hall WD (2011) Academic doping or Viagra for the brain? The history of recreational drug use and pharmacological enhancement can provide insight into these uses of neuropharmaceuticals. EMBO Rep 12:197–201

Macpherson CC (2004) To strengthen consensus, consult the stakeholders. Bioethics 18:283–292

Maher B (2008) Poll results: look who's doping. Nature 452:674–675

McCabe SE (2008) Misperceptions of non-medical prescription drug use: a web survey of college students. Addict Behav 33:713–724

McCabe LL, McCabe ERB (2004) Direct-to-consumer genetic testing: access and marketing. Genet Med 6:58–59

Mehlman MJ (2004) Cognition-enhancing drugs. Milbank Q 82:483–506

Mehlman MJ, Berg JW, Juengst ET, Kodish E (2011) Ethical and legal issues in enhancement research on human subjects. Camb Q Healthc Ethics 20:30–45

Mehta MA, Owen AM, Sahakian BJ, Mavaddat N, Pickard JD, Robbins TW (2000) Methylphenidate enhances working memory by modulating discrete frontal and parietal lobe regions in the human brain. J Neurosci 20:65

Mohamed AD, Sahakian BJ (2011) The ethics of elective psychopharmacology. Int J Neuropsychopharmacol 15:559–571

Outram SM (2010) The use of methylphenidate among students: the future of enhancement? J Med Ethics 36:198–202

Outram SM, Racine E (2011) Developing public health approaches to cognitive enhancement: an analysis of current reports. Public Health Ethics 4:93–105

Parens E (2005) Authenticity and ambivalence: toward understanding the enhancement debate. Hastings Cent Rep 35:34–41

Parens E (2006) Creativity, gratitude, and the enhancement debate. In: Illes J (ed) Neuroethics: defining the issues in theory, practice and policy. Oxford University Press, Oxford, pp 75–86

Pellegrino E (2006) Balint lecture. Paper presented at the ASBH summer conference: bioethics and politics, Albany, 2006

Potter VR (1971) Bioethics: bridge to the future. Prentice-Hall, Englewood Cliffs

President's Council on Bioethics (2003) Beyond therapy. President's Council on Bioethics/Harper Collins, Washington, DC

Racine E (2003) Discourse ethics as an ethics of responsibility: comparison and evaluation of citizen involvement in population genomics. J Law Med Ethics 31:390–397

Racine E (2010) Pragmatic neuroethics: improving treatment and understanding of the mind-brain. MIT Press, Cambridge, MA

Racine E, Forlini C (2009) Expectations regarding cognitive enhancement create substantial challenges. J Med Ethics 35:469–470

Racine E, Forlini C (2010) Cognitive enhancement, lifestyle choice or misuse of prescription drugs? Ethical blindspots in current debates. Neuroethics 3:1–14

Racine E, Bar-Ilan O, Illes J (2005) fMRI in the public eye. Nat Rev Neurosci 6:159–164

Repantis D, Schlattmann P, Lainsey O, Heuser I (2008) Antidepressants for neuroenhancement in healthy individuals: a systematic review. Poiesis Prax 6:139–174

Repantis D, Laisney O, Heuser I (2010a) Acetylcholinesterase inhibitors and memantine for neuroenhancement in healthy individuals: a systematic review. Pharmacol Res 61:473–481

Repantis D, Schlattmann P, Laisney O, Heuser I (2010b) Modafinil and methylphenidate for neuroenhancement in healthy individuals: a systematic review. Pharmacol Res 62:187–206

Sabini J, Monterosso J (2005) Judgments of the fairness of using performance enhancing drugs. Ethics Behav 15:81–94

Simonson P (2002) Bioethics and the rituals of media. Hastings Cent Rep 32:32–39

Teter CJ, McCabe SE, Cranford JA, Boyd CJ, Guthrie SK (2005) Prevalence and motives for illicit use of prescription stimulants in an undergraduate student sample. J Am Coll Health 53:253–262

Teter CJ, McCabe SE, LaGrange K, Cranford JA, Boyd CJ (2006) Illicit use of specific prescription stimulants among college students: prevalence, motives, and routes of administration. Pharmacotherapy 26:1501–1510

White BP, Becker-Blease KA, Grace-Bishop K (2006) Stimulant medication use, misuse, and abuse in an undergraduate and graduate student sample. J Am Coll Health 54:261–268

Zimmerman E, Racine E (2012) Ethical issues in the translation of social neuroscience: a policy analysis of current guidelines for public dialogue in human research. Account Res 19:27–46

Part III
Sociological, Political and Legal Aspects of Cognitive Enhancement

Chapter 16
The Biopolitics of Cognitive Enhancement

Peter B. Reiner

Abstract The ongoing discussion regarding the propriety of cognitive enhancement is, at its core, biopolitical. In an attempt to understand the underlying issues, it is instructive to view the subject through the perspectives of those at the extremes – the Transhumanists who enthusiastically support the development of cognitive enhancement and the Bioconservatives who view the enterprise with disapprobation. I argue that many of the positions undertaken by proponents of either view are influenced to a considerable degree by a variety of pre-existing political positions as much as, if not more so than, by deeper philosophical reflection, a phenomenon very much in keeping with Jonathan Haidt's social intuitionist model of moral reasoning (Haidt, Psychol Rev 108:814–834, 2001). Despite the vocal nature of the extremes, there also exists a broad middle ground which I term The view from reasonableness which is much less of a polemic and more of a conversation. Given that the ultimate outcome rests more on how the populace responds than the views of experts, I suggest that supplementing normative claims driven by biopolitical intuitions with empirical data on public attitudes towards cognitive enhancement might allow us to move beyond polemics to deepen our appreciation of how to move forward.

Keywords Cognitive enhancement • Transhumanists • Bioconservatives • Society • Public discourse

---

P.B. Reiner (✉)
National Core for Neuroethics, University of British Columbia Hospital, Vancouver, BC, Canada
e-mail: peter.reiner@ubc.ca

E. Hildt and A.G. Franke (eds.), *Cognitive Enhancement: An Interdisciplinary Perspective*, Trends in Augmentation of Human Performance 1,
DOI 10.1007/978-94-007-6253-4_16,

16.1 Historical Roots of the Enhancement Debate

The debate over the propriety of cognitive enhancement has a long and storied history. Worries about the impact of enhancement upon the quality of life go back at least as far as Socrates, who nearly 2,400 years ago warned that the use of writing would degrade memory. In a famous quote from Phaedrus, he intoned:

> At the Egyptian city of Naucratis, there was a famous old god, whose name was Theuth; the bird which is called the Ibis is sacred to him, and he was the inventor of many arts, such as arithmetic and calculation and geometry and astronomy and draughts and dice, but his great discovery was the use of letters .... It would take a long time to repeat all that Thamus said to Theuth in praise or blame of the various arts. But when they came to letters, This, said Theuth, will make the Egyptians wiser and give them better memories; it is a specific both for the memory and for the wit. Thamus replied: O most ingenious Theuth, the parent or inventor of an art is not always the best judge of the utility or inutility of his own inventions to the users of them. And in this instance, you who are the father of letters, from a paternal love of your own children have been led to attribute to them a quality which they cannot have; for this discovery of yours will create forgetfulness in the learners' souls, because they will not use their memories; they will trust to the external written characters and not remember of themselves. The specific which you have discovered is an aid not to memory, but to reminiscence, and you give your disciples not truth, but only the semblance of truth; they will be hearers of many things and will have learned nothing; they will appear to be omniscient and will generally know nothing; they will be tiresome company, having the show of wisdom without the reality (Jowett 1892).

Many of the key elements of the modern debate are evident in this passage; ironically, it is precisely because Plato memorialized the Dialogues that the apocryphal story is with us today. Indeed, enthusiasts who favor the adoption of cognitive enhancement often point out that despite apparent tradeoffs, cognitive enhancement is likely to improve the world around us in myriad ways. Moreover, Socrates notwithstanding, today there is little in the way of dissent about whether writing contributes to the quality of life in the modern world. The irony comes around full circle when we notice that much of the debate over cognitive enhancement today involves the use of drugs to improve the very same memory that Socrates warned might degrade with the introduction of the *aide de memoire* of writing (Chatterjee 2004; Farah et al. 2004; Glannon 2006; Greely et al. 2008; Erler 2011).

The issue has taken on added urgency with reports of the widespread off-label use of prescription drugs for pharmacological cognitive enhancement (McCabe et al. 2005; White et al. 2006; Franke et al. 2011) and the prospects of inexpensive and seemingly effective technologies such as transcranial direct current stimulation looming on the near term horizon (Hamilton et al. 2011; Kadosh et al. 2012). The debate is loudest at the extremes (President's Council on Bioethics 2003; Savulescu 2005; Bostrom and Sandberg 2009; Harris 2010), but there is a healthy middle ground position that neither restricts nor encourages the use of cognitive enhancement (Parens 1998; Greely 2010).

The previous decade witnessed a great deal of public interest in the issue of cognitive enhancement, with a seemingly never-ending glut of magazine and newspaper articles discussing the issue, sometimes with appropriate subtlety and other times with hyperbolic enthusiasm (Hall 2003; Garreau 2006; Schulz 2006; Baker 2009; Cascio 2009; Stix 2009). It is always hard to identify the source of a cultural meme, but one candidate catalyst was the release of the publication *Beyond Therapy* (President's Council on Bioethics 2003), a commentary on the topic of enhancement (genetic and otherwise) from an advisory body to then US president George W. Bush, which, like the president, advanced an unabashedly anti-science agenda (Mooney 2006). It is not so much that enhancement was not a topic of interest previously, but after the publication of *Beyond Therapy*, the enhancement debate became mired in biopolitics.

16.2 The Biopolitical Poles

When considering the biopolitics of cognitive enhancement, it is instructive to view the subject through the perspectives of those at the extremes – those who enthusiastically support the development of cognitive enhancement and those who view the enterprise with disapprobation. I shall argue that many of the positions undertaken by proponents of either view are influenced to a considerable degree by a variety of pre-existing political positions as much as, if not more so than, by deeper philosophical reflection.

16.3 The Transhumanists

A group of individuals who collectively call themselves transhumanists are amongst the most fervent of the enthusiasts. It was Julian Huxley, the well-respected evolutionary biologist, humanist, and internationalist, who first elaborated the concept of transhumanism. A prominent member of the British Eugenics Society (as well as brother of *Brave New World* author Aldous Huxley and half-brother of physiologist and Nobel laureate Andrew Huxley), Julian Huxley was among the first to recognize that advances in technology increasingly provide humans with the opportunity to hasten evolutionary change.

> The human species can, if it wishes, transcend itself – not just sporadically, an individual here in one way, an individual there in another way, but in its entirety, as humanity. We need a name for this new belief. Perhaps transhumanism will serve: man remaining man, but transcending himself, by realizing new possibilities of and for his human nature (Huxley 1957).

Modern day transhumanists espouse a position that calls not just for the adoption of enhancement technologies, but for radical enhancement such that we enter a new era in which at least some members of the population become *posthuman*. The transhumanists have been both activist and scholarly in their endeavours, having established the World Transhumanist Organization (which subsequently was rebranded as Humanity+); the overall objectives of the movement are described in the Transhumanist Declaration (Baily et al. 1998). It seems fair to say that transhumanism has at least achieved a modicum of maturity: by 2007, some deemed the literature on transhumanism as having reached critical mass (Agar 2007). No longer an inchoate philosophical position, transhumanist thought is well established in academic circles, most notably at Oxford University, although many contributors to the debate work from outside of academe. While proponents disagree on many details (and, to be fair, even some proponents of radical enhancement would bristle at being characterized as transhumanists), the desire to advance a program that promotes cognitive enhancement has been a prominent theme of transhumanist literature (Schneider 2009).

Most salient to the present discussion are the political underpinnings of the desire for radical human enhancement. The dominant sentiment that runs through the transhumanist literature is *libertarian*. Thus, we have the final of eight points of the Transhumanist Declaration stating:

> We favour allowing individuals wide personal choice over how they enable their lives. This includes use of techniques that may be developed to assist memory, concentration, and mental energy; life extension therapies; reproductive choice technologies; cryonics procedures; and many other possible human modification and enhancement technologies. (Baily et al. 1998)

The libertarian stance is succinctly summarized by Julian Savulescu (2006) who tells us:

> According to free-market or libertarian theories, the rich may buy enhancements which the poor cannot afford provided that their assets have been justly and legally acquired. Those with the highest levels of well-being and privilege will be able to buy even greater happiness and opportunity. This, according to some, is unfair. It is worth stressing that on libertarian accounts of justice, like that of Nozick (1974), this is not unfair.

A common way in which this argument is presented is to refer to concerns that are widespread in the enhancement debate and then reformulate them in terms of individual choice:

> Given that all medical interventions carry some risk, and that the benefits of enhancements may often be more subjective and value-dependent than the benefits of being cured of a disease, it is important to allow individuals to determine their own preferences for tradeoffs between risks and benefits. (Bostrom and Sandberg 2009)

The libertarian perspective on enhancement presented by Savulescu, Bostrom, Sandberg and their colleagues appeals to one of the most cherished of Enlightenment values: autonomy. At the same time, these proponents often (but not always, see below) minimize other values that we have inherited from enlightenment thinkers. Chief among those is communitarianism, and here James Hughes, former executive director of the World Transhumanist Association, offers an alternative: originally termed democratic transhumanism and more recently techno-progressive thought, Hughes argues for a more inclusive perspective on radical enhancement, one that balances responsibilities to society at large with individual rights (Hughes 2009).

Hughes has further reviewed this relationship, pointing out that transhumanism is, in many ways, a step-child of the enlightenment with its heavy reliance upon the primacy of rationality (Hughes 2010). He also points out several unresolved challenges: the question of universal application of rights in a hypothetical world where posthumans coexist with (mere) humans and that the very technologies that transhumanists seek may challenge the notion of the discrete self on which Enlightenment political theory is based.

Despite the strong vein of libertarian thought that runs through the transhumanist literature, it often seems as if transhumanists are more wedded to the objective – radical enhancement – than to any given political philosophy. Thus we see the normally libertarian-minded Julian Savulescu (2006) comfortably adopting an argument from fairness when it serves the interests of promoting radical enhancement: "Justice/Fairness requires we get as many people as possible up to the minimum IQ necessary for a decent chance of a decent life. Fairness thus requires enhancement."

Note the shift from adulation of individual choice, the hallmark of libertarianism, to a kind of Rawlsian argument of justice; individual autonomy is sacrificed when a political position is found that compels us to accept enhancement. A different stance, utilizing hyperbole that brooks no opposition, is seen in the following comment from John Harris: "Enhancements *per se* are not ethically problematic: they are unequivocally good, clearly ethical. Unless the downside can be demonstrated and is significant, enhancement has the moral high ground" (Harris 2010).

These examples reflect the fervor that regularly accompanies transhumanist rhetoric. Indeed, it is not unreasonable to suggest that the transhumanists approach their topic with something akin to religious zeal. This is precisely what is seen when the debate begins with the conclusion that enhancement is good, and then proceeds to develop arguments in support of the position. Moreover, to suggest otherwise is to risk being branded as a Luddite or worse. Notably, this phenomenon (which I shall discuss in further detail below) is remarkably similar to what is seen at the other pole of the debate.

16.4 The Bioconservatives

Perhaps no stance on cognitive enhancement has been as public, as political, and as rigidly bioconservative as that of George Bush's President's Council on Bioethics (PCB). Led by Leon Kass, the PCB was in many ways a lightning rod for the debate over cognitive enhancement. With the publication in 2003 of their broadside, *Beyond Therapy: Biotechnology and the Pursuit of Happiness* (President's Council on Bioethics 2003), the PCB set the stage for a decade of debate on the topic. The position adopted by the PCB was couched in an appeal to naturalness but a sophisticated one: the argument that was presented was one that declared enhancement an affront to human dignity.

Both the resistance to and the impetus for the PCB's perspective was its inherent founding on religious precepts: the argument from naturalism derives from the notion that if God wanted us to be enhanced, we would already be so and that any attempt to develop enhancements is an affront to the wisdom of the Creator. This view returns the place of man to a pre-Enlightenment stage of development, characterizing rational thinking about man's place in the modern world as hubristic and as an affront to God.

This bioconservative view, drenched in religious fundamentalism, found itself in alliance with another group of bioconservatives who similarly rebelled against enhancement, but for somewhat different reasons. These are the environmentally sensitive cultural critics best characterized by the writer Bill McKibben, famous for being among the first to bring the notion of global warming to the public's attention. For McKibben and his ilk, the modern world is increasingly estranged from Nature, and enhancing ourselves beyond what is normal is just the latest symptom of how disconnected we have become from our 'humanity' (McKibben 2003).

Notice the difference. For Kass it is reverence for, and even supplication to a deity that warrants capitalizing God's name. For McKibben, the deification is implicit rather than formalized, but it is there nonetheless: the wisdom of Nature again warrants capitalization, thereby seamlessly and yet subtly transforming the natural world into divinity. The appeal to naturalness via God on the one hand and Nature on the other resulted in an alliance that Jeremy Rifkin characterized as strange bedfellows (Rifkin 2002).

Critiques of the bioconservative view have been most eloquently provided by James Hughes (2009) who, to his credit, also lambasts some ideas of the radical transhumanists. Even more forceful critiques of Kass and his ilk are provided by Jonathan Moreno, a philosopher and historian who has carefully documented the ebb and flow of American thinking on this issue (Moreno 2011).

Of course, even before his tenure as Chair of the PCB, Kass was famous for another argument from naturalness, his 1997 article *The Wisdom of Repugnance* (Kass 1997). Published in the widely-read popular magazine *The New Republic*, Kass suggested that there is something unsettling about enhancement (he was specifically referring to genetic enhancement, but his arguments extend to other forms of enhancement rather easily), and that this something is repugnance, or more colloquially, the yuck factor. While many people might agree that there *is* something disquieting about enhancement (an observation about which we shall have more to say below), his critics latched on to the intellectual error in his argument: the suggestion that this sentiment, whatever its cause, represents a form of wisdom. For Kass, paying close attention to what we find repugnant is a way to arrive at morally acceptable outcomes because revulsion is "*prima facie* evidence of foulness and violation."

Kass' invocation of repugnance as a source of wisdom was roundly criticized. Notably, Martha Nussbaum pointed out that repugnance was the basis of many prejudices over the millennia (Nussbaum 2004). Indeed, there is now a large body of scholarship that provides empirical evidence that strongly questions Kass' attribution of wisdom to repugnance, and this line of scholarship has much to say about biopolitics. The most well-developed theory is Jonathan Haidt's social intuitionist model of moral decision-making (Haidt 2001). According to this view, people arrive at moral conclusions not through reasoning, but rather through rapid responses that are primarily driven by emotional considerations. Once the moral conclusion is arrived upon, rational thinking is used to justify the moral principle; hence, Haidt's characterization of the phenomenon as an emotional dog wagging a rational tail. Although the hypothesis is not without its detractors, in the decade since Haidt introduced the social intuitionist model, the evidence in support of this mode of thinking has grown markedly (Sunstein 2005; Rozin et al. 2009; Mercier and Sperber 2011).

One example cogently illustrates the issue, and simultaneously sheds light upon one of the key sources of moral repugnance: disgust. The idea of eating chocolate that has been molded into the shape of feces evokes strong disgust responses. If people were responding rationally, of course, they would just gobble up the chocolate, smiling merrily as they do so. More commonly, people take tentative bites, grimacing as if the chocolate was a bitter pill to swallow: even when they fully understand the absence of potential harm (it is chocolate, after all), there lingers some vestige of disgust (Rozin et al. 1986). In this particular example, it is hard to argue that there is 'wisdom' in the response to feces-shaped chocolate.

Of course, that is not to say that there is *no* wisdom in disgust, at least when viewed through the lens of evolutionary psychology. Indeed, there is a perfectly reasonable (I might even say rational!!) explanation for why disgust and morality are often intermixed: stimuli that readily evoke disgust – be it rotten flesh, maggots, or feces (readers can add their own 'favorites' to the list) – all share the same feature: contamination. The hypothesis then is that disgust is an evolutionary adaptation to contamination and that the rapid, emotional response to such stimuli has served us well throughout evolutionary history by preventing us from consuming foods that might make us ill.

It is not just experiments such as the fashioning of chocolate into the shape of feces that demonstrate the illogical nature of repugnance as wisdom. Consider the case of moral dumbfounding: when individuals are challenged with compelling rational arguments that debunk their initial moral conclusions (based, at least in part, upon repugnance), they routinely construct additional arguments to support their obviously irrational views. Finally, when these have all been exhausted, they declare, "Well, it's just wrong. I can't tell you why, it just is" (Haidt 2001). We shall return to the issue of moral dumbfounding below; for now, it is sufficient to observe that such responses hardly occasion crowning individuals as exhibiting wisdom.

These observations provide a solid basis for rejecting the thrust of Kass' argument that we should allow our moral compass to be set by the direction and intensity of our sentiments of repugnance. Indeed, in all but the most culturally conservative circles, the wisdom of repugnance has itself become a disgraced meme, with Kass himself conceding that repugnance is, "at most a pointer, and of course the objects of disgust are to a considerable extent and in many cases culturally malleable" (cited in Jones 2007).

Had Kass been but a bit more modest in his claims, he might be garnering plaudits rather than scorn. For the basis of his argument was sound: there does appear to be something about enhancement that is problematic to many. Indeed, unease with enhancement in general, and cognitive enhancement in particular, seems to be a persistent sentiment, irrespective of whether one is a cultural conservative like Leon Kass, an environmentally-sensitive cultural critic such as Bill McKibben, or, as we shall see below, one who holds a more balanced view of the issue.

16.5 The View from Reasonableness

Sometimes it seems as if the biopolitics of cognitive enhancement is dominated by the two poles: the transhumanists, who are often seen as clever but self-serving libertarians, and the cultural conservatives, who are commonly perceived as religious zealots finding yet another venue for their fundamentalist views on the world. With the waning of the influence of the Kass-led PCB, the cultural conservatives have in many ways lost their voice. The transhumanists too seem to have realized that their approach was neither making friends nor influencing people: first the World Transhumanist Organization was rebranded as Humanity+, and at least a few adherents (Douglas 2008; Faust 2008; Persson and Savulescu 2008) have moved their focus from memory enhancement to moral enhancement (the cynical view might be that they calculated this to be more acceptable to the populace than their dreams of radically enhancing memory). Loud as the noises that emanate from the poles may be, there exists a substantial group of thinkers who espouse what might be termed *the view from reasonableness*.

The proponents of this view are many (Parens 1998; Chatterjee 2004; Farah et al. 2004; Riis et al. 2008; Schermer 2008; Racine and Forlini 2010; Singh and Kelleher 2010; Hyman 2011; Mehlman et al. 2011; Partridge et al. 2011; Sahakian and Morein-Zamir 2011; Outram 2012), but I shall focus my comments on the highly influential paper by Hank Greely and his colleagues (Greely et al. 2008), a benchmark in the field that provides practical guidelines for the responsible use of cognitive enhancers. I shall not reprise the arguments put forward, nor provide a critique, but rather I will note that the paper by Greely et al. espouses the essence of the view from reasonableness: recognizing that concerns exist, the paper offers pragmatic solutions to moving forward in measured fashion.

What is most relevant to the biopolitical landscape is not the paper itself but the response: as recounted by Hank Greely 2 years later (Greely 2010), the overwhelming majority of correspondents thought that he and his colleagues had either lost their minds or been co-opted by the pharmaceutical industry. In his deconstruction of what might have motivated those sentiments, Greely suggests that the current off-label use of methylphenidate and similar agents provides an association with drug culture writ large, as well as that the enhancement debate is often confused with cheating in sports. These are well-worn tropes, but I am more intrigued by the other issue that Greely raises: people feel threatened by cognitive enhancements. Whether because of status-quo bias (Bostrom and Ord 2006) or the rapidity with which modernity is changing the world around us (Taylor 1989), this concern has a ring of truth, and brings us back to Leon Kass and his introduction of repugnance to the discussion.

It seems to me that what Hank Greely was observing when people wrote to him was a kind of moral dumbfounding (Haidt 2001), one that is based upon the persistent *angst* that people have when considering the issue of cognitive enhancement. As much as the field of psychology has explored the origins of moral behavior in contamination and disgust, there has been less attention paid to the sources of angst in the cognitive enhancement debate. Greely's assessment of the threat experienced by many gets at part of the answer, but there is a fair bit more terrain to explore.

One argument that has at least a modicum of empirical support is that people are reluctant to modify traits that are fundamental to self-identity (Riis et al. 2008). Niemelä recently suggested that there exists a kind of psychological essentialism that the folk perceive as being violated by cognitive enhancement (Niemelä 2010). Rather than being a standard argument from naturalness, this argument for the origins of the angst over cognitive enhancement is predicated upon the notion that people are, by and large, at least soft neuroessentialists (Reiner 2010); that is to say, on some level they subscribe to the view that 'I am my brain'. In this view, anything that changes the brain in a manner that is not seen as 'normal' is a threat not just to naturalness but also to the constitution of the self (Bolt and Schermer 2009; Kraemer 2010; Levy 2011). One reason that I find this explanation intriguing is that even philosophers who have long considered the pluses and minuses of cognitive enhancement will admit, when pressed, that while they cannot provide rational reasons for resisting enhancement, neither can they wholeheartedly endorse the prospect, concluding with sentiments akin to those provided by Hank Greely at the end of his review (Greely 2010) of the response to their 2008 paper: it depends.

In many ways, this comment exemplifies the distinction between *The view from reasonableness* from that of the *Transhumanists* and the *Bioconservatives*: a reliance upon reason in the best sense of the word. It is worth recalling the social intuitionist model of moral reasoning (Haidt 2001) in which people come to the debate with well-established views and then filter the information through the lens of their extant moral reasoning. My claim is that such a phenomenon is very much at work at the poles of the debate, where enthusiasts proclaim the virtues of radical enhancement and bioconservatives thunder in return with concerns dripping with fundamentalist religious sentiment. In contrast, the view from reasonableness is much less of a polemic and more of a conversation. Given that the debate ultimately rests on how the populace will respond, I suggest that supplementing normative claims with empirical data might allow us to move beyond polemics to obtain an understanding of how the public perceives the issues: Is the view from reasonableness the majority view? Are concerns about violations of selfhood at the core of lingering angst regarding cognitive enhancement? Answers to these questions and more are sure to come as we begin to unravel public attitudes towards cognitive enhancement (Nadler and Reiner 2010; Nadler et al. submitted).

Acknowledgement Supported by a grant from the Canadian Institutes of Health Research.

References

Agar N (2007) Whereto Transhumanism? The literature reaches a critical mass. Hastings Cent Rep 37:12–17

Baily D, Sandberg A, Alves G, More M, Wagner H, Vita-More N, Leitl E, Staring B, Pearce D, Fantegrossi B et al (1998) Transhumanist declaration. http://humanityplus.org/philosophy/transhumanist-declaration/

Baker S (2009) Building a better brain. Discover 30:54–59

Bolt I, Schermer M (2009) Psychopharmaceutical enhancers: enhancing identity? Neuroethics 2:103–111

Bostrom N, Ord T (2006) The reversal test: eliminating status quo bias in applied ethics. Ethics 116:656–679

Bostrom N, Sandberg A (2009) Cognitive enhancement: methods, ethics, regulatory challenges. Sci Eng Ethics 15:311–341

Cascio J (2009) Get smarter. Atlantic 304:94–100

Chatterjee A (2004) Cosmetic neurology: the controversy over enhancing movement, mentation, and mood. Neurology 63:968–974

Douglas T (2008) Moral enhancement. J Appl Philos 25:228–245

Erler A (2011) Does memory modification threaten our authenticity? Neuroethics 4:235–249

Farah MJ, Illes J, Cook-Deegan RM, Gardner H, Kandel ER, King P, Parens E, Sahakian B, Wolpe PR (2004) Neurocognitive enhancement: what can we do and what should we do? Nat Rev Neurosci 5:421–425

Faust H (2008) Should we select for genetic moral enhancement? A thought experiment using the MoralKinder (MK+) haplotype. Theor Med Bioeth 29:397–416

Franke AG, Bonertz C, Christmann M, Huss M, Fellgiebel A, Hildt E, Lieb K (2011) Non-medical use of prescription stimulants and illicit use of stimulants for cognitive enhancement in pupils and students in Germany. Pharmacopsychiatry 44:60–66

Garreau J (2006) A dose of genius: "smart pills" are on the rise. But is taking them wise. Washington Post, June 11, 2006, D01

Glannon W (2006) Psychopharmacology and memory. J Med Ethics 32:74–78

Greely HT (2010) Enhancing brains: what are we afraid of? Cerebrum. http://dana.org/news/cerebrum/detail.aspx?id=28786

Greely HT, Sahakian B, Harris J, Kessler RC, Gazzaniga MS, Campbell P, Farah MJ (2008) Towards responsible use of cognitive-enhancing drugs by the healthy. Nature 456:702–705

Haidt J (2001) The emotional dog and its rational tail: a social intuitionist approach to moral judgment. Psychol Rev 108:814–834

Hall S (2003) The quest for a smart pill. Sci Am 289:54–65

Hamilton R, Messing S, Chatterjee A (2011) Rethinking the thinking cap: ethics of neural enhancement using noninvasive brain stimulation. Neurology 76:187–193

Harris J (2010) Enhancements are a moral obligation. Wellcome Sci 1:16–17

Hughes J (2009) Technoprogressive biopolitics and human enhancement. In: Moreno JD, Berger S (eds) Progress in bioethics: science, policy, and politics. MIT Press, Cambridge, pp 163–188

Hughes J (2010) Contradictions from the enlightenment roots of transhumanism. J Med Philos 35:622–640

Huxley J (1957) New bottles for new wine. Chatto & Windus, London

Hyman SE (2011) Cognitive enhancement: promises and perils. Neuron 69:595–598

Jones D (2007) The depths of disgust. Nature (Lond) 447:768–771

Jowett B (1892) The dialogues of Plato: translated into English, with analyses and introductions. Oxford University Press, London

Kadosh RC, Levy N, O'Shea J, Shea N, Savulescu J (2012) The neuroethics of non-invasive brain stimulation. Curr Biol 22:R108–R111

Kass LR (1997) The wisdom of repugnance: why we should ban the cloning of humans. New Repubic 216:17–26

Kraemer F (2010) Authenticity anyone? The enhancement of emotions via neuro-psychopharmacology. Neuroethics 4:51–64
Levy N (2011) Enhancing authenticity. J Appl Philos 28:308–318
McCabe S, Knight J, Teter C, Wechsler H (2005) Non-medical use of prescription stimulants among US college students: prevalence and correlates from a national survey. Addiction 100:96–106
McKibben B (2003) Enough: staying human in an engineered age. Times Books, New York
Mehlman MJ, Berg JW, Juengst ET, Kodish E (2011) Ethical and legal issues in enhancement research on human subjects. Camb Q Healthc Ethics 20:30–45
Mercier H, Sperber D (2011) Why do humans reason? Arguments for an argumentative theory. Behav Brain Sci 34:57–74
Mooney C (2006) The republican war on science. Basic Books, New York
Moreno JD (2011) The body politic. Bellevue Literary Press, New York
Nadler RC, Reiner PB (2010) A call for data to inform discussion on cognitive enhancement. Biosocieties 5:481–482
Nadler RC, Fitz N, Manogaran P, Chong E, Reiner PB (submitted) Public attitudes towards cognitive enhancement
Niemelä J (2010) What puts the "yuck" in the yuck factor? Bioethics 25:267–279
Nozick R (1974) Anarchy, State, and Utopia. Blackwell, Oxford
Nussbaum M (2004) Hiding from humanity: disgust, shame, and the law. Princeton University Press, Princeton
Outram SM (2012) Ethical considerations in the framing of the cognitive enhancement debate. Neuroethics 2:172–184
Parens E (1998) Is better always good? The enhancement project. Hastings Cent Rep 28:S1–S17
Partridge B, Lucke J, Finnoff J, Hall W (2011) Begging important questions about cognitive enhancement, again. Am J Bioeth 11:14–15
Persson I, Savulescu J (2008) The perils of cognitive enhancement and the urgent imperative to enhance the moral character of humanity. J Appl Philos 25:162–177
President's Council on Bioethics (2003) Beyond therapy: biotechnology and the pursuit of happiness. President's Council on Bioethics, Washington, DC, p 373
Racine E, Forlini C (2010) Cognitive enhancement, lifestyle choice or misuse of prescription drugs? Neuroethics 3:1–4
Reiner PB (2010) The rise of neuroessentialism. In: Illes J, Sahakian B (eds) Oxford handbook of neuroethics. Oxford University Press, Oxford, pp 161–175
Rifkin J (2002) Fusion biopolitics. Nation-New York 274:7–23
Riis J, Simmons JP, Goodwin GP (2008) Preferences for enhancement pharmaceuticals: the reluctance to enhance fundamental traits. J Consum Res 35:495–508
Rozin P, Millman L, Nemeroff C (1986) Operation of the laws of sympathetic magic in disgust and other domains. J Pers Soc Psychol 50(4):703–712
Rozin P, Haidt J, Fincher K (2009) Psychology. From oral to moral. Science 323:1179–1180
Sahakian BJ, Morein-Zamir S (2011) Neuroethical issues in cognitive enhancement. J Psychopharmacol (Oxf) 25:197–204
Savulescu J (2005) New breeds of humans: the moral obligation to enhance. Reprod Biomed Online 10:36–39
Savulescu J (2006) Justice, fairness, and enhancement. Ann N Y Acad Sci 1093:321–338
Schermer M (2008) Enhancements, easy shortcuts, and the richness of human activities. Bioethics 22:355–363
Schneider S (2009) Mindscan: transcending and enhancing the human brain. In: Schneider S (ed) Science fiction and philosophy: from time travel to superintelligence. Wiley, Chichester, pp 241–256
Schulz K (2006) Brave neuro world. The Nation 282:11–16
Singh I, Kelleher KJ (2010) Neuroenhancement in young people: proposal for research, policy, and clinical management. AJOB Neurosci 1:3–16

Stix G (2009) Turbocharging the brain. Sci Am Mag 301:46–55
Sunstein C (2005) Moral heuristics. Behav Brain Sci 28:531–573
Taylor C (1989) Sources of the self: the making of the modern identity. Harvard University Press, Cambridge
White BP, Becker-Blease KA, Grace-Bishop K (2006) Stimulant medication use, misuse, and abuse in an undergraduate and graduate student sample. J Am Coll Health 54:261–268

Chapter 17
Are We Heading Towards an 'Enhancement Society'?

Armin Grunwald

Abstract The fields of cognitive and 'neuro-enhancement' that aim to enhance our cognitive capabilities have attracted increasing awareness over recent years. By far, most of the reflections on human enhancement technologies refer to ethical questions and criteria, which usually focus on the individual level and on speculative issues. In this chapter, however, I will examine the questions: Are we witnessing a historical change from a performance society to an enhancement society with an inherent and infinite spiral of enhancement? Does such a shift also include increased self-exploitation and self-instrumentalization? The main motivation of this chapter is to focus on aspects of the enhancement debate that might tell us something about ourselves at the present time rather than about speculative future developments. I will pursue the hypothesis that what we can learn from the ongoing debate on human enhancement is about ourselves, society and our contemporary perceptions.

Keywords Cognitive enhancement • Human enhancement technologies • Society • Competition • Self-exploitation

17.1 Introduction and Focus

A controversial international debate about human enhancement technologies has been going on for about 10 years in the wake of an influential publication by the National Science Foundation (Roco and Bainbridge 2002). This controversial debate enriched an older debate about the future of human nature by adding further insights, particularly those concerning the role of technological advance and the relationship between humans and technology. This debate built on discussions about

A. Grunwald (✉)
Institute for Technology Assessment and Systems Analysis (ITAS),
Karlsruhe Institute of Technology (KIT), Karlsruhe, Germany
e-mail: armin.grunwald@kit.edu

E. Hildt and A.G. Franke (eds.), *Cognitive Enhancement: An Interdisciplinary Perspective*, Trends in Augmentation of Human Performance 1,
DOI 10.1007/978-94-007-6253-4_17,

perfecting humans by means of genetic engineering and the philosophical criticism of this kind of manipulation (e.g. Habermas 2001). The fields of cognitive and 'neuro-enhancement' that aim to enhance our cognitive capabilities have attracted specific awareness (e.g. Farah et al. 2004; Galert et al. 2007).

While the philosophical debate focuses on ethical aspects of possible and partially-speculative *future* developments (e.g. Siep 2006; Schöne-Seifert et al. 2009), it is assumed in other parts of the debate that the acceptance of enhancement techniques and even their use are already widespread:

> Today, on university campuses around the world, students are striking deals to buy and sell prescription drugs such as Adderall and Ritalin – not to get high but to get higher grades, to provide an edge over their fellow students or to increase in some measurable way their capacity for learning (Greely et al. 2008, 702).

Opponents of human enhancement also frequently acknowledge that enhancement is already happening. Michael Sandel (2004), for instance, analyzes the mainstream behavior of the contemporary US middle class. He concludes that there is a widespread tendency, particularly with regard to education, training, and child rearing, that displays aspects of enhancement and, in his words, the desire for perfection. Many parents employ all available means in order to maximize their children's abilities. He writes of parents who are willing to pay tens of thousands of dollars in order to increase their child's height in the hopes that he will become a better baseball player, without any guarantee of success. We also see acceptance and willingness to use performance enhancing drugs at the workplace (DAK 2009) and at the university (see the quote by Greely et al. above).

Consequently, the question arises: Are we witnessing a historical change from a performance society to an enhancement society with an inherent and infinite spiral of enhancement including, as critics assume, increased self-exploitation and self-instrumentalization (see Sect. 17.3)? Coenen and colleagues point out:

> One could argue that there is growing evidence for the hypothesis that we are witnessing at the moment a transition from a performance-oriented society, in which the fulfilment of predefined tasks is rewarded, to a performance-enhancing society, in which tasks in work life, and even private life, are ever harder to calculate and foresee, and therefore the most pressing task for individuals is the competitive improvement of bodily preconditions and requirements for successful performance (Coenen et al. 2009, 45).

Thoughts of this kind have been expressed so far in few publications (e.g. Coenen 2008; Wolbring 2008; Coenen et al. 2009); by far most of the reflections on human enhancement technologies refer to ethical questions and criteria, which usually focus on the individual level and speculate about the future. While those explorations and reflections do have their own value (Grunwald 2010), it might be that they only address parts of the entire picture. The main motivation of this essay is to look for possible other relevant issues involved. In particular, I would like to focus on aspects of that debate that might tell us something about our society and contemporary perceptions, attitudes and concerns. My hypothesis is that what we can learn from the ongoing debate on human enhancement *is about ourselves today rather than about the future of the human nature.*

This hypothesis is at odds with most of the current writings on cognitive enhancement. Therefore I will have to put good arguments on the table. I will take them from a recent analysis of the very nature of 'the future' such as visions and scenarios which allows for the conclusion that those future scenarios might tell us more about ourselves than about the future they pretend to anticipate (Sect. 17.4). This insight then advances an argument in favor of a shift in the debate from ethical issues bound to the liberal assumption of individual judgments involving freedom of choice to a more communitarian perspective that addresses issues of contemporary society (Sect. 17.5).

17.2 Enhancement: An Open Process Without *Telos*

Before beginning my discussion, I would like to lay some groundwork by clarifying some terms and basic distinctions. While the notion of 'enhancement' is often synonymous with those of 'optimization' and 'perfection', I would like to put emphasis on a deep-ranging difference that is mostly ignored: optimization and perfection include a *telos* but enhancement does not (cf. Grunwald 2008, 2012 for this section).

Enhancement represents an activity through which an object is changed in a particular direction: there are *actors* (the subjects of enhancement) who enhance *something* (the object of enhancement) according to *criteria*. In accordance with this, enhancement necessarily includes three semantic dimensions:

1. A *starting point*. Any change is only plausible relative to a starting point of change; therefore, the starting point has to be identified.
2. A *criterion* of enhancement. A criterion relative to which some change might be classified as enhancement must be given. It consists of the declaration of a *parameter* (quantitative or qualitative) and the *direction* in which the parameter will be altered to constitute an enhancement.
3. A *measure* of enhancement. Measuring the *size* of an enhancement is primarily significant in weighting processes if enhancement in one place is offset by deterioration in another, and balancing is necessary.

Whereas *enhancing* relates to the change compared to a starting point in the intended direction, *optimizing* and *perfecting* are different subjects. Enhancing entails only the *direction* of a change, while optimizing and perfecting are oriented to an envisaged *final or target status* and therefore involve a *telos*. Although enhancing is bound up with a direction, it is open in measure and has no defined end, while an optimization is at an end when the optimum is reached. Optimization is a teleological approach while enhancement opens up an infinite step-by-step process during which criteria and direction of enhancement might change. In this sense, enhancement is close to what Sir Karl Popper (1957) and Charles Lindblom (1973) called 'piecemeal engineering' and 'muddling through,' respectively. These processes should start with a deficit analysis of the present state and then ask for incremental improvement, in opposition to the classical planning approach which departs by setting goals and then asks for instruments to reach them (see Camhis 1979). Probably it would be promising to analyze the relation between (1) the current debate on human enhancement, (2) the dynamics of scientific and technological advance and (3) older ideas of an open, yet ever-developing society.

The current debate on human enhancement is accompanied by controversies about its definition. The borderline between healing in the classical understanding within medical ethics and enhancement, which goes beyond healing, has been particularly subject to debates. Frequently, premises and points of departure of proposed clarifications are not really transparent. The application of the aforementioned understanding of enhancement as an action that changes particular performance indicators relative to certain starting points allows for the introduction of the following distinctions (see Grunwald 2012, chapter 9), which should add a greater degree of clarity:

- *Healing*: The elimination or compensation of an individual's deficits relative to the accepted standards of an average healthy human being as a starting point.
- *Doping*: An increase in an individual's performance potential without there being a deficit in terms of the accepted standards of an average healthy human being and without the individual's performance exceeding what still appears as conceivably normal, i.e., within the spectrum of usual human performance, whether in sports or in normal life.
- *Enhancement*: An increase in performance that goes beyond abilities that are regarded to be 'normally' achievable by humans who are healthy, capable, and ready to perform under optimal conditions.
- *Alteration* of the human composition that exceeds increasing the performance of existing functions (Jotterand 2008), for example, by implementing new organs.

Consider, for the purpose of illustration, the following example and assume that a person lost his/her legs in a traffic accident, and was given prostheses in order to compensate for the loss of his/her legs. Restoring the capabilities of that person according to the status before the accident would obviously fall under the category of 'healing'. If, on the contrary, completely new mechanisms of mobility would become possible via the prostheses – for example, that enable this person to jump like an Australian kangaroo – then this would be an 'alteration' according to the list above. If we consider the highest speed available in short-distance running as the relevant parameter, we could make this distinction: if the prostheses would allow the person to finish a 100-m sprint in 5 s, we would agree that this is a super-human capability. In this sense we should speak of a real enhancement as mentioned above. If the prostheses would allow one to finish a 100-m sprint in about 10 s, then things would become really complicated – as the 'real-life' case of Oscar Pistorius has already illustrated (Wolbring 2008). In this case a debate would arise about whether the technical prostheses would give the owner an advantage over his competitors – and this is precisely the issue at hand in the well-known debate of doping in sports.

This proposal for a more precise usage of terms in the enhancement debate operates with the difference between existing and new human functions. In particular, it closes a gap in the usual debate about the distinction between healing and enhancement. The semantic gap between healing and enhancement is filled by putting 'doping' at this place: when the outcome of an improvement exceeds the average performance of a healthy person but does not exceed a customary human level, we will speak of doping as a term that goes beyond the world of sports. Doping is something in between healing and going beyond abilities that are humanly possible – and its main theme is the tension between fairness and the wish to be better than all other competitors.

In cases of doping in this sense (going beyond the narrow field of sports), the topics that dominate the enhancement debate, such as man's self-image and the future of human nature, are not touched upon. Even for new forms of doping, normative frameworks exist, which can be taken into account (see Gerlinger et al. 2008 for the field of sports); whereas, they do not exist at all for human enhancement, i.e., in terms of the production of 'superhuman' or 'trans-human' capabilities and features. There is hardly any overlap between considerations of how we should handle the consumption of drugs to improve brain performance and questions about the direction a society would take in which the ability to discriminate between humans and technology would be largely removed in the future.

The distinction between doping as performance-improving by a measure we are familiar with and enhancement that allows for super-human achievements is essential in the interest of a clear use of language. It is of use also with regard to the respective normative frameworks that I have touched upon, and, thus, with regard to the necessities and issues of philosophical and ethical reflection. While doping is characterized above all by the violation of two normative demands on human action, namely, the demand for fairness, and that people act responsibly toward their own bodies, enhancement gives rise to much broader ethical and anthropological debates.[1]

The next step in a semantic clarification concerns the notion of a *technical* enhancement. A technical improvement of humans, if it is not to be meant merely metaphorically, needs modeling of humans following models of technology through the declaration of 'performance parameters' that will be enhanced. For instance, assume that a sensory organ like a human eye could be technically rebuilt in a way which is functionally equivalent to the natural eye (Grunwald 2008, 2012). An artificial eye of this kind would – as is customary in technical development and production – be given a version number by its manufacturer: this would be 'eye 1.0'. Surely, version 1.0 will not be the last one because as soon as version 1.0 has been developed and tested, engineers and physicians will be thinking of the next version: continuous improvement of what has been achieved already is a technological imperative in modern technology. A technical enhancement of

[1]By the way, the analysis in the next section will show that the most relevant part of the enhancement debate with respect to my hypothesis is not based on enhancement but on doping issues.

humans, thus, is revealed as a consequential step of a technical restoration of failed or deficient bodily functions. The technological imperative *necessarily* moves from healing to enhancing, if it is not guided or restricted normatively by arguments of a different type.

Enhancement has no intrinsic limits or measures but opens up infinite possibilities (see above). Once a status has been achieved in human enhancement, the enhancement process does not necessarily stop in the sense of a target being reached; rather, it serves as the starting point for the next enhancement, and so on. This feature radically distinguishes healing from enhancement: healing comes to an end when the patient is healthy, while enhancement does not come to an end even if it is successful but is driven ever onwards by the relentlessness of the technological imperative.

17.3 The Debate on an 'Enhancement Society'

The debate on human enhancement still focuses on ethical issues at the individual and collective level. The notion of an enhancement society has only seldom been used so far. In this section, I will tell the story the other way around and will look at what the debate could tell us about ourselves and our contemporary society.

17.3.1 Enhancement Technologies Entering Society

According to my observations, no strong ethical arguments have been raised against human enhancement so far. "Strong arguments" refer to those that work on the consequences that will necessarily follow the introduction of enhancement technologies and that are therefore not dependent on uncertain developments in the future (Grunwald 2008, 2012). If human enhancements are not carried out on people who are unable to give their consent, an informed consent could and would have to be created. This informed consent takes on a central role in ethical argumentation and prevents the possibility of "strong" ethical arguments.

The ethical debate has instead focused in a consequentialist manner on the unintended side effects, in particular, on questions of distributive justice (Siep 2006):

> The ability divide will develop between the poor and the rich within every country. It will be bigger between low and high income countries than it will be within any given country. Not everybody will be able to afford enhancement of their body, and no society will be able to enhance everyone, even if they wished it (Wolbring 2006, 31).

Also researchers who are more positive about human enhancement are concerned about this issue (Galert et al. 2009). Arguments of this type are, in terms of validity, rather weak because they have to make a series of more or less uncertain assumptions about the future (Grunwald 2012, chapter 10). Furthermore, arguments based on the possible manifestation of unintended side effects do not have to be understood as arguments against the new technology per se but, rather, as arguments that something must be done politically and socially, and perhaps also technically, to limit, prevent, or compensate in some suitable manner for the manifestation of these unintended consequences. Reflections of this kind enable society to think about possible regulations in early stages of development, e.g. in the fields of liability, and can be regarded as precursory deliberations made in preparation for the advent of new technology rather than as indicators of a "strong" ethical objection.

Viewed from this perspective, it does not seem implausible for enhancement technologies to be introduced according to a market model in the absence of strong ethical arguments (Grunwald 2012). It is not only possible but even probable that there might be an emerging demand for enhancement technologies caused by increasing competition or other various developments:

> [...] several market pressures leading to rapid development of HE [human enhancement] technologies: 1) global competitiveness; 2) brain drain/depopulation economics; 3) national security concerns; and 4) quality of life/consumer life-style demands. (Williams and Frankel 2006, 3)

The market model means that enhancement technologies could be offered as a kind of service analogous to aesthetic surgery today. Consumers willing to enhance themselves would be informed about the available enhancement services, their costs, implications and possible risks. Then they could decide within the framework of informed consent. This scenario would exemplify the libertarian ideal of autonomous persons deciding about their personal issues (e.g. Greely et al. 2008).

17.3.2 The Omnipresence of Competition

However, the libertarian perspective might not cover the whole story. The question is whether individuals are "really" autonomous or whether they are subject to external pressures and forces. If we look to possible driving forces of an emerging marketplace for enhancement technologies, we soon become aware of the crucial role of competition. This model strongly resembles that of competition within the world of sports: for example, the stronger the competition and the corresponding pressure of the system, the larger the willingness of individual sportsmen and -women to use drugs for doping.

A society that has accepted the idea of competition as its central motor at nearly every level, from the economy to the military to life style, might feel a need to achieve human enhancement if this were available. Phrased differently, competition and enhancement are inextricable (Grunwald 2012). Since, from a technical point of view, doping and enhancement are only different in degree and because the technological imperative leads directly from the former to the latter (Sect. 17.2), the pressure of competition will result in developments that first are of the "doping type" and then will lead in the direction of enhancement.

In the United States, the creation of an enhancement society as a possible political goal has already become an item of debate (Coenen 2008). Michael Sandel (2004) offered a diagnosis of the US middle class showing that the idea of enhancement is already part of its self-understanding and everyday acting, in particular in the field of parenting (see Sect. 17.1). The appearance of an attractive world market for enhancement technologies and procedures has already been predicted. In particular, in view of the fact that societies are aging, enhancement could be an appropriate means to create competitive advantages for companies and economies. The question of social diagnosis is whether we stand at a transition from a capitalist achievement society to an enhancement society. Consider the following:

> In a political-analytical and sociological line of reasoning of this kind, social structures are considered which favour the spread of HET [human enhancement technologies] as are new tendencies which may boost their use. The pathologisation and medicalisation of more and more emotional and physical states, the commodification of the human body, its use as an improvable tool for competition, and the prospects of radically changing the human body by means of secondstage HET are only some of the aspects relevant here (Coenen et al. 2009, 44).

The changes in the world of employment and disruptions in recent decades presumably have also been accompanied by changes in our understanding of achievement. It often no longer appears sufficient for an individual to perform in a context defined by salaried employment. More and more people feel compelled to constantly improve their performances and possible actions in an increasingly flexible world of employment as well as in other areas of life. Rankings and ratings, constant evaluations, the necessity of presenting oneself and beating the competition, whether at work, in the context of one's love life, whether in the attempt to become the next top model or to use some crazy idea to get the media's attention or to be featured in the *Guinness Book of Records* – competition is ubiquitous, and competitiveness is measured by these abilities. In this way, culture becomes an aspect of competitiveness or a location factor; education becomes an article of competition, and soft skills are required to increase one's own competitiveness (Grunwald 2012).

Lifelong learning comprises one element in this ongoing challenge. Such concepts oriented towards education could be supplemented or replaced in the future by technical enhancements. An above-average and constantly improving performance at work, a beautiful and strong body, high resistance to stress: these abilities (Wolbring 2008) are moving to the top of the agenda of many people. They are both expressions of an atmosphere where constant enhancement is necessary just to main one's quality of life, not to mention to improve it; in addition, enhancement serves as a motor for maintaining and strengthening this atmosphere (Grunwald 2012). Competition and abilities are inseparable; thus, the improvement of one's abilities becomes part of the dynamics of development under thepressures of competition.

This is because every success in competition is only for a finite period of time and is constantly threatened by the possibility that others might catch up or pull ahead. Enhancement as an infinite process without a *telos* (Sect. 17.2) opens up an infinite spiral of further and further improvement. In this manner, the idea of competition and human technical enhancement indeed are inseparable.

At the level of philosophy, political liberalism is deeply related to the driving force of competition. The grounding of ethical statements on the idea of informed consent with regard to an individualistic approach could be a door-opener of human enhancement:

> A more mundane vision in a similar vein, presents us with a society in which "morphological freedom" and "cognitive liberty" are core values [. . .]. In such a society, every individual would have the right to treat his or her own body as fully malleable object. While some critics have denounced this as a reduction of the body to a commodity or a fashion accessory [. . .], the promoters of morphological freedom argue for a new notion of individual freedom which allows for aesthetic self-realisation, overcoming the "genetic lottery", free experimentation with all kinds of mood- and mind-altering drugs and technologies, and transcendence of the merely human (Coenen et al. 2009, 44).

In a liberal market model, regulation would be limited to compensation for the side effects of a market failure (e.g., from the clarification of liability issues, such as what would happen if an enhancement does not succeed) and to ensuring distributive justice and access:

> Governments may also need to protect the level playing field for consumers, so that the already-enhanced do not act, through non-market means, to protect their status by preventing others from becoming enhanced. (Sarewitz and Karas 2006, 10)

Consequently, it seems that there are, through a combination of liberalism, ubiquitous competition, and the imperative of technology (Sect. 17.2), strong forces that support human enhancement. These forces are part of our contemporary social reality and are already at work. But what about the reality of cognitive enhancement? Does it work?

17.3.3 Do Cognitive Enhancers Really Enhance Cognition?

Interestingly, there are serious doubts about whether cognitive enhancement drugs such as Ritalin really have enhancing functions in healthy persons. A recent study for the German Bundestag concluded (see also Lieb 2010):

> There is some reason to believe that presently available substances have shown performance-relevant effects – insofar as they have done so at all – only in cases in which the subjects concerned suffered from some kind of deficit at baseline. Furthermore, some indications suggest that in subjects with a high level of alertness at baseline any additional activation of general wakefulness or increase in neurotransmitter concentrations leads rather to a deterioration in cognitive performance (Gerlinger et al. 2011).

It seems that the expectations concerning the effectiveness of enhancement drugs are much greater than what can be measured empirically:

> Overall, it can be asserted that there is no proof that any presently available substance can enhance human performance without at the same time causing significant side effects. All that can be demonstrated are effects on single cognitive abilities (e.g. alertness, reaction time) that are to some extent thought to be of special relevance to present-day occupational training and working environments (Gerlinger et al. 2011).

In drastic contrast to these more or less disappointing results – disappointing from the standpoint of promoters of human enhancement – expectations in parts of the general public are high, as is the willingness to use enhancement drugs if they were available. In particular, groups facing stress in education and at the workplace show significant willingness to use drugs to enhance their cognitive capabilities in order to be better able to meet the challenges of their lives.

In my opinion, it is an interesting fact per se that there seems to be a clear demand for cognitive enhancement without a similarly clear technological or pharmacological offer. In contrast to many other fields of new science and technology, the field of human enhancement is not governed by the "technology push" model where new technology first is developed and then, after or upon entering the marketplace, creates its own demand. Here, a social demand seems to be a reality before pharmacology can offer appropriate enhancement drugs.

This seems to be interesting also in another respect. While people often reject the general idea of human enhancement for principled reasons – which is my impression after a lot of public presentations and debates about the ethics of enhancement – they seem to be willing to accept and welcome enhancement technologies in specific contexts; namely, in situations of high competition where people are afraid of lagging behind their competitors and hope for a boost from enhancement. Stories and visions of cognitive enhancement seem to be able to exert real power and influence in spite of the absence of real effects of cognitive enhancers:

> Even where such HET are non-effective, their widespread use changes social interaction, promotes the establishment of new norms, and raises a variety of questions such as those relating to distributive or procedural fairness, to the fabrics of society, to the relationships between the self-image of individuals and their social roles, and to social key concepts such as competition, ability and disability, happiness, self-realisation, and individual freedom (Coenen et al. 2009, 44).

At this point, it is a good occasion to remind the reader of the hypothesis guiding this chapter: what we can learn from the ongoing debate on human enhancement *is about contemporary perceptions of ourselves rather than about the future of human nature*. In order to understand the intensity of the debate and the great controversies involved, as well as the high engagement of many researchers and thinkers, it seems to be promising not to look at what the future visions of human enhancement tell us but at controversies in place today. In order to explain this diagnosis, I will briefly present some thoughts about the future in a more general, philosophical direction.

17.4 Stories of the Future as Parts of the Present

The future in its many illustrations such as visions, expectations, scenarios or predictions does not exist per se, nor do such illustrations arise of their own accord. On the contrary, the future is "made" and linguistically constructed in a more or less complex manner (Grunwald 2011). The designing of the future is a specific type of action, intended often to provide orientation. The future, regardless of whether it is manifested in terms of forecasts, scenarios, plans, programs, visions, speculative fears or expectations, is "produced" using ingredients of different types such as the available knowledge, value judgments, emotions and suppositions. This constructed character of the future, i.e., that it is the result of a construction process and does have authors, is essential.

Self-evidently, the construction of the future has to take place at a certain point of time, in a particular social and cultural context, and is, therefore, bound to this context. Forecasters, scenario-builders and visionary writers cannot break out of their present time and their contexts. They are always making their projections and predictions on the basis of *present* knowledge, *present* assessments, *present* diagnoses and perceptions, and *present* values – because they do not have any other kind.

Future facts or processes can be neither *logically deduced* (Goodman 1954) nor *empirically investigated* but rather remain constructions out of the ingredients stemming from a particular point of time and context. The only issues we really can talk about are the *images* which we make of the future, but not the future itself that will at some time become the present. For this reason, we can talk about a *possible* future, about alternative possibilities for imagining the future, and about the justification with which we can expect something in the future. These are always *present futures* and not *future presents* (Picht 1971). In spite of the trivial nature of this insight it is frequently overseen, or its implications are not taken seriously. The most radical formulation might be: the future cannot escape from the present but are bound to a specific immanence of the present (Grunwald 2006) – the future is part of the respective present.

If we talk about cyborgs or far-ranging human enhancements perhaps becoming possible in the future, we are not talking about how these developments will "really" occur and what their implications for human nature and society might be but how we *imagine them today*. The future is thus something that is always contemporary and changes with the changes in each present. *A future is thus not something separate from the present, but a specific part of each present* (Grunwald 2006). The future expresses specific insights, expectations, values, fears, hopes, attitudes, diagnoses and problem perceptions of a particular present time.

The future is created in accordance with available knowledge, but also with reference to assessments of relevance, value judgments, and interests. Because of the socially constructed character of the future, its "quality" – it might be difficult to explain what the quality of a future would mean and how it could be made operable (Grunwald 2011) – will strongly depend on what was put inside its construction. The question arises as to which ingredients are invested in building a future and in which way these ingredients have been assembled and composed. In a rough approximation, the following gradation of knowledge and non-knowledge components can initially be made (Grunwald 2011, 2012):

- *Present knowledge*, which is proven according to accepted criteria (e.g., of the respective disciplines) to be knowledge (e.g., according to the issue at stake from the field of nanotechnology, engineering, economics etc.)
- *Estimates* of future developments that do not represent current knowledge but that can be substantiated by current knowledge (e.g., demographic change, energy needs etc.)
- *Ceteris paribus* ("*All other things being equal*") *conditions*, whereby certain continuities – business as usual in some sense or a lack of disruptive changes – can be assumed as a framework for the prospective statements
- *Ad hoc suppositions*, which are not substantiated by knowledge, but 'given' (e.g., the future validity of a German phase-out of nuclear energy, or the nonoccurrence of a catastrophic impact of a comet on the Earth)
- *Implicit and underlying convictions, world views, and possible elements of 'Zeitgeist'*.

All types of ingredients have to be identified at the point of time and within the social and cultural context in which the respective story of the future shall be made and communicated.

This view on the future allows us to pose the usual question of what we can learn from stories of the future in a different way. It turns out that because the future is parts of a particular present time, we could possibly learn to better understand a particular present time by investigating the stories of the future it produces and debates. In the case of human enhancement: what does the strong demand for cognitive enhancement mean for our contemporary society? What does this tell us about ourselves?

17.5 Conclusion: What Does This All Mean?

Debating about the future serves different purposes; one of them is to prepare society for emerging developments. An example of this approach to making use of the future can be found in mainstream nano-ethics and in the ethical inquiry of human enhancement. These lines of thought identify ethical issues and explore ways to deal with them responsibly. This is, in the sense of an explorative philosophy (Grunwald 2010), a legitimate means of reflection.

However, in this essay, I follow a different rationale, in accordance with the preceding section. It has been convincingly shown in Sects. 17.2 and 17.3 of this chapter that competition and enhancement are interrelated and that they possibly reinforce each other. If, taking the immanence of the future seriously, the enhancement debate is not about the future of human nature but about present social uneasiness, then the perception of the role of competition in contemporary society will be at the heart of this view regarding the future. This would mean that many people feel uneasy with the ever-increasing and dominant role of competition in many fields of life, particularly, within education and work-life. Moreover, there also might be hidden criticisms against capitalism and concerns about more and more self-exploitation in many fields of life. The increase of efficiency under the rationale of meeting challenges of increasing (mostly economic) competition is seemingly regarded as an increasing burden on individuals. Cognitive enhancement, put in this context, could be an indicator for an increasing discomfort with existing economic structures, pressures, and forces. If the dominant imperative of sports – you have to be better than your competitor – holds true for large parts of society, and if people are concerned about this situation, then the story of cognitive enhancement is not about creating superhumans in the future but about competition in our contemporary societies.

The thought that the possible developments toward an enhancement society – in which cognitive enhancement technologies develop according to a market model and then spread – might be associated with the overarching social system, in particular, with a capitalist economy, has hardly been expressed so far. If this connection could be substantiated, then this would say something about the location and direction of the social debate that would have to be conducted about this situation. It would then not simply be about ethical issues, which could be answered in one direction or another, but also about the type of society in which we live and about its implications. The politically explosive nature of this question is obvious. It includes the question: do we approach a new type of society characterized by crude social Darwinism and the possible end of the welfare state based on solidarity according to the Western European Model?

If this interpretation would address, at least, parts of the debate on human enhancement, the world of sports could serve as an illustration. Analogies often are misleading, but it seems promising to take a look at this world under the perspective of doping and enhancement as exemplified in a recent TA report to the German Bundestag:

> The principal social and political relevance of the topic «Enhancement» arises not because enhancement is perceived as contributing towards a scientifically and technically based «improvement of human beings», but rather because pharmacological interventions to improve performance form part of the «medicalization of a performance (enhancement)-oriented society». The social and political debate about this issue should therefore focus on the likely future status of pharmacological and other (bio)medical strategies and measures for coping with performance targets and demands in a globalized educational and working environment, and on the consequences of demographic change. To this end, rather than assuming at the outset that adoption of strategies designed to maximize individual and collective performance is inevitable, we need to look into conditions in secondary and tertiary education and at the workplace, and where appropriate adjust performance indicators. Commercial and economic considerations also favor such an approach, at least in the medium and long term. In this regard the example of doping in sport shows how a system of competition could potentially self-destruct as a result of unlimited expectation of ever-improving performance (Gerlinger et al. 2011).

Independent of whether this analogy might shed light on the "immanence of the present" of the debate on cognitive enhancement, we should be aware that stories about the future, such as visions, can exert real power and influence on the respective present (Grunwald 2007):

> In foresight and technology assessment as in any other reflection on the future of HET, one has to be aware of the fine line between taking a broad look at the future and feeding the hype. As has been shown for nanotechnology (Paschen et al. 2004), the enthusiasm which optimistic futuristic visions can evoke is often deliberately utilised as a means of promoting technology development. Such a strategy of "hype and hope" always appears to be precarious. This strategy can have both positive effects (e.g. incentives for young scientists, or arousing and sustaining political and business interest) and adverse effects as well. Among the latter is the danger that expectations will be set too high, making disappointment inevitable. Second, it may popularise the reverse of the optimistic futurism – a pessimistic futurism involving apocalyptic fears and visions of horror, which itself is being increasingly used to raise attention for nascent fields. Third, the focus on far-reaching visions may hinder a sober discussion of the potentials of technologies (Coenen et al. 2009, 38)

Perhaps the most relevant challenge today is not a possibly-coming enhancement society with problems of access and justice, one with cyborgs entering our life-worlds, but more that many people today do believe in it or are afraid of it – possibly because of the aforementioned uneasiness with increasing and omnipresent competition.

However, my thoughts in this essay are qualitative and speculative. They bring together issues of the current debate on cognitive enhancement and observations of current society with an analysis of what we can learn by investigating the future. These thoughts need an in-depth analysis and the development of indicators of how the emergence of an "enhancement society" could be measured. Results of those "measurements" then could be used to underpin or to falsify my hypothesis by empirically-based research.

There is another way in which my thoughts could be developed further. In a sense, I doubt in this chapter whether the purely individualistic approach of liberal ethics would be able to tell and assess the full story of cognitive enhancement. If my diagnosis of deep-ranging relations between the debate on cognitive enhancement, economic competition, technological progress, and uneasiness in large parts of the population with increasing self-exploitation are correct, at least to a certain degree, then statements of this type,

> Like all new technologies, cognitive enhancement can be used well or poorly. We should welcome new methods of improving our brain function. In a world in which human workspans and lifespans are increasing, cognitive enhancement tools – including the pharmacological – will be increasingly useful for improved quality of life and extended work productivity (Greely et al. 2008, 705),

are either naïve, because they ignore the fact that individuals are not free but subject to pressure and external forces. Or statements of this type might be ideological – which is not much better – because they intentionally ignore those pressures and forces.

References

Camhis M (1979) Planning theory and philosophy. Tavistock Publications, London

Coenen C (2008) Von der Leistungs- zur Leistungssteigerungsgesellschaft? TAB-Brief 33. Büro für Technikfolgen-Abschätzung, Berlin, pp 21–27

Coenen C, Schuijff M, Smits M, Klaassen P, Hennen L, Rader M, Wolbring G (2009) Human enhancement. European parliament (IP/A/STOA/FWC/2005-28/SC32 & 39), Brussels

DAK (Deutsche Angestellten Krankenkasse) (2009) DAK Gesundheitsreport 2009. http://www.dak.de/content/filesopen/Gesundheitsreport_2009.pdf. Accessed 7 Apr 2011

Farah MJ, Illes J, Cook-Deegan R, Gardner H, Kandel E, King P, Parens E, Sahakian B, Wolpe PR (2004) Neurocognitive enhancement: what can we do and what should we do? Nat Rev Neurosci 5:421–425

Galert T, Bublitz JC, Heuser I, Merkel R, Repantis D, Schöne-Seifert B, Talbot D (2009) Das optimierte Gehirn. Ein Memorandum zu Chancen und Risiken des Neuroenhancements. Gehirn&Geist 11:40–48

Galert T, Merkel R, Boer G, Fegert J, Hartmann D, Nuttin B, Rosahl S (2007) Intervening in the brain. Changing psyche and society. Springer, Berlin

Gerlinger K, Petermann T, Sauter A (2008) Gene doping. Scientific basis – gateways – monitoring. Technology assessment report 3. Office of Technology Assessment at the German Bundestag, Berlin (http://www.bundestag.de/htdocs_e/bundestag/committees/a18/translations/gene_doping.pdf)

Gerlinger K, Sauter A, Petermann T (2011) Pharmakologische Interventionen zur Leistungssteigerung als gesellschaftliche Herausforderung. TAB-Arbeitsbericht Nr. 143. TAB, Berlin. English summary available at: www.tab-beim-bundestag.de

Goodman N (1954) Fact fiction forecast. Harvard University Press, Cambridge, MA

Greely H, Sahakian B, Harris J, Kessler R, Gazzaniga M, Campbell P, Farah M (2008) Towards responsible use of cognitive-enhancing drugs by the healthy. Nature 456:702–706

Grunwald A (2006) Nanotechnologie als Chiffre der Zukunft. In: Nordmann A, Schummer J, Schwarz A (eds) Nanotechnologien im Kontext. AKA, Berlin, pp 49–80

Grunwald A (2007) Converging technologies: visions, increased contingencies of the conditio humana, and search for orientation. Futures 39:380–392

Grunwald A (2008) Auf dem Weg in eine nanotechnologische Zukunft. Philosophisch-ethische Dimensionen, Freiburg

Grunwald A (2010) From speculative nanoethics to explorative philosophy of nanotechnology. NanoEthics 4(2):91–101

Grunwald A (2011) Energy futures: diversity and the need for assessment. Futures 43:820–830. doi:10.1016/j.futures.2011.05.024

Grunwald A (2012) Responsible nanobiotechnology. Philosophy and ethics. Pan Stanford Publishing, Singapore

Habermas J (2001) Die Zukunft der menschlichen Natur. Suhrkamp, Frankfurt

Jotterand F (2008) Beyond therapy and enhancement: the alteration of human nature. Nanoethics 2:15–23

Lieb K (2010) Warum wir nicht alles schlucken sollten. Artemis & Winkler, Düsseldorf

Lindblom CE (1973) The science of "muddling through". In: Faludi A (ed) A reader in planning theory. Pergamon Press, Oxford, pp 151–170

Paschen H, Coenen C, Fleischer T, Grünwald R, Oertel D, Revermann C (2004) Nanotechnologie. Forschung und Anwendungen. Springer, Berlin et al

Picht G (1971) Prognose Utopie Planung. Klett, Stuttgart

Popper K (1957) Die offene Gesellschaft und ihre Feinde. Mohr Siebeck, Tübingen

Roco MC, Bainbridge WS (eds) (2002) Converging technologies for improving human performance. National Science Foundation, Arlington

Sandel M (2004) The case against perfection. Harvard University Press, Boston

Sarewitz D, Karas TH (2006) Policy implications of technologies for cognitive enhancement. Consortium for Science, Policy & Outcomes, Arizona State University, Tempe

Schöne-Seifert B, Ach JS, Talbot D, Opolka U (eds) (2009) Neuro-enhancement. Ethik vor neuen Herausforderungen. Mentis, Paderborn

Siep L (2006) Die biotechnische Neuerfindung des Menschen. In: Abel G (ed) Kreativität. Akten des XX. Deutschen Kongresses für Philosophie, Hamburg, 2006, pp 306–323

Williams E, Frankel MS (2006) Good, better, best: the human quest for enhancement. Summary report of an invitational workshop. Convened by the scientific freedom, responsibility and law program. American Association for the Advancement of Science, Washington, DC, 1–2 June 2006

Wolbring G (2006) The choice is yours. Ableism and NBICS. http://innovationwatch-archive.com. Accessed May 13, 2011

Wolbring G (2008) Oscar Pistorius and the future nature of Olympic, Paralympic and other sports. Scripted 5(1):14–60 (wolbring.wordpress.com/2008/05/15/oscar-pistorius-and-the-future-nature-of-olympic-paralympic-and-other-sports/). Accessed May 13, 2011

Chapter 18
Leveling the Playing Field: Fairness in the Cognitive Enhancement Debate

Greta Wagner

Abstract One controversy within bioethics regards whether the use of potential cognitive enhancements would decrease fairness in society by giving advantages to the already privileged or increase fairness by enabling the worst off to compete with others. Both positions are based on a notion of society as competitive, in which fairness is the purpose of a certain degree of institutional intervention. In the context of cognitive enhancement, this intervention would be the regulation or deregulation of the use of certain drugs by schools, universities, employers or legislators with the goal to protect competition. In the context of neo-liberalism the state intervenes in the market to protect competition. In both contexts, the aim is to level the playing field and to provide equal opportunities to compete. Within the framework of Michel Foucault's terminology, I will argue that the bioethical debate on fairness forms part of a neoliberal governmentality.

Keywords Cognitive enhancement: fairness • Competition • Neoliberalism • Michel Foucault

> Where Foucault analyzed biopolitics, we now must analyze bioeconomics and bioethics, for human capital is now to be understood in a rather literal sense – in terms of the new linkages between the politics, economics and ethics of life itself. (Rose 2003, 58)

The competition-based allocation of work, income and opportunity is regarded as a normative foundation of modern societies. In the last two decades, competition has become a dominant mode of interaction in more and more spheres of life beyond economics. Competitions are regarded as legitimate modes of resource distribution, assuming everyone has the same opportunity to compete. This, of course, can never

G. Wagner (✉)
Department of Social Sciences, Institute of Sociology, Goethe-University Frankfurt am Main, Frankfurt, Germany
e-mail: greta.wagner@soz.uni-frankfurt.de

E. Hildt and A.G. Franke (eds.), *Cognitive Enhancement: An Interdisciplinary Perspective*, Trends in Augmentation of Human Performance 1,
DOI 10.1007/978-94-007-6253-4_18,

be the case. Level playing fields are always distorted by numerous inequalities that social life brings along. Nonetheless the question of whether cognitive enhancement has an impact on level playing fields and fairness in competition is an important issue in bioethical literature on cognitive enhancement.

The reason is that most bioethical papers on cognitive enhancement do not deal with the empirical practice of self-medication with prescription drugs intended for patients of Attention Deficit and Hyperactivity Disorder (ADHD) in order to improve alertness and ability to focus. Rather, they examine ethical problems of a future scenario for which some hope and others fear: The invention of a 'smart pill' without any side effects.

Both issues raise very different ethical questions but are nonetheless often dealt with together under the term, cognitive enhancement. So far, there is no drug on the market that is capable of making a patient smarter or increasing one's IQ. Besides, drugs without any side effects usually have no effect at all (cf. Quednow 2010). On the other hand, there is a certain prevalence[1] of students without a diagnosis for ADHD taking drugs like Ritalin® and Adderall® to be able to study longer and overcome distractions. In this sense, they take stimulants as second-order cognitive enhancers. These substances would better be called vigilance or motivation enhancers, as Boris Quednow emphasizes (Quednow 2010, 153). Clinical trials have shown that those who profit from drugs are people who have cognitive deficits either caused by lack of sleep or by constitution. "Hence, an already optimally tuned brain can hardly be enhanced" (Quednow 2010, 153). After performing twelve qualitative interviews in Germany and in the U.S. with users of Ritalin®, Adderall® and modafinil, it is my impression that users are aware that they are not taking smart drugs and that by taking medication, they are not writing better papers. They do it either to be able to write their papers on time or to ease their struggle with writing a paper. The results of these interviews gave me the impression that taking pills to improve working performance is something many people desire. In fact, I was asked countless times by colleagues if these pills actually work and where to get them. On the other hand, almost everyone seems to reject a society where most people take pills to enhance their cognition.[2] In the ongoing media debate on cognitive enhancement, one gets the impression that the phenomenon is far more prevalent than it really is. Many journalists even state that taking Ritalin® has already been a common practice among students for some time. In a study that explored how the newsprint media portrays neuroenhancement, Bradley J. Partridge et al. found out that 94 % portrayed it as common, increasing or both (Partridge et al. 2011, 1). Many articles are titled with terms like 'brain doping' or 'mind doping'

[1] In the US, the life-time prevalence of non-medical prescription stimulant use in 2004 was 6.9 %, according to McCabe et al. (2005). In Germany a recent study from the University of Mainz surveyed a lifetime prevalence non-medical prescription stimulant use of 0.78 % (Franke et al. 2011).

[2] To further explore the fantasies and underlying normative assumptions on cognitive enhancement, I conducted focus group discussions in Germany and the US with students who do not necessarily take medication themselves. Results are not published yet.

thus drawing a parallel from studying to high-performance sports where doping is the illegitimate use of drugs for a desired effect.

In the bioethical debate, the question of fairness is one of the key issues regarding cognitive enhancement. Fairness is one issue that affects society as a whole as opposed to ethical objections that pertain to the individual, such as safety and authenticity. Related topics are social inequality and distributive justice. The analogy to doping in sports, which is drawn by using terms like 'brain doping,' conceptualizes the context in which cognitive enhancements are being used as competitive. I will, after presenting the conceptions of fairness in the cognitive enhancement debate, shed light on the question of competition and how to keep it 'fair.' The focus on fairness that bioethicists deal with, whether they regard fairness as being undermined or not, is based on competition as a major mode of interaction. I will argue that the focus on fairness in the cognitive enhancement debate can be seen as a sociological diagnosis of time because this focus provides that a field is structured by modes of competition. The theoretical framework that I will use for my thesis is Michel Foucault's term of neoliberal governmentality, a term he uses to analyze the intertwinement of governmental goals with strategies of self-conduct. In order to clarify this approach of governmentality, I will briefly outline the neo- and ordoliberal concepts of fairness. They disagreed about the level of state intervention required to protect competition and free markets; these arguments are analogous to those of bioethicists regarding the issue of competition between enhanced and non-enhanced individuals.

18.1 Bioethics

Cognitive enhancement is one of the core issues of the very young discipline of neuroethics; signs of its successful establishment within academia during the last decade include: institutionalization at universities, the founding of academic journals devoted to this topic and the procurement of funding.[3] Progress in the neurosciences as well as a focus on the connection between a person's brain and her identity have been cited as the major reasons for establishing a new discipline beyond bioethics (Farah and Wolpe 2004). This 'neuroethics exceptionalism' has been criticized, for example, by Fernando Vidal as being part of an 'ideology of brainhood' (Vidal 2009).

In the following, I will use the term bioethics in general because not all scholars dealing with the ethical implications of cognitive enhancement regard themselves as neuroethicists.

[3]Foundation of the "Neuroethics Society" (2006), Foundation of the "European Neuroscience and Society Network" (2007). Two neuroethical journals: *Neuroethics* (2008) and *American Journal of Bioethics – Neuroscience* (2010).

Scholars' positions within the field of bioethics can be classified by the terms, liberal and conservative. There are two aspects that I want to emphasize with regard to that division: First, the distinction between liberal and conservative positions in bioethics is done in German- and English-language publications, although the sense of the term "liberal" differs significantly between Germany and the U.S. Within the German political sphere, the term "liberal" is associated with economic liberalism/laissez-faire, while in the U.S., it is associated with progressive or left-leaning cultural political attitudes. Additionally, the term "progressive" tends to imply a linear view of history and is, therefore, problematic since not all progressives favor, for example, all forms of technological progress.

Second, I would like to point out that the liberal/conservative-distinction in bioethics is not in line with political camps as we know them. For example, Francis Fukuyama and Jürgen Habermas are both strongly opposed to enhancement biotechnology, but, at the same time, they hold completely opposite political positions. Francis Fukuyama, famous for diagnosing "the end of history" in 1990 and for participating in the President's Council on Bioethics under George W. Bush, is an advocate of liberalism and free market capitalism on the one hand. On the other, Jürgen Habermas is a social democrat who regards himself as following in the tradition of the Frankfurt School, which is known for its critical stance towards capitalism. Despite their divergent political views, they have both been labeled "bioconservative" by Ian Hacking – a label that Habermas happily accepted as he wrote in a letter to Ian Hacking: "I never thought that any version of 'conservatism' would apply to me, but 'bioconservatism' is a wonderful term!" (Hacking 2009, 14)

James Hughes, a transhumanist who calls himself a 'technoprogressive,' has outlined that in the twentieth century there have been progressives and conservatives on two axes: cultural and economic politics. Hughes argues that in the twenty-first century, however, a third axis called biopolitics needs to be added (Hughes 2010). Such a model can exemplify how various combinations are possible, as in the case of culturally and economically progressive but biopolitically conservative Habermas. But of course the problem remains that what is regarded as progress depends on normative assumptions.

In the following, my classification into liberal and conservative arguments in bioethics is only based upon policy recommendations that either call for restriction or for liberalization. In my effort to outline the position of conservatives, I limit myself to the President's Council on Bioethics, appointed by George W. Bush and chaired by Leon Kass. Their book, *Beyond Therapy* (President's Council on Bioethics 2003), is probably the most widely-quoted, bioethically conservative position in the debate on cognitive enhancement. As I outline liberal positions in the debate, I will first give a broader overview and then go into more detail regarding the conception of fairness that Julian Savulescu proposes.

18.1.1 Bioconservatives

In *Beyond Therapy*, the President's Council on Bioethics (2003) evaluates the ethical implications of different forms of enhancement technologies (genetic and pharmacological ones). The authors' rejection of any form of enhancement stems from their assumption that achievement, performance and competition have a great impact on society. Leon Kass and his colleagues see the intrinsic value of effort as endangered by the popularization of cognitive enhancing drugs. They hold that "the true dignity of human activity" would be undermined by biotechnological enhancement of muscles, genetic codes or the brain.

"What is at stake here is the very meaning of human agency, the meaning of being *at-work* in the world, being at work as *myself* and being at-work in a *humanly excellent* way" (The President's Council on Bioethics 2003, 141). Success must not be detached from authentically made efforts. In the authors' view, competitive deeds lose their value when the respective actors use performance enhancers. They underline the importance of striving for "our own best possible performance, to be sure, but our best performance as human beings, not animals or machines" (The President's Council on Bioethics 2003, 153). What do animals and machines have in common, in contrast to humans? The absence of will, volition and reason?[4] In *Beyond Therapy*, a strong character who is able to resist the temptations of shortcuts is celebrated over and over. Although improving performances by taking Ritalin® or modafinil is tempting, the authors state (drawing an analogy to doping in sports) that the essence of a competition is worthless if achievements were not carried out naturally. By only valuing the results of an action and divorcing them "from any idea of what is humanly superior," the council depicts a dark dystopia, a pessimistic picture of a society, where human agency has lost its significance. "We would become a society of spectators and our activities a mere spectacle. Or a society of parasites, needing and taking, but never doing or acting" (The President's Council on Bioethics 2003, 155f). The intrinsic value of effort as a normative source of our social order that is seen as endangered by pharmacology has polemically been called "pharmacological calvinism" by Gerald Klerman (1972). Indeed, this approach strongly alludes to what Max Weber analyzed in *The Protestant Ethic and the Spirit of Capitalism* (Weber 1976) as the influence of Calvinist work ethics on the development of capitalism. The combination of an intrinsic value of effort with a description of society as competitive in almost all spheres leads to the point where taking cognitive enhancing drugs is seen as nothing but cheating:

> Still, in those areas of human life in which excellence has until now been achieved only by discipline and effort, the attainment of similar results by means of drugs, genetic

---

[4]Bruno Latour who explores the dualistic distinctions made by modernists between nature and society, between humans and non-humans, asks (Latour 1993, 124): "Are not most ethicists busy with those two opposite but symmetrical tasks: defending the purity of science and rationality from the polluting influence of passions and interests; defending the unique values and rights of human subjects against the domination of scientific and technical objectivity?".

> engineering, or implanted devices looks to many people (including some Members of this Council) to be 'cheating' or 'cheap'. Many people believe that each person should work hard for his achievements. [...] This matter of character – the merit of disciplined and dedicated striving – is surely pertinent. (The President's Council on Bioethics 2003, 291)

In sum, the Presidents' Council on Bioethics extols endeavor, regardless of its result. In their view, endeavor has a value in itself, which would be undermined by enhancement technologies that ease achievements. Allowing people to pharmacologically enhance their cognition would provide a shortcut to the endeavor that others make through self-discipline and would, therefore, be unfair.

18.1.2 Bioliberals

Liberal bioethicists also deal with the question of fair competition although they do not regard it as being undermined by cognitive enhancers *per se*. For instance, in "Towards responsible use of cognitive enhancing drugs by the healthy," Henry Greely and colleagues (Greely et al. 2008) do admit that cognitive enhancement could undermine fairness by giving the example of a math test where some students are allowed to use a calculator while others are not. However, they then state that this kind of unfairness already exists in private tutoring, preparatory courses and so on, which give some students advantages over others.

The authors consider "mitigate[ing] this inequity by giving every exam-taker free access to cognitive enhancing drugs, as some schools provide computers during exam week to all students" (Greely et al. 2008, 704). This, they argue, could prevent cognitive enhancers from becoming the exclusive domain of the rich. They propose to develop policies governing the use of cognitive enhancers in competitive situations to minimize enhancement-related socioeconomic disparities.

Arthur Caplan also argues for the liberalization of enhancement technologies. According to him, market societies encourage their members in their efforts to enhance their capacities. The capital to do so is not equally distributed, but this should not lead to a restriction of those means. He gives the example of his son who has the privilege to go to a private school. A parent from a less wealthy neighborhood would probably not blame him for providing such an advantage to his son but would also wish he were able to do the same for his son. Thus, the solution is not to stop enhancement but to provide fair access to it, whether by way of private teachers or via biotechnological enhancement devices (Caplan 2003). But the question remains here as well as in other approaches: What does fair access mean? Is it fair to provide access to a good through the market? And in which way are market-based resource allocations fair? Or is access to cognitive enhancement fair if public institutions provide it? Should it then be distributed equally or according to need? And what would constitute a need for cognitive enhancing drugs?

Another influential article is Martha Farah and colleagues' "Neurocognitive Enhancement: what can we do and what should we do?" (Farah et al. 2004).

This article deals with the fairness problem in a similar way by raising the question of implicit coercion that arises from the pure fact that others are taking cognitive enhancers: "Merely competing against enhanced co-workers or students exerts an incentive to use neurocognitive enhancement, and it is harder to identify any existing legal framework for protecting people against such incentives to compete" (Farah et al. 2004, 423).

Farah et al. admit, that cognitive enhancers, like most other goods, will not be fairly distributed, that there will be cost barriers and social barriers that might compound already-existing disadvantages of groups with low socioeconomic status in education and employment. But these inequities would actually exist and still no one would restrict advances in health or quality of life because these goods potentially would not be distributed equally. The authors discuss the possibility that cognitive enhancement could help to equalize opportunity as cognitive enhancing drugs can more easily be distributed fairly than other goods that contribute to gaps in socioeconomic achievements (Farah et al. 2004, 423).

But, above all, they reject the restriction of the use of cognitive enhancement because this *a fortiori* would be coercive.

In 2009, a group of German scholars published a programmatic paper titled "Das optimierte Gehirn" (The Optimized Brain), which raised a lot of media attention because it was the first attempt to initiate a debate about liberalization of cognitive enhancement in Germany.[5] They also admit the risk of coercion, but highlight the freedom to choose and refer to non-competitive situations where cognitive enhancement could improve quality of life such as learning languages faster, listening to music more intensely and so on.

Julian Savulescu, an Oxford bioethicist, who can also be placed in the liberal camp has published one of the few papers that explicitly examines the term 'fairness' in the cognitive enhancement debate by relating it to philosophical discussions. He argues that fairness is not undermined by pharmacologically enhancing performance but that fairness even requires enhancement. He proposes a "Welfarist Definition of Human Enhancement" and defines this term as "Any change in the biology or psychology of a person which increases the chances of leading a good life in circumstances *C*" (Savulescu 2006, 324). Defined this way, enhancement is either a medical treatment for disease or an increase in capabilities within or beyond the typical range of the human species. "When enhancement is understood as an intervention which increases the chances of a person having a good life, it is hard to see how there could be any objections to trying to make people's lives go better" (Savulescu 2006, 326). In his article, Savulescu neither examines the question of what constitutes a good life, nor does he deal with the risks and benefits of enhancement; rather he explores the objections dealing with the introduction of inequality, injustice and unfairness. Since fairness-based objections would refer

[5] Two of the co-authors of the memorandum, Jan Christoph Bublitz and Dimitris Repantis, are also contributors to this book.

to enhancement of positional goods[6] in a competitive society, Savulescu takes the example of unfairness in sports. He argues that allowing safe performance enhancers would reduce unfairness because everyone would have the option to use them, while to ban them would give a relative advantage to 'cheaters' (cf. Savulescu 2006, 326). Cheating in this sense conveys the same notion as it does in *Beyond Therapy*: breaking a rule; however, Savulescu conceptualizes rules as arbitrary because they are defined by human beings, while the Presidents' Council on Bioethics has a somewhat metaphysical notion of rules based on an essentialist concept of human nature.

In the second part of his paper, Savulescu deals with the objections of injustice caused by increasing inequality through enhancement. He argues that the application of different theories of justice lead to the conclusion that enhancement increases justice rather than injustice. A libertarian theory of justice would not consider it unfair if the rich can buy enhancements and thus increase their opportunities "provided that their assets have been justly and legally acquired" (Savulescu 2006, 332). Utilitarian approaches also would require enhancement, given that enhancement is a good that increases well-being. Savulescu quotes Jeremy Bentham, who advocates the "greatest good to the greatest number" (Savulescu 2006, 332). Furthermore, an egalitarian approach in theories of justice, like John Rawls' concept of Justice as Fairness, would, according to Savulescu, require enhancement as well: "According to Rawls' Justice as Fairness, we should distribute enhancements so that the worst off in society are as well off as they can be" (Savulescu 2006, 332; Rawls 1971).

Savulescu argues, that with regard to a "fair go" for everyone and assuming an IQ of at least 70 is needed to live a good life, Justice/Fairness would require that, "we get as many people as possible up to the minimum IQ necessary for a decent chance of a decent life. [. . .] But where we set the minimum threshold for treatment or enhancement is up to us. It depends on how we define a decent chance of a decent life in the way society and the world are likely to be" (Savulescu 2006, 334).

Savulescu does not deal at all with the problem of implicit coercion, nor does he address the problem that, probably, the worst off in society would not be the only ones to benefit from an efficient enhancement. Therefore, it seems to be highly questionable whether a Rawlsian approach of Justice as Fairness can really justify a moral obligation to enhancement in order to provide equal opportunity.

Both liberal and conservative ethicists agree that fair competition has an enormous normative impact on the legitimacy of social order. However, they offer different answers to the question of how to govern freedom and how to protect fairness in competition. Liberals and conservatives differ profoundly on the question of whether cognitive enhancement poses a threat to a fair, competitive society and whether the state should restrict the use of drugs or distribute drugs to preserve level playing fields. Moreover, important questions remain as to whether cognitive enhancement equalizes opportunities for those who did not benefit in 'the lottery of nature' or whether it increases the socioeconomic gap by providing

[6]Positional goods are goods that are valued for their scarcity alone (cf. Hirsch 1977).

advantages to already privileged individuals. Much discussion centers on whether state intervention is necessary to protect fair competition or if such interventions are too coercive in a liberal society. Conservatives tend to accentuate the integrity of human nature as a normative source[7] to prohibit enhancement technologies. Liberals, instead, criticize an essentialist term of naturalness and accentuate the freedom of choice to enhance one's own capabilities and maintain that it is still oneself making the effort, even if one is pharmacologically enhanced.

Nonetheless, bioliberal and bioconservative positions seem to be in agreement regarding the significance of fair competition as a key source of the normative order of our society. Interestingly neoliberals and ordoliberals argued similarly on the basis on this normative assumption. In the following, I will outline their concepts of a political and economic order to shed light on the implications pertaining to this bioethical discourse, in terms of the history of ideas.

18.2 Neoliberalism

The relation between market and state has been a controversial subject in the debate of ordo- and neoliberals. Michel Foucault analyzed the rationale of governing in ordo- and neoliberalism in his lectures on governmentality (Foucault 2008). He did not analyze neoliberalism in the way this term has been used by leftists in quite an inflationary way to criticize almost any undesirable effect of contemporary capitalism. Foucault looked at the politics of those scholars and politicians of the 1940s and 1950s who called themselves neo- or ordoliberals. Those neo- and ordoliberals sought to distinguish themselves from proponents of nineteenth-century laissez-faire liberalism, as they did not conceptualize a free market as a natural order but rather as one that has to be governed and protected by the state. The state's legitimacy in this sense depends on its success to create wealth by protecting the free market. In order to preserve fair competition, two developments mainly have to be prevented: the emergence of monopolies and the influence of lobbyists (be it unions or business lobbyists).

Neoliberals and ordoliberals also dealt with the legitimate level of coercion to protect fairness. Debates over the level of state intervention with the goal of protecting fair competition were key questions within the *Colloque Walter Lippmann*, the birthplace of neoliberalism. Friedrich August von Hayek, for example, emphasized that fairness basically means to comply with the rules of the game. Fairness requires neither breaking the rules nor demanding that exceptions are made (cf. Vanberg 2006). He favored a minimum of state intervention and rejected all kinds of policy

[7] Ian Hacking points out, that although Habermas put "Human Nature" into the title of his book it is not human nature's integrity itself that forms the basis of his argument, but human dignity. Fukuyama, instead, uses this phrase in itself as the basis of his entire argument (cf. Hacking 2009, 16).

action with the purpose of balancing unequal starting conditions. Although Hayek also rejected laissez-faire liberalism, his conception of state interventions was much narrower than that of ordoliberals. He conceptualized a spontaneous order, which is the result of "human action but not of human design" (Hayek 1978). The role of the state was to only maintain the rule of law since Hayek conceptualized the economy as so contingent and unpredictable that the consequences of state intervention in the end would always be harmful and lead to authoritarian rule. In contrast, ordoliberals like Alexander von Rüstow, for example, called for much stronger state interventions to provide equal opportunities (to compete), for instance by means of an extremely high inheritance tax. Rüstow, who originally invented the term 'neoliberalism' in order to distinguish it from nineteenth century laissez-faire liberalism, later called himself ordoliberal because he saw the term already occupied by more market-oriented thinkers like Hayek.

As Graham Burchell puts it, both ordoliberals and neoliberals still deal with the question of to what extent "competitive and optimizing market relations and behavior can serve as a principle not only for limiting governmental intervention, but also rationalizing government itself" (Burchell 1996, 23). Criticism of neoliberalism, misunderstood as laissez-faire, often refers to any kind of economic state intervention as being progressive; we have seen that neoliberal politics cannot be seen as simply "governing less." Another critique of neoliberalism, in the sense of market fundamentalism, refers to its lack of moral values. Especially since the recent financial crisis, many have bemoaned the fact that moral commitment has been replaced by pure materialism. But, in fact, it has a strong normative impact: an "ethic of competition," as Bruno Amable puts it. As neoliberals reject the natural character of the market order, it is up to everyone to adopt an ethos of individual responsibility, which means to be responsible for one's own existence as well as for one's integration into the market (Amable 2010, 13). With this highly normative notion of individual responsibility, 'fairness in competition' is the key source of legitimacy and, thus, the only reason for state intervention in neoliberalism.

> It [neoliberalism] is based on the idea, that the ideal world order should be a 'free' and 'fair' competition between individuals. This competition is always under threat by groups who try to protect themselves from its rigour and consequences and seek to obtain more than their due share. Public intervention is thus legitimated, when it tries to restore the conditions of fair competition and 'level the playing field'. (Amable 2010, 3)

Here we see that the core of neoliberal thought shares its assumptions with the protagonists of the bioethical debate on cognitive enhancement. Bioconservatives discuss the temptation to bypass the rigor of competition and speak of enhancements as 'shortcuts'; whereas, bioliberals tend to accentuate the freedom to choose one's means to compete and to enhance one's competitiveness. Or they, instead, favor enhancement as a means to level the playing field.

In the following, I will outline how Foucault took the thoughts of Neo- and Ordoliberals as a starting point to develop what he called neoliberal governmentality.

18.3 Neoliberal Governmentality and Technologies of the Self

Ethical guidelines developed by bioethicists refer to both policy recommendations as well as individual ethics. Foucault analyzed how forms of guidance for collective goals and forms of self-conduct are intertwined. He therefore created the term, governmentality: "Foucault defines government as conduct, or, more precisely, as 'the conduct of conduct' and thus as a term which ranges from 'governing the self' to 'governing others'" (Lemke 2001, 191). The term governmentality refers to the (self-)production of subjectivity, to the production of technologies of the self that can be entangled with governmental goals (cf. Bröckling et al. 2000, 29).

In neoliberalism, governmental goals are the products of entrepreneurial behavior and the establishment of competitive situations also within spheres that previously have not been market-based. Social scientists call this process "marketization." The perspective of governmentality reveals that self-determination and free choice are not the boundaries of governmental actions but, rather, constitute an instrument of governing oneself, i.e. a form of subjectification.

> Liberalism, particularly its modern versions constructs a relationship between government and the governed that increasingly depends upon ways in which individuals are required to assume the status of being the subjects of their lives, upon the ways in which they fashion themselves as certain kinds of subjects, upon the ways in which they practice their freedom. (Burchell 1996, 29)

Neoliberal governmentality, which is based on the predominance of market mechanisms, engenders the construction of auto-regulated or auto-correcting selves (Lemke 2001). With the term 'technologies of the self' Foucault describes techniques of governing the self and operations that form such auto-regulated selves. Technologies of the self are defined as strategies that "permit individuals to effect by their own means or with the help of others a certain number of operations on their own bodies and souls, thoughts, conduct, and way of being, so as to transform themselves in order to attain a certain state of happiness, purity, wisdom, perfection, or immortality" (Foucault 1988, 18). Technologies of the self are always embedded in power relations, but they cannot be reduced to their mere effects. When we analyze technologies of the self, we have to take into account the contemporary power relations by which the individual is governed via individualization. Nicolas Rose explains this process of constant self-monitoring which 'neurochemical selves' have to engage in:

> The active and responsible citizen must engage in a constant monitoring of health, a constant work of modulation, adjustment, improvement in response to the changing requirements of the practices of his or her mode of everyday life. These new self-technologies do not seek to return a pathological or problematic individual to a fixed norm of civilised conduct through a once-off programme of normalisation. Rather, they oblige the individual to engage in constant risk management, and to act continually on him or herself to minimise risks by reshaping diet, lifestyle and now, by means of pharmaceuticals, the body itself. The new neurochemical self is flexible and can be reconfigured in a way that blurs the boundaries between cure, normalisation, and the enhancement of capacities. And these pharmaceuticals offer the promise of the calculated modification and augmentation of specific aspects of self-hood through acts of choice. (Rose 2004, 28)

How do Foucault's theoretical framework and his terms, "governmentality" and "technologies of the self," now refer to cognitive enhancement? When we conceptualize taking cognitive enhancing drugs as a technology of the self and when we further apply Foucault's method to analyze the history of the present, then we can draw two lines: the first one is the history of self-techniques to improve one's own mental capabilities, whether by meditation, nutrition, or practice, which has been a human goal for a long time. Foucault analyzed ancient technologies of the self in *The Care of the Self* (Foucault 1986). With this approach we would rather conceptualize cognitive enhancement as a technique of self-constitution than as a technique of domination. When we instead put our focus on techniques of domination, we can draw a second line: the history of neuro-interventions, i.e. the various forms by which clinical institutions have tried to modify the human brain. What we see from this perspective is a line from coercion to request.[8] Early neuro-interventions turned patients into passive objects with brain surgery such as lobotomy. Psychiatric treatments were aimed at controlling the population as well as disciplining criminals and deviant subjects. Nowadays, brain interventions can be understood as forms of self-conduct. They aim to help individuals to lead autonomous lives and to make decisions for themselves rather than to oppress their individual freedom. So, one can argue that in cases of brain interventions, coercion conceptually has been replaced by request (Schaper-Rinkel 2007, 96). One example that fits this category of request for psychopharmacology is the desire of many to be diagnosed with adult ADHD so they will in turn be able to use methylphenidate legitimately. The distinction between normal and deviant subjects gets more and more irrelevant. Every brain is now considered as a site of potential improvement. The normal and the deviant in every subject is the object of neuroscientific research and intervention (cf. Schaper-Rinkel 2007, 104). Potentially, everybody could be aided by neuroscientific technology.

18.4 Conclusion

In the first part of my chapter, I have analyzed the bioethical discourse on cognitive enhancement. I have argued that the bioethical debate has an implicit conception of 'fair competition' that can be understood as analogous to debates on fair competition between ordo- and neoliberals. Ordo- and neoliberals both agree on the core meaning that competition has for the society as a whole. While ordoliberals were more skeptical that markets would successively undermine fair competition and, therefore, argued for more state intervention to protect competition, neoliberals argued for a minimal degree of state intervention into the market to avoid a distortion of competition.

[8] This doesn't mean that today's psychiatric neuro-interventions are not coercive any more.

Likewise, both bioconservatives and bioliberals regard fairness of competition as the normative source of social order. One could argue that they are both based on the assumption that competition is the dominant mode of social interaction. They disagree on how to protect fairness in competition. Bioconservatives are concerned about level playing fields; they favor competition on the merits and therefore reject enhancements. Bioliberals are also concerned about level playing fields but regard enhancements as a possible means to level the playing field.

I have analyzed this analogy in the context of Foucault's work on neoliberal governmentality to emphasize the societal context within which this debate on fair competition is located. Neoliberal governmentality is a certain rationale of governing by implementing and protecting market mechanisms.

The medical anthropologist Nicolas Langlitz also criticizes the analogy of doping in sports that is frequently mentioned in the cognitive enhancement debate. The exaggeration of prevalence of cognitive enhancers would unwittingly advertise products of the pharmacological industry, whose long-term effects are widely unknown. The discussion would increase feelings of a permanent competitive struggle, and although the working environment is broadly determined by competition, a society would also form itself by the images it draws of itself. The achievement of level playing fields in biology, as well as in society, is a fiction. Langlitz questions whether it is even desirable to view our life as analogous to high-performance sports, as the brain-doping debate has been framed (Langlitz 2010, 52).

Even the fact that we substitute the term 'cognitive enhancement' for what we referred to as 'substance abuse' in previous decades reveals an orientation towards competition in more and more societal spheres. It does not mean that the term, "substance abuse," which has a pejorative connotation, would be more accurate. But, we should carefully examine where qualities actually are being enhanced and where the drive for competitiveness is being advanced.

Acknowledgements I would like to thank Nicolas Langlitz for many very inspiring discussions that have enriched this chapter and Thomas Biebricher, Georg Fischer and Isabell Trommer for valuable editing and comments.

References

Amable B (2010) Morals and politics in the ideology of neo-liberalism. Socio-Econ Rev 8:1–28. doi:10.1093/ser/mwq015

Bröckling U, Krasmann S, Lemke T (2000) Gouvernementalität der Gegenwart. Studien zur Ökonomisierung des Sozialen. Suhrkamp, Frankfurt am Main

Burchell G (1996) Liberal government and techniques of the self. In: Barry A, Osborne T, Rose N (eds) Foucault and political reason. The University of Chicago Press, Chicago

Caplan AL (2003) Is better best? A noted ethicist argues in favor of brain enhancement. Sci Am 289:104–105

De Jongh R, Bolt I, Schermer M, Olivier B (2008) Botox for the brain: enhancement of cognition, mood and pro-social behavior and blunting of unwanted memories. Neurosci Biobehav Rev 32:760–776. doi:10.1016/j.neubiorev.2007.12.001

Farah M, Wolpe P (2004) Monitoring and manipulating brain function. New neuroscience technologies and their ethical implications. Hastings Cent Rep 34:35–45. doi:10.2307/3528418

Farah M, Illes J, Cook-Deegan R, Gardner H, Kandel E, King P, Parens E, Sahakian B, Wolpe P (2004) Neurocognitive enhancement: what can we do and what should we do? Nat Rev Neurosci 5:421–425. doi:10.1038/nrn1390

Forlini C, Racine E (2009) Disagreements with implications: diverging discourses on the ethics of non-medical use of methylphenidate for performance enhancement. BMC Med Ethics 10:9. doi:10.1186/1472-6939-10-9

Foucault M (1986) The care of the self. Pantheon, New York

Foucault M (1988) Technologies of the self. In: Martin LH, Gutman H, Hutton PH (eds) Technologies of the self. University of Massachusetts Press, Amherst

Foucault M (2008) The birth of biopolitics: lectures at the collège de France, 1978–1979. Palgrave Macmillan, Basingstoke

Franke AG, Bonertz C, Christmann M, Huss M, Fellgiebel A, Hildt E, Lieb K (2011) Non-medical use of prescription stimulants and illicit use of stimulants for cognitive enhancement in pupils and students in Germany. Pharmacopsychiatry 44:60–66. doi:10.1055/s-0030-1268417

Galert T, Bublitz C, Heuser I, Merkel R, Repantis D, Schöne-Seifert B, Talbot D (2009) Das optimierte Gehirn. Gehirn und Geist 11:40–48

Greely H, Sahakian B, Harris J, Kessler RC, Gazzaniga M, Campbell P, Farah M (2008) Towards responsible use of cognitive enhancing drugs by the healthy. Nature 456:18–25. doi:10.1038/456702a

Hacking I (2009) The abolition of man. Behemoth J Civil 3:5–23. doi:10.1524/behe.2009.0017

Hayek FA (1978) The results of human action but not of human design. In: Hayek FA (ed) New studies in philosophy, politics, economics, and the history of ideas. Routledge & Kegan Paul, London/Henley

Hirsch F (1977) The social limits to growth. Routledge & Kegan Paul, London

Hughes J (2010) TechnoProgressive biopolitics and human enhancement. In: Moreno J, Berger S (eds) Progress in bioethics. MIT Press, Cambridge

Klerman GL (1972) Psychotropic hedonism vs. pharmacological calvinism. Hastings Cent Rep 2:1–3. doi:10.2307/3561398

Langlitz N (2010) Das Gehirn ist kein Muskel. Frankfurter Allgemeine Sonntagszeitung. http://www.faz.net/aktuell/wissen/medizin/neuro-enhancement-das-gehirn-ist-kein-muskel-1912020.html

Latour B (1993) We have never been modern. Harvard University Press, Cambridge

Lemke T (2001) The birth of bio-politics: Michael Foucault's lectures at the College de France on neo-liberal governmentality. Econ Soc 30:190–207. doi:10.1080/713766674

McCabe SE, Knight JR, Teter CJ, Wechsler H (2005) Non-medical use of prescription stimulants among US college students: prevalence and correlates from a national survey. Addiction 100:96–106. doi:10.1111/j.1360-0443.2005.00944.x

Partridge BJ, Bell SK, Lucke JC, Yeates S, Hall WD (2011) Smart drugs: "as common as coffee": media hype about neuroenhancement. PLoS One 6(11):doi:10.1371/journal.pone.0028416

Quednow BB (2010) Ethics of neuroenhancement: a phantom debate. BioSocieties 5:153–156. doi:10.1057/biosoc.2009.13

Rawls J (1971) A theory of justice. Harvard University Press, Cambridge

Rose N (2003) Neurochemical selves. Society 41:46–59. doi:10.1007/BF02688204

Rose N (2004) Becoming neurochemical selves. In: Stehr N (ed) Biotechnology, commerce and civil society. Transaction Publishers, New Brunswick

Sabin J, Daniels N (1994) Determining "medical necessity" in mental health practice: a study of clinical reasoning and a proposal for insurance policy. Hastings Cent Rep 24:5–13. doi:10.2307/3563458

Savulescu J (2006) Justice, fairness and enhancement. In: Sims Bainbridge W, Roco MC (eds) Special issue: progress in convergence: technologies for human wellbeing. Ann N Y Acad Sci 1093:321–338. doi:10.1196/annals.1382.021

Schaper-Rinkel P (2007) Die neurowissenschaftliche Gouvernementalität. Re-Konfigurationen von Geschlecht zwischen Formbarkeit, Abschaffung und Re-Essentialisierung. In: Dölling I, Dornhof D, Esders K, Genschel C, Hark S (eds) Transformationen von Wissen, Mensch und Geschlecht. Transdiziplinäre Interventionen. Ulrike Helmer Verlag, Königstein/Taunus

The President's Council on Bioethics (2003) Beyond therapy: biotechnology and the pursuit of happiness. Dana Press, Washington, DC

Vanberg VJ (2006) Marktwirtschaft und Gerechtigkeit. Zu F.A. Hayeks Kritik am Konzept der "sozialen Gerechtigkeit". In: Held M, Kubon-Gilke G, Sturn R (eds) Jahrbuch Normative und institutionelle Grundfragen der Ökonomik, Bd. 5, Soziale Sicherung in Marktgesellschaften. Metropolis-Verlag, Marburg

Vidal F (2009) Brainhood, anthropological figure of modernity. Hist Hum Sci 22:5–36. doi:10.1177/0952695108099133

Weber M, The agrarian sociology of ancient civilizations (trans: Frank RI, 1976). Verso Classics, London

Weber M (2001) The protestant ethic and the spirit of capitalism. Routledge, London

Chapter 19
My Mind Is Mine!? Cognitive Liberty as a Legal Concept

Jan-Christoph Bublitz

Abstract This chapter explores some of the legal issues raised by mind-interventions outside of therapeutic contexts. It is argued that the law will have to recognize a basic human right: cognitive liberty or mental self-determination which guarantees an individual's sovereignty over her mind and entails the permission to both use and refuse neuroenhancements. Not only proponents but also critics of enhancements should embrace this right as they often ground their cases against enhancement on precisely the interests it protects, even though critics do not always seem to be aware of this. The contours and limits of cognitive liberty are sketched, indicating which reasons are good (or bad) grounds for political regulations of neurotechnologies.

Keywords Neuroenhancement • Cognitive liberty • Law • Human rights • Political regulations

19.1 Preliminaries: Some Observations on the German Enhancement Debate

As an introduction, let me share some observations on the German debate on neuroenhancement (NE). In 2009, I was among a group of German scholars from various disciplines who concluded a research project by publishing a "Memorandum on Neuroenhancement" in a popular science magazine with the aim of spurring a public debate. In short, we suggested that "the principled objections leveled against

J.-C. Bublitz (✉)
Faculty of Law, University of Hamburg, Hamburg, Germany
e-mail: christoph.bublitz@uni-hamburg.de

E. Hildt and A.G. Franke (eds.), *Cognitive Enhancement: An Interdisciplinary Perspective*, Trends in Augmentation of Human Performance 1,
DOI 10.1007/978-94-007-6253-4_19,

pharmaceutical improvement of the mind are not convincing. NE is the continuation of humankind's quest to enhance cognitive capacities by different means" (Galert et al. 2009, 47, transl. J.-C. B.).

By "principled objections" we meant arguments against NE not based on empirical issues such as negative side-effects on health or personality but those grounded on more fundamental, normative considerations. To us, neither the goals pursued nor the means employed warrant categorical objections against NE. They do not necessarily undermine the authenticity of persons, nor corrupt the value of achievement or endanger other important common goods. Nevertheless, specific substances or consumption practices may indeed lead to undesired results for both the individual and society. They should be carefully observed, prevented and remedied by appropriate measures. And quite certainly, traditional means to achieve goals may sometimes be preferable to pharmaceutical shortcuts, particularly when they promote secondary virtues such as endurance and self-confidence or when they confer self-knowledge (Kipke 2010). Yet, after all, and without blindness to the perils and pitfalls they may pose, if there were substances with significant enhancing effects and tolerable risk-profiles, we should, in principle, welcome rather than condemn them. Therefore we called for:

> An open and liberal, but by no means uncritical or incautious approach to NE [...]. While the arguments of NE opponents are not strong enough to warrant blanket prohibitions, some of them are worthy of further consideration and raise important questions about what is desirable for society and the individual [...]. NE prompts each and every one of us to reconsider what is meaningful in our lives. Moreover, they reflect problematic tendencies of modern times, especially the orientation towards performance and competition that increasingly pervades society. (Galert et al. 2009, 47, transl. J.-C. B.)

Additionally, project members published the results of systematic reviews of potential enhancers. In a conservative evaluation, they conclude that, at the moment, there is not any reliable data proving significant enhancement effects, mostly due to a lack of controlled studies designed to capture them (Repantis et al. 2010a, b).

Two years later, the claims of this memorandum still sound quite moderate to me. The public response, nonetheless, ranged from suspicion to outright rejection. Lifting the taboo on the desire to support one's psyche – if necessary by pharmaceutical means – was, in the eyes of many, too far a step, amounting to nothing less than a "declaration of war against mental health" as one of Germany's biggest newspapers put it (FAZ, 13.10.2009). The memorandum struck a nerve, not surprisingly, perhaps, in light of the highly ideological debates over drug policies in general.

In this chapter, I shall not defend the memorandum against the (remarkably few) substantive criticisms nor reiterate the extensively exchanged pro and con arguments. Rather, I shall explore a fundamental background assumption of the memorandum: the right to cognitive liberty (CL). It is, or rather, should be the central legal principle guiding the regulation of neurotechnologies, guaranteeing the right to alter one's mental states with the help of neurotools as well as to refuse to do so. Most legal systems, however, have yet to acknowledge such a right. The public debate has demonstrated that not only the law but also the wider public has been reluctant to embrace the ideal of cognitive liberty. Thus, I shall explore its meaning, scope and limits, especially in light of what I take to be the strongest legally-relevant argument *against* NE: worries over social pressure.

Very likely, a widespread use of enhancements will create pressures on persons preferring to abstain from using them. By raising standards of cognitive fitness in competitive job markets or by subtly shifting ideas of mental normality, nonusers may be confronted with the social expectation to "take a pill." This might lead to the "paradoxical perspective of some people expanding their freedom of action by restraining the freedom of will of equally numerous others" (Merkel 2007, 289). Critical arguments along this line implicitly draw on the idea of cognitive liberty.

In a broader perspective, the desire and demand for NE seem to be related to economic developments. In a thought-provoking paper, Hess and Jokeit suggest that today's "neurocapitalism" adapts "the innate neurobiological capacity of humans as a productive force to the technologies of globalization" (Hess and Jokeit 2009, 6). As the western world reduces accumulation of wealth through physical labor and industrial production, it is transformed into a "mental economy" (Franck 2005). From scientific innovations, creative work, cultural and media productions to financial markets, economic progress depends on the generation of novel ideas, knowledge capital and intellectual property. In mental economies, minds are the places of production. With information processing as its foundation, mental labor creates specific demands on workers and employees, primarily with respect to their cognitive capacities. And as we can witness in our own lives, the exploding amount of data constantly flowing into our minds easily exceeds our ordinary capacities. It does not come as a surprise then that the inability to cope with information overload characterizes the mental *conditio humana* in the age of neurocapitalism:

> Attention deficit disorder probably encapsulates the key symptoms of mental illness in the twenty-first century. Just as the repression of past centuries gave rise to the silent drama of neurotic symptoms, and the apparently boundless excess of the second half of the twentieth century created a breeding ground for the desireless state of depression, so the elevation of pre-selective attention skills and emotional intelligence to decisive competitive advantages could, in the event of failure, be very harmful to precisely these [. . .]. Wriggling helplessly between a dearth and an excess of stimuli, unable to escape the ubiquitous flood of signals, the relaxation mechanisms impaired and experience of emotions brutalised – all of these are symptoms that in the collective consciousness go under the general heading of ADD. (Hess and Jokeit 2009, 6)

Since historically, capitalism has always succeeded in producing the "scientific and technological wherewithal to [. . .] mitigate the self-generated 'malfunctioning' to which its constituent subjects are prone" (Hess and Jokeit, ibid), cognitive and mood enhancements might be the latest and most intrusive means to mold the minds of the workforce to cater to capitalist needs. And when mental skills become commodities, optimizing the brain is optimizing the means of production.

Relatedly, sociologists diagnose the acceleration of many parts of social life. "Fast food" and "speed dating" are the surface symptoms of a thoroughly sped-up pace of life. Without eternity, a good life for secularized subjects consists in making the most out of their spare time, creating a subtle but constant feeling of scarcity, of never having enough time (Rosa and Scheuerman 2009). Then, not incidentally, sleep-reducing drugs as Modafinil and alertness-increasing amphetamines (street name "speed") are becoming popular, not only for work but also for leisure activities.

These observations help to understand people's motives for enhancements. And while contextualizing technologies within the dynamics of societies is indispensable for their assessment, the normative ramifications often remain unclear. For instance, in many public debates on NE, variants of a "better-world" argument can be found, roughly running like this: In a better world, e.g. free from capitalist working conditions, no one would desire to take NEs; hence, it is preferable not to take them. Normative principles such as cognitive liberty and other arguments in favor of the right to take NE are then dismissed as misguided, apologetic and insensitive to flawed societal conditions. Yet, by such reasoning, contextualizing technologies leads to conflating different levels of analysis. While one may rightly criticize e.g. working conditions and their impact on mental well-being, it does not undermine the importance of the normative concept of cognitive liberty. On the contrary, concerns about negative mental effects of substances as well as social conditions should lead critics to support mind-protecting rights such as cognitive liberty.

More generally, any regulation of technologies for mind-interventions is confronted with defining legal principles that guide policies over changing minds. Thus, every call for legislative action over neurotechnologies has to provide proposals which cannot merely consist in *ad hoc* suggestions tailored to suit one's view on NE. Rather, they have to be aligned with and embedded in the broader framework of rights and duties that constitute legal orders. Even though cognitive liberty is not among the rights found in positive law, it is, so I claim, among the implicit assumptions of any liberal democratic state. It may have been neglected, even fallen into oblivion in legal thinking, but, nonetheless, it cannot be ignored by anyone who formulates policy recommendations. Cognitive liberty's main claim, the right to self-determine what is on (and in) one's mind, can be inferred from general and widely-accepted ideas of the relation between the individual and the state, granting persons wide ranging liberties in self-regarding matters. Historically, the idea of cognitive liberty can be found in the works of authors such as Kant and Mill (Sect. 19.3), and traces can be found in some provisions of today's positive law (Sect. 19.4). Nonetheless, the law has yet to define scope, contours (Sect. 19.5) and limits of mental freedom (Sect. 19.6). Only then, more substantive arguments about neuropolitics can be made. But before we turn to the law, some words about the relation between neuroethical and legal arguments are in order.

19.2 From Neuroethics to Neurolaw

The differences between neuroethics and neurolaw can be illustrated by one of the objections leveled against the memorandum's presumption of liberty to take NE:

> [The] initial point of our inquiry is the liberty of every competent person to find and define for herself the paths to a good life which entails self-determination over her body and her mind. This perspective is neither arbitrary nor negotiable: It is the foundational presumption of any liberal democratic state and of the German Constitution. Hence, it is not the potential consumers of NE who are in need of justifying their cause, but, on the contrary, those who seek to restrict the liberty to change one's mind [...]. Although strict prohibitions are currently not warranted, soft ethical recommendations against NE might well be, especially in light of considerations over what constitutes a flourishing life. (Galert et al. 2009, 40, transl. J.-C. B.)

The presumption of liberty implies that restricting freedoms needs stronger justification than exercising freedoms. Critics were quick to contend that the Memorandum is hence based on a *petitio principii* with the liberal conclusion being assumed in the premise. Even if counterarguments to NE were truly unconvincing, critics claim, a liberty to take NE cannot be inferred, at least not without further argument. To them, the presumption of liberty appears as a rhetorical trick instead of a substantial argument, shifting the argumentative burden on opponents of NE (Hoppe 2009). Apart from the fact that I hold it to be self-evident that, *other things being equal*, improved cognitive capacities are valuable and worthy to pursue, the critics' objection points to differences between neuroethics and neurolaw.

In order to define policies and legal regulations regarding NE, the abundant (neuro-)ethical arguments have to be translated into legal, rights-based arguments. Not all legitimate ethical concerns are automatically legitimate bases for policy decisions since the perspectives of ethics and law differ (Basl 2010). While ethicists tackle questions such as "What should I do?" or "In what kind of a society do we want to live?" and may, in answering, resort to whatever metaphysical, political or spiritual conception they deem favorable, legal philosophers are concerned with justifying coercion and infringement of legally-protected interests.

This change in perspective has normative consequences: In light of the law, some of the common issues in the debate, such as the treatment/enhancement distinction or the issue of authenticity, bear different argumentative weight. While some interests are protected by strong rights, others may not enjoy any legal protection at all. We will see how some ethical arguments reappear in a legal context in a moment. Here, it is important to note that every prohibitive or restrictive legal norm *prima facie* infringes on rights and, hence, needs to be justified by prevailing interests of others or society. Thus, any call for curbing access to or use of NE is in need of justification. It should be borne in mind that it is a main feature of liberal constitutional orders to protect individuals against state interferences – not only against the power of despots, but also against what may be called the ethical tyranny of the "moral majority." Rights are, by their very nature, constraints of public power. So, ethicists may devise kinds of society worth living in; yet, in enforcing them (against the wills of affected persons), democratic governments have to observe constitutional limits, particularly the following two:

19.2.1 Liberty: Question Begging or Fundamental Value?

First, rights confer on their holder "spheres of freedom" that restrict the scope of legitimate governmental interferences. Constraints follow roughly this idea: The more measures affect primarily the interests of the individual, the less others – and the state – should have to say about them, and *vice versa*. Issues exclusively concerning the individual are to be decided by her alone; whereas, others are *res publica*, issues of public interests proper. In regulating the latter, elected governments have a wide margin of appreciation and room for political decisions. The tension between the individual and society is the foundational conflict that democracies face and which legal constitutions are designed to appease and adjust. The problem in delineating the private from the public sphere is that *everything* a person does might in some way or another affect others and society. Thus, *judgments* about private and public domains have to be made, and their constant re-arrangement sets the background to many current controversies from internet anonymity to counterterrorism. Reformulated in legal terms, some (but few) rights are absolute and inviolable which means that governments cannot interfere with them for any reason. Other rights are strong but restrictable, placing high demands of justification on infringing measures, and some, such as the basal freedom of action in continental jurisdictions, can be limited quite easily.

Let me illustrate this with the body. It is contested whether the right to life is absolute – many constitutions and human rights treaties have limitation clauses allowing governments to take the life of citizens in special situations. Human dignity, by contrast, is often regarded as "inviolable".[1] Thus, killing another person might be permissible in exceptional circumstances, whereas, humiliating and degrading treatment (e.g. torture) never is. Logically, the right to bodily integrity has to be weaker than the right to life, but it nonetheless ranks among the strongest rights in many constitutions and human rights treaties. Therefore, what individuals do to their bodies is left to them to a large extent. From tattooing and cosmetic surgery to extreme sports, potential (statistically, even lethal) threats to health and bodily integrity are within the domain of personal decisions. In these cases, state regulations primarily concern safety standards, duties to inform about risks (of e.g. surgical interventions) and issues of consent, regardless of the fact that such bodily modifications may in one way or another relate to the social sphere. Plastic surgery, for instance, supposedly has already influenced collective ideas of beauty and aesthetics and is very likely a contributing factor to the prevalence of dissatisfaction with one's physical appearance. Some persons may even feel "pressured" to undergo surgery themselves. Even if this were to be proven empirically, society may – and

[1] Art. 1 European Charter of Fundamental Rights (ECFR); Art. 1 I German Constitution; Art. 3 European Convention on Human Rights (ECHR).

should – criticize the superficial values expressed in cosmetic surgery but not ban it. Self-determination over one's body finds limits in extreme cases (such as the amputation of healthy limbs in Body Identity Disorder).[2]

At least a similar scope of self-determination has to apply to the mind. Bodily self-determination is not an idea inherently bound to the physical part of persons, but rather a manifestation of the general principle of far-ranging autonomy in primarily self-regarding matters, itself a pre-condition for an effective and full enjoyment of many other, more specific human rights. From these broad observations, the presumption of liberty (to take NE) follows. Concededly, in a strict sense, the presumption of liberty is not a moral, but a legal or political argument. Legal permissions for actions do not imply that they are ethically advisable; they do not provide orientation on *what* one ought to do, only *who* should have the power to decide. And when it comes to the mind, it has to be the affected person herself. Surely, presumptions can be rebutted and liberties restricted if opposing interests are stronger. This depends on the weight assigned to cognitive liberty and where the boundaries of the individual sphere are drawn in respect to the mind. These are the novel and not-yet-fully-addressed challenges for neurolaw. Moreover, there is another limit to be observed by democratic majority rule: Restrictions have to be based on (somewhat) *neutral* reasons.

19.2.2 State Neutrality and Authenticity

The principle of state neutrality roughly commits governments to neither favor nor discriminate against particular worldviews, at least in regard to their different conceptions of a good life (Rawls 2005, 191). The finer details of neutrality are subject to ongoing debates; accounts differ in their orientation on the effect of measures or its justification and more generally on the role of the state as such. Some, especially European, states traditionally take a more pro-active stance towards fostering social values while others, e.g. the US, have traditionally lower state involvement in matters of a good life. Prime examples of the divergent views within Europe can be found in court judgments on religious head-scarves or crucifixes, which demonstrate that legal orders do not adhere to a firm neutrality doctrine.[3] Sometimes it seems as if the neutrality thesis has gathered more attention in political philosophy than in constitutional practice. Nonetheless, its main claim

[2] Some peculiarities of legal provisions concerning the body should be noted. Feminists point to the fact that as soon as social interests are at stake, the uniquely personal body becomes highly political, e.g. restrictions on abortion, prostitution, organ selling, surrogate motherhood (cf. Fabre 2006). These limits to self-determination are probably best understood as (arguably too restrictive) dignity-based attempts to not commodify the most intimate aspects of persons.

[3] See the European Courts of Human Rights (ECtHR) recent decision allowing crucifixes in Italian Schools (Lautsi v. Italy; App. 30814/06) compared to the ban by the German Constitutional Court (BVerfGE Vol. 93, 1).

can hardly be pretermitted. While states can en- or discourage citizens' conduct through a variety of measures, if freedoms are curbed by legal, i.e. binding-for-all, prohibitions, it is plausible to demand that the interests and values justifying the prohibition should, in principle, be acceptable to everyone affected by it.

Some central arguments in the enhancement-debate evoke the suspicion of not being neutral in this sense. The controversy over authenticity, for instance, is caught between two opposing poles (Parens 2005). On the one side, essentialist conceptions imbued with ideas of a pre-given, rather static "true self," promote self-discovery via an introspective journey and reject any artificial alterations of "who one really is." On the other side of the spectrum, existentialist views deny any such pre-given structures since "existence precedes essence." Without a predestined purpose to be found in an inner essence, persons have to actively develop and shape their personality, creating, modeling and choosing how they want to be. For this, NEs could be valuable tools. Both views are reasonable, yet mutually irreconcilable (Bublitz and Merkel 2009). After all, how to argue about authenticity e.g. vis-à-vis a Buddhist denying the existence of selves (also see Metzinger 2003)? In cases like this, governmental regulations should not be grounded on one particular conception.

The same suspicion applies to the closely-related argument that enhancements express an "improper disposition toward the naturally given world: the failure to properly appreciate and respect the 'giftedness' of the world" (President's Council on Bioethics 2003, 288; cf. Sandel 2007, who, of course, is skeptical about neutrality). In this view, artificial alterations of one's self do not only violate obligations vis-à-vis oneself, but also against nature or a divine creator. However, I may (and do) personally feel indebted to some entities, say, my parents, ancestors or teachers, but simply fail to feel indebted to a creator. Even though Sandel (2007, 93) holds that reverence for giftedness does not require a giver, an entity indebted to, it is hard to see where the reverence should come from if one does not share it intuitively. And with regard to nature, it would be equally (im-)plausible to revere the powers of evolution or the Quantum Universe and hence dynamic and transformative processes. Nowhere, nature is aptly characterized as the mere conservation of the status-quo. At any rate, the neutrality doctrine speaks against the use of state power to impose a lifestyle of giftedness on those who find it incomprehensible while, of course, no one should be restrained from following such stronger personal moral convictions. On the other hand, the neutrality constraint should not set impossibly burdensome standards on state action where decisions have to be made and cannot, by the nature of their subject matter, but favor one side (Dees 2010, 54).

In a sense, the gist of the foregoing is that between neuroethics and neurolaw stands political philosophy. Anyone making policy recommendations or calling for legal regulations of NE has to acknowledge that passing bills and enforcing regulations is only possible within the framework of the legal order, a cornerstone of which is the presumption of liberty. And as neutrality – in whichever exact sense – is a plausible constraint of majority power, a double onus is placed upon prohibitive proposals: They need to demonstrate that protected interests are substantial enough to rebut the presumption of liberty and provide additional arguments for why their view should be binding for all.

19.3 Toward a Legal Concept of Cognitive Liberty

Now let us turn to genuine legal considerations.[4] Cognitive liberty or a right to mental self-determination guarantees individuals sovereignty over their minds. As said, such a right is not enshrined in constitutions, human-rights treaties or legal textbooks. To jurists content with describing positive law, cognitive liberty does not have much of an appeal. Currently enacted drug-regulations might even appear to refute the thesis of cognitive liberty as a fundamental principle of law. Yet, as Kant once remarked, a "merely empirical doctrine of right is a head that may be beautiful to look at, but unfortunately it has no brain" (Kant 1797, 230). In a sense, brainlessness also fittingly describes the current state of positive law. While legal orders have detailed rules over permissible conduct with bodies, there are hardly any criteria for permissible ways of interfering with brains and minds. Legal principles pertaining to the body cannot simply be transferred to the mind (or the brain): We do know, for instance, what constitutes illegitimate injury to other bodies; whereas, it is quite unclear what constitutes illegitimate *mental* harm – don't we hurt each other all the time? Legal norms relating to mental injuries are often scattered and incoherent, and, at any rate, cannot be equal to those relating to bodily injury. Also, consider manipulative interferences. While persons seek to influence and manipulate each other in almost every aspect of social life, from family matters to public affairs, there seems to be a qualitative difference between these ordinary influences and e.g. covertly administering psychoactive substances. Traditional legal categories such as lying and deception are insufficient to capture the latter kind of manipulations on the level of synapses and neurotransmitters (Bublitz and Merkel 2012). Therefore, and without presupposing an ontological mind-brain dualism (at least, of a stronger kind), legal protection of the mind cannot be identical to the protection of the body, but requires distinct and yet-to-be-worked-out criteria. Elsewhere, I have suggested that some jurisdictions should even consider introducing a criminal offence penalizing grievous interventions into other minds (Bublitz and Merkel 2012). The point is this: the lack of a theoretical framework of negative interventions into *other* minds is entwined with the lack of considerations on positive, self-determined alterations of *one's own* mind. Normatively, both are two sides of the same coin: Cognitive liberty.

For many reasons and in many ways, the mind is still a *terra incognita* for the law. In pre-neuroscience days there neither seemed to be a practical necessity for mind-protecting norms, nor were there any ways to meaningfully incorporate what were perceived as intangible, immaterial and inviolable mental states into legal doctrines. This has changed, and legal thinking should change accordingly.

Furthermore, the legal premise of free will seems to obstruct clear thinking about cognitive liberty and mental self-determination. Lawyers often entertain an overly simplistic understanding of free will. It is presupposed that persons have free will (in whichever exact sense), or at least, that the law has to treat persons as if they had free

[4] As national legal systems differ, the following remarks are rather general legal observations.

will, so that, in the eyes of the law, persons are quite ideally self-controlled agents. This picture does not leave much room for deficient self-determination over mental phenomena. If the law were to acknowledge the vulnerabilities and manipulability of the mind, it might be suspected of contradicting its own premises. Although rarely made explicit, background reasoning along these lines seems to cloud the view on a right to mental self-determination.

But on a closer look, the assumed tension between these legal premises vanishes. On the contrary, the fact that the law presumes that persons possess quite strong mental powers even supports calls for a right to cognitive liberty. In a nutshell: If the law treats persons as self-determined over actions and antecedent mental states by ascribing to them mental powers which, in reality, they may only rarely have, and if it holds them accountable for consequences of mind-states (in criminal and contract law, "meeting of the minds") *as if* they had free will, then, as a corollary, it has to grant them the legal powers of self-determination. Responsibility entails self-determination. Cognitive liberty is, in a way, the right to free will, protecting the conditions of possibilities of "free" actions and therewith of blame and retribution. In light of this, cognitive liberty is not merely a political claim that one may favor or reject. Rather, it is an implicit assumption of any legal order based on individual self-determination and responsibility.

19.3.1 The Notion of a Legal Subject

This thesis finds support in another line of reasoning. The very first step in setting up legal orders is to define the entities constituting it; namely, the legal subjects entering into a social contract and a state of rule of law. In a subsequent step, the content of the contract (i.e. the rights and obligations these subjects owe to each other) can be deliberated upon. In the prior "original position," the body of a person is considered to "belong" to her; bodily self-determination is not a right to be assigned in the course of negotiation but assumed from the outset. But even more constitutive of a subject, *cogito ergo sum*, is her mind. It is not just one aspect among many, but, arguably, it is what defines subjects, and hence, legal subjects. In fact, it is hard to conceive any conception of a legal subject in which the mind and mental capacities (e.g. acting from reasons, deliberation) are not among its necessary constitutive conditions.[5] Thus, I submit, the claim "my mind is mine" is not based on property rights nor on a legally established relation of ownership between entities, not subject to distribution of *meum* and *tuum*, but rather the point from which any legal order originates, intrinsic to the very notion of legal subjects. The "innate" – not acquired – right of everyone is the right to one's person, to make use of one's mental and bodily powers and to remain free from interferences.

[5]To include non-conscious humans such as *nascituri*, requirements may be lowered to potentiality for mental processes. Also, acceptance of corporate legal personhood does not necessarily refute the above claim, but I must leave this issue aside here.

These arguments demonstrate that cognitive liberty is deeply anchored in the foundations of the law. Unfortunately, legal thinking has never thoroughly explored its meaning. The term cognitive liberty has only recently been put forward by US legal scholars and civil rights activists (e.g. Boire 2000; Sententia 2004; Blitz 2010).[6] As Boire correctly observes:

> The right to control one's own consciousness is the quintessence of freedom. If freedom is to mean anything, it must mean that each person has an inviolable right to think for him or herself. It must mean, at a minimum, that each person is free to direct one's own consciousness; one's own underlying mental processes, and one's beliefs, opinions, and worldview. This is self-evident and axiomatic. (Boire 2000, 8).

The idea behind cognitive liberty, however, is anything but new and can be found in the works of some of the intellectual "founding fathers" of modern constitutional theory, Kant and Mill.

19.3.2 Historical Traces of Cognitive Liberty

19.3.2.1 Kant's Doctrine of Right

In Kant's doctrine of right, the distinction between internal and external actions plays a pivotal role. To him, the concept of right "has to do, first, only with the external and indeed practical relation of one person to another, insofar as their actions, as deeds, can have (direct or indirect) influence on each other." Juridical laws are those "directed merely to external actions and their conformity to law" (Kant 1797, 230, 214). Kant restricts the purview of the law to the regulation of actions in the external world. In his view, the law's function is to ensure and enforce equal freedoms of everyone, i.e. independence from being subjected to other people's choices. Different freedoms can only collide with each other in the real world, where actions of one come in practical conflict with those of others. Mediating this conflict legitimizes law. Therefore, legal obligations can proscribe external conduct, but as freedoms of others are not constrained by events internal to agents (mind-states), legal coercion to modify them is never justified. In this view, mental duties are *ultra vires*, outside of the legitimate mandate of the law, as long as a person's outward behavior conforms to the law (Ripstein 2009; Kersting 2007, 83). Kant never clearly laid out where the boundaries between internal and external actions run, an issue still being discussed today (von der Pfordten 2007). Yet, by positing the mental as the legally private realm and the external as the public sphere, Kant formulates a key ingredient of cognitive liberty, severely limiting state powers over minds.

---

[6] See the Journal of Cognitive Liberties at http://www.cognitiveliberty.org, particularly "On Cognitive Liberty I–IV", to which this chapter is indebted.

19.3.2.2 Mill: On Liberty

In his "On Liberty", Mill writes:

> [T]here is a sphere of action in which society, as distinguished from the individual, has, if any, only an indirect interest; comprehending all that portion of a person's life and conduct which affects only himself, or, if it also affects others, only with their [...] consent. When I say only himself, I mean directly, and in the first instance: for whatever affects himself, may affect others through himself [...]. [T]he appropriate region of human liberty [...] comprises, first, *the inward domain of consciousness*; demanding liberty of conscience, in the most comprehensive sense; liberty of thought and feeling; absolute freedom of opinion and sentiment on all subjects [...]. Secondly, the principle requires liberty of tastes and pursuits; of framing the plan of our life to suit our own character; of doing as we like, subject to such consequences as may follow; without impediment from our fellow-creatures, so long as what we do does not harm them even though they should think our conduct foolish, perverse, or wrong. (Mill 1859, 82–83)

In view of the states' interest in the "whole bodily and mental discipline of every one of its citizens" (Mill 1859, 81), Mill forcefully argues that any notion of liberty implies restrictions of governmental powers in self-regarding domains, first and foremost, in respect to body and mind over which the individual has to be sovereign. Another interesting suggestion is that even other-affecting actions should be considered self-regarding if they affect others only "through" the individual herself – an idea we will return to.[7]

Generally, one does not have to subscribe to Kantian ideas or the Millian harm-principle – in fact, the following argument allows for greater restriction of cognitive liberty than both might have approved of – in order to acknowledge the main point: The idea of strictly limited state powers in matters of the mind is conceived as a prime principle in the founding age of modern democracies, reinforcing my claim that cognitive liberty has been a neglected precondition of the legal order. Furthermore, traces of cognitive liberty can be found in today's positive law.

19.4 Rights in the Proximity of Cognitive Liberty

Let us briefly take a look at some rights in the proximity of cognitive liberty. Of course, as legal systems differ, rights accepted in one jurisdiction may be absent in another. My focus here is on European Human Rights and German Constitutional law.

---

[7]Interestingly, "On Liberty" was written during a time in which alcohol was prohibited in some parts of the UK and the US (1859, 151; Boire 2003). Alcohol, the most widespread (social and communicative) enhancer illustrates that persons have always had an interest in changing their minds, and despite all the problems it causes, a new prohibition is unthinkable in the western world.

19.4.1 Freedom of Thought

"Thought is free" is not only the main line of a famous German political folk-song, but also a fundamental legal principle. Freedom of thought is one of the strongest existing rights, enshrined in every human rights treaty,[8] but not explicitly enumerated in most (European) national constitutions.[9] Nevertheless, the European Court of Human Rights (ECtHR) and the German Constitutional Court (BVerfG) have repeatedly proclaimed its significance for any democratic state.[10] Freedom of thought is an absolute right, i.e. there are no limitation clauses allowing restrictions. Whatever falls within the ambit of freedom of thought is hence off-limits for state regulations.[11]

Its theoretical importance, however, is contrasted by its practical insignificance. There are no court cases defining meaning, scope and limits of this fundamental freedom (Blitz 2010).[12] Not even the outspoken and critical legal commentaries define its contours in more detail. Most agree that freedom of thought protects the *forum internum*, understood as a person's inner sphere in which opinions and thoughts are formed and revised (in contrast to the outward manifestation of beliefs in the *forum externum* – protected by freedom of speech). Commonly cited violations of freedom of thought are practices such as "brainwashing" or "indoctrination" (Vermeulen 2006, 851), but these are quite vague notions themselves (Taylor 2004). Even in philosophy, the very discipline of free thought, the idea has received little and perhaps insufficient attention (at least in comparison to free will; Pettit and Smith 1996).

In light of its absolute (unrestrictable) nature, freedom of thought has to be construed narrowly, but, I suggest, not void of any practical application. The right has to guarantee basic capacities required for performing mental acts such as thinking, rational reflection or revision of arguments and has to protect against manipulations. Therefore, it has to encompass the brain processes that underlie thinking and decision-making, including their modulation to both detrimental and beneficial effects. However, while negative interferences with other persons' thinking processes may violate their free thinking, it does not follow that banning tools to enhance one's own thinking does likewise. After all, this might imply that persons with ordinary cognitive capacities cannot think freely, a misguided contention. So, freedom of thought seems to be interfered with only if capacities fall below a threshold. Accordingly, the right to free thought may e.g. mandate states

---

[8]Art. 9 ECHR; Art. 10 ECFR, Art. 18 Universal Declaration of Human Rights (UDHR).

[9]The US Supreme Court apparently referred to freedom of thought in some decision, yet it is not recognized as part of the 1st amendment protection in the US (cf. Blitz 2010).

[10]ECtHR: Kokkinakis v. Greece (App. 14307/88), 25.05.1993, § 31; Decisions of the German Constitutional Court (BVerfGE) Vol. 80, 367 (381 – dissenting vote).

[11]E.g. UN General Comment No. 22, 1993: Art. 18 UDHR does "not permit any violation whatsoever on the freedom of thought."

[12]I have yet to find one European case in which freedom of thought played a decisive role.

to provide children with school education, but not oblige them to provide access to NE. Yet, these are tentative first approximations. Formulating a modern-day concept of freedom of thought, informed by empirical sciences and philosophy of mind, is an open task for the law. Its ambiguities notwithstanding, freedom of thought and the protection of the *forum internum* are reminiscent of Kant's distinction between internal and external actions and underscore that the "inward domain of consciousness" is a highly sensitive area for legal regulations.

19.4.2 Personality and Privacy Rights

Another set of rights relating to legal regulations of NE are personality or privacy rights. Art. 2.1 of the German Constitution guarantees:

> Everyone shall have the right to the free development of their personality provided that they do not interfere with the rights of others [...] or the public order.

Art. 22 of the United Nations' Universal Declaration of Human Rights (UDHR) stipulates:

> [E]veryone [...] has the right to social security and is entitled to realization [...] of the economic, social and cultural rights indispensable for his dignity and the free development of his personality.

Other constitutions and human right treaties guarantee "privacy" or "private life" (e.g. Art. 8 ECHR; Art. 7 ECFR) protecting a space of seclusion against unwanted intrusions – "be that the head or the home" (Marshall 2009, 3). By drawing spatial limits, privacy rights exemplify the law's approach to delineating spheres of individual and public concern in quite a literal way. Privacy rights have been steadily expanded and often overlap with personality rights. The judicature of the ECtHR has evolved from protecting privacy to personal autonomy (Marshall, ibid). In the words of the Court:

> For numerous [...] authors the right to respect for private life is the right to privacy, to live protected from publicity [...]. In the opinion of the Court, however, it does not end there. It comprises also the [...] development and fulfillment of one's personality.[13]

Notably, the right even guarantees the development of one's personality in *social interaction*, which may, at first glance, appear as the very opposite of privacy. Privacy is particularly visible in US and French law, and legal scholars from these traditions have proposed to further develop the concept of privacy to offer protection against mind-reading and -interventions ("mental privacy", cf. Halliburton 2007; Tovino 2007). Here, I shall concentrate on personality rights in German law.

---

[13]ECtHR: X v. Iceland (App. 6825/75), 1976.

Interestingly, the right to develop one's personality bears some relation to the ethical debate on authenticity. German Constitutional Law distinguishes between the protection of a person's "core personality" on the one hand and actions contributing to its development on the other. Whereas the core personality is inviolable and enjoys absolute protection against state-interventions, mere personality- forming actions enjoy weaker protection. By defining the scope of the personality core, legal reasoning resembles the diverging ethical views on what constitute authentic personalities. In legal practice, only few aspects of the person are considered part of the core of the personality (e.g. gender and sex issues) while many interventions (e.g. coerced medication) are regularly not regarded as core-interferences. Presumably, the underlying picture of what a personality core consists of is permeated with normative and essentialist assumptions that, unfortunately, have never been articulated openly. Similar developments can be observed in the judicature of the ECtHR (Marshall 2009). Here, legal reasoning could be informed by the ethical discourse.

Conversely, the legal situation points to an underappreciated *normative ambivalence* of authenticity: While it is hard to draw on authenticity to formulate an objection against a person's deliberate transformations of herself without presupposing "true selves," it can meaningfully be used against alterations of other persons' personalities. "Do not severely interfere with another person's character" is a normative claim without much metaphysical baggage, and it is precisely what personality (and privacy) rights guarantee. They express the principle that others (and the state) regularly do not have a legally enforceable interest that an individual's character should be a certain way or exhibit particular traits and, consequently, bar state-interference with existing personality structures.

By contrast, actions by individuals to actively develop and transform their personalities are protected to a considerably lesser degree. In German law, this freedom is synonymous with the basic and easily limitable right to free action. For example, the ban on cannabis has been tested against this weak right only. In a constitutional challenge against its prohibition, the German Constitutional Court held that cannabis consumption falls under the right to develop one's personality, but only in this weaker form, hence it is limitable by reasonable public interests.[14] If that judgment were a precedent for other mind-altering substances, the level of protection for NE is marginal as interferences with free action are easily justifiable. As a result, personality transformations via NE would enjoy the same degree of protection as most other, even mundane activities such as "horseback riding in the woods" or "feeding pigeons."[15] While feeding animals is surely beneficial to the development of one's personality, (German) courts have yet failed to appreciate the difference between these trivial actions and practices which aim primarily at individuals' inner aspects and alter personalities in a much more straightforward way.

---

[14]Decisions of the German Constitutional Court (BVerfGE) Vol. 90, 145.

[15]These are two well-known German cases invoking constitutional protection for trivial activities (BVerfGE Vol. 54, 143; Vol. 80, 137).

The reason for the weak protection of freedom of action is that actions in the external world, just as Kant pointed out, interfere with freedoms of others or public interests in innumerous ways and have to be restricted all the time. However, this does not seem to be the case with all personality-transforming actions; some, especially the consumption of NE, are different. The enhancing action, e.g. ingesting a pill, does not *per se* interfere with reasonable interests of others. Rather, it is the *effects* of the substance on the person that raise concerns. However, the altered personality as such does enjoy full and strong legal protection, irrespective of its genesis. A thought experiment: If there were permanent enhancements, say magic pills turning users into cognitive super-humans, the state would not have the competency to order super-humans to reverse their transformations. Just as any other personality, an enhanced personality is off-limits for state interventions. So here is the catch: If the result of an action is nothing that others can (legally) complain about, and the action itself does not interfere with others' legally-protected interests, it is hard to see where the legitimacy of the state's power to prohibit the action comes from. At least, the usually adduced reasons for banning actions do not capture the normative peculiarities of self-transformations. Therefore, the threshold for restricting such actions has to be considerably higher than for those directly infringing on the freedoms of others.

I suspect this is what Mill hinted at by referring to actions affecting others only "through" the individual herself. Nevertheless, *pace* Mill, I do not think that such actions can be considered entirely self-regarding. As said, this requires normative judgments about the limits of state power. Given the fact that neurotechnologies may not only transform the psyche of the individual but also society at large, some regulations seem justifiable.

19.4.3 Mental Integrity and the Treatment/Enhancement Distinction

Finally, let me briefly note another currently existing right: mental integrity, as protected e.g. by Art. 8 ECHR and Art. 3 ECFR. Again, meaning and scope of the right to respect for mental integrity are unclear. Presumably, it is meant as a right to mental health; some commentators suggest that it could imply a right to mental self-determination (Höfling 2006), which, in my understanding, *is* a right to cognitive liberty. At the very least, these European provisions show that the mind has to enjoy strong human-rights protection.

The scope of the right to mental health relates to the often challenged distinction between treatment and enhancement. While there are certainly grey areas between illness and health, open to cultural and social variances, the claim that *any* distinction is arbitrarily drawn is hard to accept. At least, there is quite a significant legal difference between health and illness: No state can deny patients' access to necessary (and effective) forms of therapy.[16] This does not, however, imply corresponding obligations of states to provide patients with medical care since governments have discretion over – and limited financial resources to meet – such positive obligations. Morally, however, if there is any positive obligation of solidarity towards each other, it is support in times of tragedy and illness. Thus, patients have a strong legal right against being barred access to existing treatments, which is supported by a moral obligation of the state to provide for necessary therapeutic resources. Similarly, the right to health outweighs e.g. concerns over fairness: states cannot deny persons a form of therapy just because others would not be able to afford it. Obviously, these normative considerations do not equally apply to enhancements, and, therefore, despite gray areas and factual problems in delineating treatment from enhancement, their difference is relevant. Another consequence is that public funds and resources should be *prioritized* for treatment and research.

19.4.4 Gaps in Current Law

To sum up: There are some rights that are closely related to the idea of cognitive liberty, yet in their current state, they fail to adequately cover the peculiarities of mind-interventions. There are no systematic approaches to define permissible and impermissible ways to change minds, so that legal theory has yet to develop more fine-grained doctrines dealing with the mind and mental states. The ramifications of the theoretical underappreciation of the mind reverberate in several areas of the law (e.g. tort-law damages for mental injuries are highly controversial and may be reformed in light of brain imaging evidence; Grey 2011). Nonetheless, one cannot interpret the fundamental ideas behind liberal constitution other than a guarantee of protection of the essential elements of a person. With respect to novel technologies, the law has to buttress the individual against invasions of her inner domain, even if interventions fall short of violating the "core" of a personality or obstructing free thinking; likewise, the law has to ensure self-determination over one's own mind, at least *prima facie*. Limits are, of course, subject to discussion. What is missing then is a mind-protecting right and principled criteria over means of changing minds from which more concrete regulations can be deduced. Logically, developing such a right is a step prior to defining regulations of particular NE substances. Because of this, pointing to existing drug-prohibitions conveys the false impression that self-determination over one's mind enjoys weak legal protection only. Technically, the

[16]Of course, states can regulate markets to avoid exploitation of patients, secure good-practices, assess risk-benefits, etc., but they cannot, in my view, outlaw effective therapies. Therefore, the ideologically motivated ban on the use of psychedelics in (psycho-)therapy has to be lifted, provided substances are effective and relatively safe (currently, the first LSD study for more than 30 year. is conducted by Gassner, www.maps.org/research; regarding MDMA see Mithoefer et al. 2011).

right to cognitive liberty could be construed either by blending the just mentioned ill-defined existing rights into one novel unified right or by further developing existing provisions guided by the idea of mental freedom. Establishing such a right raises a multitude of questions for various areas of the law. Here is a rough sketch of its possible scope and limits in regard to regulations of NE.

19.5 Cognitive Liberty

19.5.1 Scope of the Right

The protection of an individual's self-determination over her mind should comprise the entire *forum internum*, i.e. all mental states or capacities and therewith cognitive as well as emotional (potentially even unconscious) phenomena. Surely, speaking of mental self-determination or freedom is always entangled with metaphysical assumptions and intriguing questions of what freedom might mean in regard to both brains made up of neurons and governed by natural laws, and mental phenomena supposedly following hard-to-describe psychological dynamics.[17] In the absence of a firmer understanding of empirical and metaphysical aspects of these matters of the mind, defining contours of its legal protection could be conceived as futile. But it is not. Most importantly, one should recall the function of rights. Unlike in the free will debate, a right of mental self-determination is *not* so much concerned with the (perhaps deterministic) relation between mind and brain but between different persons. Rights primarily regulate inter-personal relations, not intra-personal psychological conditions and external, not internal impediments. Thus, mental self-determination is to be understood in contrast to heteronomy, neither requiring nor presupposing self-determination in a strong sense.

Just as any other right, cognitive liberty has several dimensions: First, the *liberty to* change one's mind, permitting to attain or discard any mental state and exercise one's mental capabilities. It also comprises the choice of whether and by which means to change one's mind, i.e. it entails a *prima facie* permission to use mind-altering tools like NEs as well as to refuse them. Secondly, it protects against interventions into other minds to preserve mental *integrity*. Moreover, the right protects against other interferences with mental self-determination, most notably

[17] A right to mental self-determination does not rely on a particular view on the mind-brain relationship. While dualists won't object to mind-brain distinctions, reductionists may agree with the protection of physical processes *as identified* by their (reducible) mental properties. All that needs to be accepted is that protection of mind- (or brain-)states cannot follow the same normative rules as the protection of the integrity of other parts of the body. Unlike the latter, the mind (and its correlative neuronal processes) is highly dynamic; negative changes in mental phenomena are hardly describable as detrimental on the physical level. An analogy might be drawn to data-protecting provisions. Erasing a computer's hard disk does not damage the disk itself but the (supervening) information, and hence, stand-alone data-protecting provisions are needed.

"mental duties" such as prohibitions of having particular mind-states (e.g. Orwellian thought crimes).[18] Therefore, both prohibitions of NE as well as mandatory NE constitute interferences with cognitive liberty.

These interrelated but not identical dimensions pertain primarily to the negative freedom from interferences and stipulate duties on others to refrain from mind-changing interventions.[19] But thirdly, there is even a positive side that obliges states to *promote* cognitive liberty. Constitutions vary widely in the extent of positive duties and usually grant governments wide margins of appreciation. However, in view of the absolute protection of freedom of thought and the core personality, states might be obliged to provide necessary resources to persons lacking a bare minimum of necessary capacities, e.g. patients in minimally conscious states or suffering from Alzheimer's disease. But I shall leave these more specific questions aside here and, instead, try to convince NE opponents why they too should recognize a right to cognitive liberty.

19.5.2 Why Critics Should Embrace Cognitive Liberty

Critics worried about mounting social pressure ground their case against NE on precisely the interest guaranteed by cognitive liberty. By calling for protection against soft 'coercion,' they recognize that there is a sphere worthy and in need of legal protection: The mind. Let us consider this argument more formally. Critics will endorse some weak claims without hesitation:

1. No one shall expose other people to psychoactive substances without consent. For instance, secretly spraying Oxytocin, a neuropeptid that modulates interesting behavioral properties (Fehr et al. 2005), or pouring NE into someone else's coffee should be illegitimate. Therefore, critics will concur with the idea that there should be an obligation toward everyone to refrain from intervening into other people's minds in these ways.
2. This obligation is of legal nature, i.e. if someone intervenes into another person's mind, he ought to be stopped, not only by appealing to his moral standards, but

[18]Whether (and to which extent) enhancements raise the "standards of reasonable care" is currently being discussed (Vincent 2012; Chap. 21 by Danaher, this volume). As standards of care are not empirical facts but normative judgments, they have to observe the right to cognitive liberty. Regularly, CL should prohibit stipulating legal expectations that others (e.g. pilots) take NE in order to discharge their duties (at least, without consent). Greater factual powers do not automatically lead to greater normative responsibilities. Exceptions might apply in severe, life-threatening circumstances (e.g. military).

[19]For the sake of argument, it is assumed that fundamental rights apply not only to the state-citizen, but also the citizen-citizen relationship. Positive and negative liberties in this sense do not precisely match Isaiah Berlin's famous distinction.

also, if necessary, by resorting to coercive state measures. Thus, there should be a legally enforceable obligation to refrain from exposing others to NE.[20]

3. At least for many legal theorists, rights and duties are correlatives, i.e. by definition, rights are entities which create duties for others, and legal duties arise if (and arguably, only if) someone else has a right.[21] In our present case, the right is a claim right, conferring to the right-holder a legally enforceable claim *erga omnes*, against everyone, to respect the protected interest (i.e. to not intervene into her mind).
4. Closely related is another claim that critics will support: No one should have a claim against another person that the latter enhances herself (i.e. a legal obligation to enhance oneself), or: No claim of anyone against another person to NE. If no one has a claim against a person to X, then the latter is not under any obligation to X, i.e. he has a liberty to not-X. Thus: No one has to enhance herself.

   While the former points should not be controversial, the following might: Who is the holder of the right and what interests does it protect?
5. The right can be held by either the individual or the state. The difference becomes obvious when asking who should be competent to consent to infringements. Should the affected individual or the legal community have the power to grant permission for mind-interventions? Particularly those concerned with societal pressures should argue for strengthening individuals' (legal) powers to protect them against (overwhelming) external forces. Even if, at the moment, the majority of people and, by extension (if representation works) parliament, might object to NE, this is anything but a stable basis to dispel worries. For one, consequentialist thinking may well prevail over current skepticism if NEs have tolerable risk-profiles and promise economic advantages. Additionally, even benevolent states seek to stabilize society and further important interests which may, in their view, outweigh objections by affected persons (e.g. "moral enhancements" inducing pro-social behavior, cf. Douglas 2008; Chap. 20 by Shaw, this volume). Placing the power of regulating minds, consciousness and neurotechnologies in the hands of the state instead of the individual creates, in the best case, a delusive sense of security and, in the worst, a carte-blanche for governmental mind-control. Therefore, it has to be a right of the individual against other persons and the state.
6. Hitherto we have an individual's right against everyone to not interfere with her mind and to not make any regulations commanding her to do so. Now we need

[20]For an introduction to a theory of rights see Thomson (1990).

[21]Concededly, there may be imperfect duties, i.e. duties without correlating rights, e.g. those owed to children, animals or future generations. However, the latter are arguably *moral* duties only, and children can be considered as fully right-bearing persons with their legal guardian(s) exercising their rights on their behalf.

to take a closer look at the content of the right. If it is an individual's right, then the right-holder can relinquish it, i.e. consent to infringements and thereby *allow* others to interfere with his mind. And if others are allowed to interfere with her mind, the right-holder herself has to have the same permission *a fortiori*. In this formulation, the right is a liberty to mental self-determination.

However, this is precisely the liberty unwillingly granted by critics. Thus, they have to put forward a different formulation of the right, encompassing only the *integrity* but not the self-determination dimension. They would need to argue that even right-holders cannot relinquish and consent to infringements of this right, without arguing that the right is not held by the individual (see point 5).

To their assistance it should be noted that from prohibitions on others to bring about a state of affairs (X), one cannot straightforwardly deduce a respective permission of the right-holder to X. For instance, from the prohibition on others to kill you, a liberty to commit suicide does not necessarily follow. A legal system prohibiting any form of ending human life, including one's own, is logically conceivable. However, rights which do not entitle the right-holder to waive them are a special kind of rights, *inalienable* rights. It is questioned whether these rights are rights or liberties in a strict sense as they only stipulate prohibitions without correlating permissions of right-holders (see e.g. Harel 2004). Sometimes courts invoke such rights with respect to supreme interests such as life or human dignity. But on closer examination, it seems that what is really protected in these cases is not the interest of an individual, but something like a collective taboo of taking life or violating dignity. If one accepts that the prohibition of unwanted mind-interventions protects primarily an individual's interest, this line of reasoning is blocked. The only remaining solution is to construe an inalienable duty of right-holders against themselves, but whether any such legal (not moral!) duties can be cogently established is highly controversial.

Moreover, even critics would concede that the mind-protecting norm for which they argue is at least sometimes waivable – when it comes to therapeutic mind-interventions. Thus, they have to allow for exceptions (and argue for their consistency). Ironically, to meet the argumentative threshold, the inalienability argument would commit critics to not only endorse cognitive liberty, but even more, to elevate it to the ranks of the strongest rights. It should have become apparent by now that whoever wants legal protection against NE does subscribe to the idea of cognitive liberty. And any formulation of such a right not entailing mental self-determination – albeit not logically impossible – faces serious obstacles. What one can argue about are its limits for paternalistic or social reasons.

19.6 Limits of Cognitive Liberty

19.6.1 Different Ways to Change One's Mind

The limits of cognitive liberty, it seems, can be drawn meaningfully only by distinguishing between means of changing minds. They fall into four normative distinct categories:

1. Mental actions – those mental actions purely directed at psychological changes – thinking, dreaming, remembering, memorizing, etc.
2. Bodily actions, from yoga to sports, in virtue of their mind-altering effects.
3. Tools and Substances, from drugs to books.
4. Interaction with the outside world, including other persons.

The reason for these distinctions is not that internal alterations (as 1 and 2) are intrinsically better or worse than external interventions (cf. Levy 2007), but that they impact others and the social sphere differently. By their very nature, actions of the last category involve interaction with the environment and potentially collide with interests of others. Thus, they have to be restrictable to a larger degree than internal alterations. After all, other persons cannot be legally coerced into interaction with someone just because it benefits the latter.[22] Current legal provisions (e.g. the right to develop one's personality) provide adequate protection for these activities. On the other end of the spectrum, mental activity (1) as such, as we have seen, never directly collides with legally-protected interests of others. If at all, it does so indirectly "through the individual" in Mill's sense. Thus, mental actions of this kind are essentially self-regarding and should enjoy strong protection (if restrictable at all). The same applies to bodily activity (2). If there is some truth to "embodied cognition", distinctions between mental and bodily activity are hard to draw. Moreover, research suggests that bodily activity may have some cognitive enhancing effects (Chap. 7 by Dietz, this volume). The only normatively relevant differences between (1) and (2) are environmental aspects. As long as the movement of the body in space does not collide with rights of others, bodily activities need to enjoy a protection similar to (1). Therefore, e.g. banning falun-gong, other meditative practices or sports in virtue of their mental effects seems illegitimate (paternalistic arguments notwithstanding). For present purposes, this leaves us with discussing the limits of (3). As said, treating (3) as similar to (4) obscures the facts that the transformative actions as such regularly do not interfere with the outside world and that their effects are normatively closer to (1) as they primarily change the inner world. Even with the concession that mind-changes via tools are not an exclusively self-regarding affair, they need to enjoy considerably stronger protection than (4).[23]

[22] As always, exceptions apply in special normative relations, e.g. parents-children.

[23] Blitz (2010) puts forward a different claim. Drawing on the extended mind thesis by Clark and Chalmers, he proposes that the protection of freedom of thought should be expanded to

19.6.2 Balancing Countervailing Rights

As with any non-absolute right, interferences with cognitive liberty can be justified by opposing rights or interests. When different rights collide, they have to be reconciled by balancing (for the German approach see Alexy 2003). Balancing rights is guided by the idea that countervailing interests should be realized to the greatest extent possible and freedoms should be curbed in the least invasive manner (principle of proportionality). In the end, this is a case-by-case decision based on a measure's positive and negative effects and the intensity of interferences. However, balancing partially depends on the abstract weight assigned to specific rights. Some are by default stronger than others, rendering them restrictable only if negative consequences to the latter are severe. In principle, cognitive liberty should be considered as one of the strongest rights. The different ways of changing minds allow for finer graduations: While mental actions (1) are hard, the use of mind-altering tools (3) is easier to restrict.

The intensity of the interference depends on various factors. Another abstract distinction can be drawn between outlawing specific mental states as such and banning the means to their attainment. The former would create mental duties proper, the legitimacy of which e.g. Kant would deny. At least, mental duties are extremely hard to justify. Regarding the ban of specific means, the intensity of the interference depends on the availability of other means to attain (sufficiently) similar states. If there aren't any (e.g. psychoactive substances yielding rare insights into subconscious phenomena), blocking the means might be tantamount to blocking the destination itself, which intensifies the interferences and calls for stronger justificatory reasons. Thus, the more peculiar the effects of substances, the higher the requirements for their ban. Additionally, the legal importance of the altered mental phenomena has to be assessed. An increased span of concentration may be useful, but less important than e.g. memory capacities (if items were otherwise lost). Eventually, any NE has to be assessed on its own. Now let's take a look at the other side of Justitia's scale: The countervailing rights potentially justifying infringements.

19.6.2.1 Paternalistic Limits

Most legal systems allow limiting rights for paternalistic reasons, i.e. restricting freedoms in order to safeguard individuals for their own good. Paternalism is a classical issue and the familiar arguments and restrictions apply to mental

"activity that is [. . .] the functional equivalent of thought" and therewith to computers, IPhones and other devices. However, this expansion eliminates the distinction between personality and property rights. While technical devices/data-storage need (and in fact, enjoy) legal protection, their protection is based on property rights. Even though machines might be functionally similar, freedom of thought can only be meaningfully construed in relation to the human mental realm.

self-determination as well. Society does not need to remain silent and turn a blind eye to people harming themselves and, hence, at least in the face of severe negative consequences, states can intervene. It is a different question whether criminal law is the right tool to enforce paternalism, i.e. whether self-harm can be penalized (cf. Feinberg 1986; Husak 1989; von Hirsch 2008). Current drug and substance-related offences (covering many potential NE substances) are often grounded in paternalistic reasoning. As long as other measures to prevent self-harm are available, however, criminalization contradicts the *ultima ratio* principle of punishment as a means of last resort. Also, if punishment is employed to avert self-harm, one should be careful that the cure does not become worse than the disease. The legitimacy of penalizing self-harm is much more problematic than current legal provisions suggest.

What may constitute severe self-harm with respect to NE? Strong side-effects of substances (like heroin) warrant restrictions. But if NEs live up to their promise, they have tolerable risk-profiles. Before substances are outlawed, other harm-reduction strategies need to be considered (e.g. restricted access under supervision by trained experts). An interesting theoretical question is whether persons lose their autonomy when acting from preferences induced by NE, i.e. whether they are responsible for their actions stemming from motives created through pills. If not, it could be argued that loss of autonomy constitutes harm in a normative sense and substances should be outlawed in virtue of this legal effect. However, usually enhancements do not undermine autonomy – rather, they will often increase it (Bublitz and Merkel 2009).

19.6.2.2 Limits: Common Good

Presumably the most challenging question is whether and which interests of the common good justify interferences with cognitive liberty. Especially transhumanists emphasize the social benefits of enhancements (Bostrom and Roache 2011). When NEs improve socially relevant mental traits (e.g. increasing IQ and productivity), the common good perspective may indeed speak in favor of their use. On occasion, e.g. in regard to delinquent behavioral dispositions, some sides of the political spectrum may even call for their mandatory use (e.g. pro-social enhancement of criminals, cf. Chap. 20 by Shaw, this volume). In other cases, however, what is in the interest of the individual may be detrimental for society. Mood enhancements, in particular, could alter the psychological foundations of culture and society. Huxley's *Brave New World* (1932) vividly depicts this dystopia: The mood enhancement Soma is free of health-risks but poison to numerous collectively held values. Consumers drug away their doubts and weaknesses, suppress their inner contradictions and likewise their ambitions; they lose their depths and difficulties out of which personalities develop. Criticism is replaced by complacency, sincerity by superficial happiness and the mental make-up of society is transformed at large. Instead of tackling problems at the social level, the origins of discontent are located and treated in the maladapted individual. What is cured are symptoms, not causes. However, reading *Brave New World* along these lines overlooks the context of an authoritarian regime that engineers society and citizens (even before conception). To demonstrate that Soma's negative effects are largely due to social context, Huxley presented a positive vision of psychedelics being used for enlightenment in his novel *Island* (1962) in which he anticipated some of the ideas more thoroughly formulated later in the 1960s. In this positive vision, mind-expanding drugs become tools for achieving a more peaceful and harmonious but by no means quietist society. In their own way, each novel demonstrates the huge influence substances can have on society, but also how pharmacological effects in turn depend on the social settings in which their use is embedded (Schermer 2007).

This relates to the interplay between technology and society in general. It would be naïve to assume that society and technology co-emerge, as it is often called, in a neutral fashion. As the philosopher Feenberg (2002, 3) remarks: "What human beings are and will become is decided in the shape of our tools no less than in the actions of statesmen and political movements." The interesting observation, then, is that such profound changes can be brought about by individuals' choices over technologies rather than by collective decisions. Neither the use of cars, computers and the World Wide Web nor their consequences have been approved of in a collective and democratic procedure. Slowly, yet powerfully, they crept into our lives and transformed them tremendously. In light of democratic ideals, this might constitute reason for concern. And NE, in particular, might be among those technologies many people would utilize for personal gain while not approving of their society-wide use. In such circumstances, the societal effects of technologies mandate democratic legislators to regulate their use. Nevertheless, enforcing societal values against the will of affected persons has to meet high standards. Detriments to society have to be severe, perhaps undermining the psychological roots of collectively valuable mental states such as guilt, empathy or solidarity. After all, society's "state of mind," the fabric from which social relations are woven, is barely understood, nor are the effects enhancers may have on it (Merkel 2007, 287). The difficulties with protection and promotion of socially desirable mental states can be illustrated by two currently discussed interferences with memory.

19.6.2.3 Blunting and Enhancing Memories

Suppose novel NEs enable soldiers to wash away their sins and pangs of remorse not by repentance, but by blocking the emotional side of recollection. This might be partly realizable in the future (Kolber 2006). There are but few things conceivably more horrendous and disgusting than human killing "machines," soldiers with a clean conscience. Perhaps this is as far as one can deviate from universally-shared conceptions of a moral being, reminiscent of the symptoms of psychopathy. In spite of this, if the killing is justified, e.g. in self-defense or a "just" war (if there is any), it is hard to see why agents should be legally obliged to face the mental consequences of their deeds. Perhaps there is a *moral* obligation to suffer from taking another's life regardless of its permissibility, but it is quite unclear whether a corresponding *legal* duty exists. How could states command soldiers to kill, yet deny them the means to come to terms with their actions in the name of a social expectation to suffer from killing? Should they bear the cost of trauma for something others have commanded? Soldiers may be deemed as having consented to mental injuries sustained in action by enlisting, but this consent can hardly take away their rights to mental health and cognitive liberty outside the battlefield. Obliging them to live through the negative mental consequences because *others* want to live in a society in which killing is accompanied by mental turmoil constitutes an act of securing collective values at the expense of the individual. Obliging culpable offenders, by contrast, to come to terms with their deeds by consciously grappling with and suffering from them can be understood as part of their sentence, as rehabilitation is among the prime peneological aims.

The same skepticism about public interests is warranted when they speak in favor of using enhancements. Klaming and Vedder (2010) have recently suggested that the common good perspective should play a stronger role in the enhancement debate and have exemplified their case by the use of memory enhancing technologies in eyewitnesses. As improving witnesses' capacities for recollection promotes public interests, they argue in favor of the use of transcranial magnetic stimulation (TMS) in police interrogations or courtroom procedures. Here, it becomes obvious that arguments based on societal benefits often stand in direct opposition to cognitive liberty. While the ordinary obligation of a witness to give testimony is a minor and justifiable infringement on her rights and although the efficacy of law enforcement is an important good in democratic states, intervening into minds to further police and state interests is quite another, much more invasive and dangerous measure – not so much because of side-effects, but because it grants governments access to the inner realm of persons for state purposes. The history of innumerous governmental attempts to change citizens' perceptions of the world throughout the ages provides a cautionary tale of what might happen if public interests are emphasized too strongly. Fortunately, under German law, the administration of "truth-sera" or the like in both suspects and witnesses are considered violations of human dignity.[24] And even if TMS turns out to be less harmful, calling for its use would signal embarking on a dangerous path.

Having said this, the vigorous appeals by some ethicists to enhance mankind in order to avert an ecological catastrophe and to tackle the great global injustices deserve attention (Persson and Savulescu 2011). Here, I can only note in passing that many such arguments exhibit a tendency to overemphasize the level of the individual and to downplay social and political conditions. It appears naïve to diagnose the cause – and respectively, the proper cure – of today's global problems in individuals' mind-sets. None of these problems seems to be due to a lack of human intelligence, nor should we hope for technological fixes for social problems. It seems quite obvious that the affluent countries have to, *inter alia*, drastically cut down overproduction and CO_2 emission and to abandon the ideological belief in incessant

[24]Cf. § 136a StPO (German Criminal Procedure Act).

economic growth (and recognize it as among the roots of the problem). The true obstacles to this, it seems, are undemocratic market economies and imbalances in power – not in brain chemistry.

19.6.2.4 Fairness

In public debates, fairness concerns are often raised. A fair and just society certainly ranks among the most foundational interests and may, in general, warrant intensive restrictions of liberties. Most fairness objections in the NE debate, however, suffer from unclear conceptions over what constitutes fairness with respect to mental capacities. This issue is beyond the scope of this chapter, but let me briefly remark that worries over NEs rendering a (by large) fair situation unfair overlook the fact that natural processes and social conditions do not distribute "mental wealth" and opportunities in accordance with any theory of justice (Savulescu 2006; Greely et al. 2008). Whether NE worsens or improves, this deficiency partially depends on empirical facts (efficacy, pricing). And of course, fairness concerns could be addressed by distributing NE in line with a theory of a fair distribution of mental capacities (Sandberg and Savulescu 2011).

The often-drawn analogy to sports also falls short of providing an adequate model of fairness. Notions of fairness in sports are inextricably linked to the "spirit of sports" which glorifies natural abilities and is founded on a system designed with the sole purpose of producing a hierarchy of winners and losers out of contestants. Sport is essentially the creation of competition for its own sake. Even though one may (unfortunately) describe some spheres of society in similar terms, the idea that sports may serve as a model of a fair and just society is to be strongly resisted. At best, some constellations are structurally and normatively comparable, e.g. exams in state universities. The right to equal treatment, at least as understood in German Constitutional Law, prohibits treating "unlike cases alike." Grading non- and enhanced students by the same standards would, arguably, violate equal treatment (Bublitz 2010). Yet, this is an exception because a legally-binding notion of equality exists for this particular case. For society at large, NE can only be assessed in reference to a general theory of justice. In capability approaches to justice (Robeyns 2011), NEs might fare well as they increase real opportunities despite potentially increasing inequalities.

19.6.2.5 Limits: Rights of Others

The strongest right to justify interferences with a person's right to cognitive liberty then is, perhaps surprisingly, the cognitive liberty of another, that is, other persons' rights to refuse NE. Quite likely, social and economic pressure will let those who refuse face severe negative consequences, and the main challenge may very well consist in realizing the demand made by Greely et al. (2008, 703): to protect unwilling persons from 'coercion.' In one of the first papers on the ethics of

NE, Farah et al. (2004, 423) asked whether it is legitimate to "den[y] people the freedom to practice a safe means of self-improvement, just to eliminate any negative consequences of the (freely taken) choice not to enhance." More concretely, the question is to which extent nonusers should be protected against overt and subtle pressure. Any answer will have to strike a balance between the liberties to use and refuse NE. As they are different dimensions of the same right, both are of the same abstract weight. The pressure on some to take NE may warrant curbing the liberty of those who freely take NE if the negative consequences are severe (e.g. side-effects). Contrary to a widespread misunderstanding, it is not the pressure alone that justifies restrictions. This becomes apparent by considering behavioral interventions for enhancement purposes. If meditation, working-memory training or other nonpharmacological enhancers were effective (Chap. 6 by Dresler, this volume; Dresler et al. 2013), persons availing themselves of these means may enjoy considerable advantages, which could potentially exert pressure on others. Still, banning working memory training or meditation seems inappropriate. Why? Because they do not have sufficiently severe negative consequences; hence, everyone can be normatively expected to make use of these techniques. To put it differently: The mere wish to not avail oneself of a technology and the resulting disadvantages cannot, by itself, justify its prohibition for everyone else. The same applies to pharmaceutical NEs: only if they have unacceptable side effects, social pressure is sufficient reason for curbing the liberty of those willing to take the risks. Thus, a threshold of acceptable risks is needed, eventually to be defined by the legislator.

One last point should be made: Increased cognitive capacities are, *other things being equal*, not a harm from which claims against others to abstain from NE could arise, or against which society should protect. Even if enhanced capabilities are not all-purpose goods and contravene some ideals of a good life (e.g. an anti-intellectual rural life), curbing others' liberty is not warranted. If there is no neutral measure as both permissions and prohibitions (dis-)favor particular conceptions of a good life, judgments have to be made. I suggest that the benefits of improved cognitive capacities are so evident that anyone pursuing other lifestyles will have to live with the consequences – becoming smarter is, by itself, not an unacceptable side-effect. By contrast, negative effects on personality, self-perception or on emotional capacities may justify restrictions. Subdued, unstable or dulled emotional capacities would be too high a price that other persons can be expected to pay for increased cognitive performance.

19.7 Conclusion

The foregoing is a legal framework in which regulations of NE have to be embedded. Surely, democratic governments have leeway in the assessment of goals and regulation of means. They can en- or discourage the use of NEs; provide or restrict access; support or impair research; define, in part, the realm of normal mental functioning, acceptable risks and the degree of permissible paternalism. These are political issues which have to be decided in light of ethical considerations over desirable mental states and the novel ways to alter them (Metzinger 2009). The law can only set outer limits and structural guidelines. Yet, it is important to recognize cognitive liberty as the basic freedom that restricts state-interferences with minds of citizens, and, hence, it cannot be ignored by anyone giving policy recommendations. As the right to both enhance and refuse to enhance one's mind, it is the central right to guide future regulations of neurotechnologies. Even though NE may primarily transform persons' inner domains, it cannot be considered as an entirely self-regarding matter, at least in a mental economy. Outside of competitive contexts, however, e.g. in psychedelic explorations of one's inner depths, there are few interests of society strong enough to justify restrictions. Harm-reduction, instead of prohibition, should be the default choice with respect to mind-altering technologies.

Thus, a liberal framework is well-suited to accommodate many worries over NE. Critics, too, should endorse a right to cognitive liberty as only strong legal protection of the mind guarantees that neither a transhumanist nor a bioconservative government can legitimately pass bills or enforce mind-invasive measures against individuals' wills. One does not have to be an apologist of a cold-hearted world of competition and capitalism in order to appreciate the idea of cognitive liberty. On the contrary, if one is worried about pressure to conform, the best answer is to strengthen rather than to curb the individual's legal as well as factual powers.

More generally, the call for a right to cognitive liberty underscores the need to re-consider the way society perceives and deals with matters of the mind. In Germany, mental problems of employees have risen to unprecedented heights (Hommel 2010). Stress, depression and burn-out are the symptoms of late-capitalist conditions, which constantly drive individuals to perform better and faster; and even outside of work-life, mental problems seem to be on the rise. Improving mental life requires taking the mind and its legal protection more seriously. This is what cognitive liberty stands for. Beyond any doubt, changing structural and societal conditions that lead to mental turmoil is necessary (albeit the extenuation of their structural causes in a globalized world is hard, perhaps impossible). Neurotools to improve one's psyche – from meditation to psychopharmacology – may be reasonable aids to cope with cognitive and emotional demands before mental disorders arise.

Nonetheless, I must add from the perspective of a criminal lawyer, the use of most substances for these purposes is currently punishable by law. The war on drugs costs thousands of lives, sends millions of persons to jail and contributes to the destabilization of states and regions, from Mexico to Afghanistan. In the end, these are the global consequences of the somewhat ideological rationale of preventing people from attaining altered mental states. At the same time, however, the use of these (and other) substances for therapeutic purposes steadily increases, which leads to two diametrically opposed societal reactions to production and consumption of similar things – condemnation and persecution for non-therapeutic use vs. state-supported therapy –, demarcated only by the thin line between therapy and enhancement. In light of cognitive liberty, drawing such strict dichotomies between what is valuable and despicable loses some of its persuasiveness. Certainly, states should counteract the negative social effects of drugs and prevent people from unnecessarily harming themselves just as they should improve the factual conditions that imperil mental freedom, from addiction to stressful working-conditions. In the age of neuroscience, reasonable, fine-grained and coherent approaches to the various challenges that our inner domain of consciousness faces are needed. Even if all these problems were to vanish in that better world often alluded to by critics, cognitive liberty is not an obstacle, but an invitation to its creation; it is not a threat, but a reminder – especially to liberal thinkers – that legal freedoms are nice ideas, beautiful to look at, but hollow inside without social conditions enabling and empowering individuals to truly develop their personalities and pursue happiness.

References

Alexy R (2003) Constitutional rights, balancing and rationality. Ratio Juris 16:131–140

Basl J (2010) State neutrality and the ethics of human enhancement technologies. AJOB Neurosci 2:41

Blitz MJ (2010) Freedom of thought for the extended mind. Cognitive enhancement and the constitution. Wisc Law Rev 2010:1049–1118

Boire RG (2000) On cognitive liberty, part I. J Cogn Lib 1:7–13

Boire RG (2003) On cognitive liberty, part IV. Mill and the liberty of inebriation. J Cogn Lib 4:15–23 (www.cognitiveliberty.org)

Bostrom N, Roache R (2011) Smart policy: cognitive enhancement and the public interest. In: Savulescu J, ter Meulen R, Kahane G (eds) Enhancing human capacities. Oxford University Press, Oxford

Bublitz JC (2010) Doping-Kontrollen im Staatsexamen? Leistungssteigernde Stimulantien und Chancengleichheit in Prüfungen. ZJS 3/2010, 306–317

Bublitz JC, Merkel R (2009) Autonomy and authenticity of enhanced personality traits. Bioethics 7:360–374

Bublitz JC, Merkel R (2012) Crimes against minds. Crim Law Philos (forthcoming). DOI: 10.1007/s11572-012-9172-y

Dees R (2010) Rawlsian "neutrality" and enhancement technologies. AJOB Neurosci 2:54

Douglas T (2008) Moral enhancement. J Appl Philos 25:228–245

Dresler M, Sandberg A, Ohla K, Bublitz JC, Trenado C, Mrozko-Wasowics A, Kühn S, Repantis D (2013) Non-pharmacological cognitive enhancement. J Neuropharmacol 1/2013, 64:529–543

Fabre C (2006) Whose body is it anyway: justice and the integrity of the person. Oxford University Press, Oxford

Farah MJ, Illes J, Cook-Deegan R, Gardner H, Kandel E, King P, Parens E, Sahakian B, Wolpe P (2004) Neurocognitive enhancement: what can we do and what should we do? Nat Rev Neurosci 5:421–425

Feenberg A (2002) Transforming technology. Oxford University Press, Oxford

Fehr E, Kosfeld M, Heinrichs M, Zak P, Fischbacher U (2005) Oxytocin increases trust in humans. Nature 435:673–676

Feinberg J (1986) Harm to self. Moral limits of criminal law, vol 3. Oxford University Press, New York

Franck G (2005) Mentaler Kapitalismus. Hanser, Munich

Galert T, Bublitz JC, Heuser I, Merkel R, Repantis D, Schöne-Seifert B, Talbot D (2009) Memorandum: Das optimierte Gehirn. Gehirn Geist 10:40–48

Greely H, Campbell P, Sahakian B, Harris J, Kessler RC, Gazzaniga M, Farah MJ (2008) Towards responsible use of cognitive-enhancing drugs by the healthy. Nature 456:702–705

Grey B (2011) Neuroscience and emotional harm in tort law: rethinking the American approach to free-standing emotional distress claims. In: Freeman M (ed) Law and neuroscience. Oxford University Press, Oxford

Halliburton C (2007) Letting Katz out of the bag: cognitive freedom and fourth amendment fidelity. Hastings Cent Law Rep 59:309–368

Harel A (2004) Theories of rights. In: Golding M, Edmundson W (eds) Blackwell guide to the philosophy of law and legal theory. Blackwell, Oxford

Hess E, Jokeit H (2009) Neurocapitalism. Orig. German in: Der Merkur 6/2009; engl. transl. www.eurozine.com/articles/2009-11-24-jokeit-en.html

Höfling W (2006) Kommentar zu Art. 3 EU-GRCH. In: Tettinger P, Stern K (eds) Kölner Gemeinschaftskommentar zur EU-GRCH. Beck, Munich

Hommel T (2010) Psychisch bedingte Fehlzeiten um 40 Prozent gestiegen. Ärzte Zeitung 20.07.2010

Hoppe C (2009) Neuro-Enhancement: Kein Verbot, aber bitte auch keine Empfehlungen. http://www.brainlogs.de/blogs/blog/wirklichkeit/2009-10-09/neuro-enhancement-kein-verbot-aber-bitte-auch-keine-empfehlung. Accessed 17 Feb 2013

Husak D (1989) Recreational drugs and paternalism. Law Philos 8:353–381

Huxley A (1932) Brave new world. Chatto & Windus, London

Huxley A (1962) Island. Chatto & Windus, London

Kant I (1797) Metaphysik der Sitten. Academy Edition: Metaphysics of Morals (trans: Gregor M (1991)), vol 6. Cambridge University Press, Cambridge

Kersting W (2007) Wohlgeordnete Freiheit. Immanuel Kants Rechts- und Staatsphilosophie. Mentis, Paderborn

Kipke R (2010) Was ist so anders am Neuroenhancement? Pharmakologische und mentale Selbstveränderung im ethischen Vergleich. Jahrbuch für Wissenschaft und Ethik 15:69–99

Klaming L, Vedder A (2010) Human enhancement for the common good – using neurotechnologies to improve eyewitness memory. AJOB Neurosci 3:22–33

Kolber A (2006) Therapeutic forgetting: the ethical and legal implications of memory dampening. Vanderbilt Law Rev 59:1561–1626

Levy N (2007) Neuroethics. Cambridge University Press, Cambridge

Marshall J (2009) Personal freedom through human rights law? Autonomy, identity and integrity under the European convention on human rights. Martinus Nijhoff, Leiden

Merkel R (2007) Treatment – prevention – enhancement: normative foundations and limits. In: Merkel R, Boer G, Fegert J, Galert T, Hartmann D, Nuttin B, Rosahl S (eds) Intervening in the brain – changing psyche and society. Springer, Berlin, pp 285–378

Metzinger T (2003) Being no one. MIT Press, Cambridge, MA

Metzinger T (2009) Ego tunnel. Basic Books, New York

Mill JS (1859) On liberty. In: Bromwich D, Kateb G (eds) (2003) edn. Yale University Press, New Haven

Mithoefer M, Mithoefer T, Wagner M, Jerome L, Doblin R (2011) The safety and efficacy of ±3,4-methylenedioxymethamphetamine-assisted psychotherapy in subjects with chronic, treatment-resistant posttraumatic stress disorder: the first randomized controlled pilot study. J Psychopharmacol 4:439–452

Parens E (2005) Authenticity and ambivalence: toward understanding the enhancement debate. Hastings Cent Rep 3:34–41

Persson I, Savulescu J (2011) Unfit for the future? Human nature, scientific progress, and the need for moral enhancement. In: Savulescu J, ter Meulen R, Kahane G (eds) Enhancing human capacities. Oxford University Press, Oxford

Pettit P, Smith M (1996) Freedom of belief and desire. J Philos 9:429–449

President's Council on Bioethics (2003) Beyond therapy. Biotechnology and the pursuit of happiness. President's Council on Bioethics, Washington, DC

Rawls J (2005) Political liberalism. Columbia University Press, New York

Repantis D, Laisney O, Heuser I (2010a) Acetylcholinesterase inhibitors and memantine for neuroenhancement in healthy individuals: a systematic review. J Pharmacol Res 61:473–480
Repantis D, Laisney O, Heuser I (2010b) Modafinil and methylphenidate for neuroenhancement in healthy individuals: a systematic review. J Pharmacol Res 62:187–206
Ripstein A (2009) Force and freedom. Kant's legal and political philosophy. Harvard University Press, Cambridge, MA
Robeyns I (2011) The capability approach. In: Stanford encyclopedia of philosophy (www.sep.org)
Rosa H, Scheuerman W (2009) High-speed society. Social acceleration, power and modernity. Pennsylvania State University Press, University Park
Sandberg A, Savulescu J (2011) The social and economic impacts of cognitive enhancements. In: Savulescu J, ter Meulen R, Kahane G (eds) Enhancing human capacities. Oxford University Press, Oxford
Sandel M (2007) The case against perfection: ethics in the age of genetic engineering. Harvard University Press, Cambridge, MA
Savulescu J (2006) Justice, fairness and enhancement. Ann N Y Acad Sci 1093:321–338
Schermer M (2007) Brave new world versus Island – Utopian and dystopian views on psychopharmacology. Med Health Care Philos 10:119–128
Sententia W (2004) Cognitive liberty and converging technologies for improving human cognition. Ann N Y Acad Sci 1013:221–228
Taylor K (2004) Brainwashing. The science of thought control. Oxford University Press, Oxford
Thomson JJ (1990) The realm of rights. Harvard University Press, Cambridge, MA
Tovino SA (2007) Functional neuroimaging and the law. AJOB-Neurosci 9:44–56
Vermeulen B (2006) Commentary on art. 9. In: van Dijk P, van Hoof F, van Rijn A, Zack L (eds) Theory and practice of the European convention on human rights, 4th edn. Intersentia, Antwerpen
Vincent N (2012) Enhancing responsibility. In: Vincent N (ed) Legal responsibility and neuroscience. Oxford University Press, Oxford
von der Pfordten D (2007) Kants Rechtsbegriff. Kant-Studien 98:431–442
von Hirsch A (2008) Direct paternalism: criminalizing self-injurious conduct. Crim Justice Ethics 1:25–33

Chapter 20
Cognitive Enhancement and Criminal Behavior

Elizabeth Shaw

Abstract Some offenders appear to have an impaired capacity to appreciate the moral significance of their acts and a limited ability to engage in effective practical reasoning. These deficits may interfere with the communicative aspects of punishment and with rehabilitation. This chapter will discuss whether it could ever be morally permissible to employ certain types of cognitive enhancements to enhance offenders' capacities for practical reasoning and moral communication as part of their rehabilitation. The text sets out the considerations that need to be balanced when deciding which types of techniques should be permissible and under what circumstances they may be used. It argues that offenders have three main interests that should be taken into account here: (1) being treated as a member of the moral community; (2) being treated as a rational human being; and (3) being spared unnecessary suffering.

Keywords Cognitive enhancement • Criminal justice system • Criminal behavior • Rehabilitation • Law

20.1 Introduction

This chapter will discuss whether it could ever be morally permissible to employ certain types of cognitive enhancements in an attempt to reform or rehabilitate criminals. Some offenders appear to have an impaired capacity to appreciate the moral significance of their acts and a limited ability to engage in effective practical

E. Shaw (✉)
School of Law, University of Aberdeen, Aberdeen, UK
e-mail: eshaw@abdn.ac.uk

E. Hildt and A.G. Franke (eds.), *Cognitive Enhancement: An Interdisciplinary Perspective*, Trends in Augmentation of Human Performance 1,
DOI 10.1007/978-94-007-6253-4_20,

reasoning. Much academic writing focuses on whether such cognitive deficits diminish the offender's criminal responsibility, or whether they should be taken into account in mitigation at the sentencing stage (see, e.g., Maibom 2008; Mitchell 2003; Penney 2012). However, it is also important to note that these deficits may pose a barrier to effective moral dialogue with the law-breaker, interfering with the communicative aspects of punishment and with rehabilitation.

Although these deficits can be classed as 'cognitive', because they result in impaired moral *understanding* and moral *reasoning*, they may stem from underlying affective abnormalities. For instance, it has been suggested that a markedly reduced ability to empathize with others can interfere with a person's ability genuinely to appreciate what is wrong with harming others. Similarly, powerful impulses towards violent aggression could 'cloud a person's mind', making practical reasoning difficult (see Douglas 2008, 231). Techniques which aim to enhance empathy or inhibit violent urges could be regarded as 'cognitive' enhancements, because their ultimate purpose is to improve the individual's capacities for moral understanding, reasoning and communication. They are classed as 'enhancements' because they are not strictly necessary to treat disease.

This chapter argues that if in the future these techniques become sufficiently refined, it may be permissible to employ them in a limited way in the criminal justice system. The purpose of this chapter is to examine whether the use of such techniques is legitimate *in principle*. It will not focus on the issues surrounding the safety and efficacy of specific interventions – that is a subject for another paper.[1] The text sets out the considerations that need to be balanced when deciding which types of techniques should be permissible and under what circumstances they may be used. It argues that offenders have three main interests that should be taken into account here: (1) being treated as a member of the moral community; (2) being treated as a rational human being; and (3) being spared unnecessary suffering.

The first section of this chapter gives some examples of potentially useful enhancements that may become available in the medium term future. The next section briefly indicates why the criminal justice system should aim to reform and rehabilitate offenders. The remainder of the chapter considers why the state must be restricted in pursuing those aims by the need to respect the offender's interests, and discusses what respecting these interests entails.

[1] For more detail on the issues connected with safety and efficacy e.g. (see Greely 2012; Farah 2004; Maletzky et al. 2006; Maletzky and Field 2003).

20.2 Examples of Potentially Useful Enhancements[2]

20.2.1 Increasing Empathy

As noted above there is some evidence to suggest that the ability to empathize is key to understanding moral norms (Blair et al. 2005).[3] For example, individuals with markedly reduced levels of empathy have exhibited difficulties in distinguishing conventional rules (such as rules of etiquette) from moral rules and in ranking wrongs in order of seriousness. Philosophers differ as to whether empathy is essential for moral understanding. However, even those who believe that it is not essential often maintain that it is indirectly helpful in moral development. If techniques were produced that increased empathy in individuals who appear to be deficient in it, then this might play a useful role in reforming offenders.

20.2.2 Decreasing Violent Urges

Certain offenders may experience repetitive violent fantasies and powerful surges of anger which they find difficult to control. As discussed in greater detail below, these factors can impair offenders' ability to think clearly about how they should act and may distort their moral judgments. Research is beginning to uncover certain neurological factors that seem to have an impact on individuals' dispositions to anger and violence. There is some evidence that selective serotonin reuptake inhibitors (SSRI's) may reduce aggression (Ferari et al. 2005; Douglas 2008; Crockett et al. 2010). It may become possible in the medium term future to develop techniques that can reduce the strength of offenders' volatile impulses or that increase their control over these impulses. In a recent article, John Harris argues that such developments may not be morally desirable (Harris 2011). He cites an example of an individual who attacked a terrorist who was about to detonate a bomb, thereby rescuing a plane full of people. According to Harris, if the rescuer had been given SSRI's to reduce his aggression he might not have managed to save the plane. Harris does highlight a genuine concern – moral understanding and morally-motivated behavior are complex phenomena. Even the well-intentioned use of biomedical interventions risks causing undesirable consequences. However, when assessing whether offenders should receive such interventions, it is important to take

[2]For a discussion of the possible negative effects and side-effects of such interventions see footnote 1 above and footnote 11 below.

[3]Studies concerning empathy and moral understanding have often been carried out on individuals with psychopathic personality disorder. There is some disagreement concerning whether psychopathy is a mental illness and hence whether interventions to increase psychopaths' emotional deficits would count as a treatment or an enhancement. For opposing views on the mental illness question (see: Mealey 1997; Kendell 2002).

into account the *likelihood* of the relevant scenarios occurring. In the case of many violent offenders, the risk that the intervention will prevent them from heroically rescuing a crowd of innocent people may seem relatively small compared with the risk that without the intervention they will reoffend.

20.2.3 Anti-libidinal Medication

Drugs have already been developed to help reduce deviant sexual urges and thoughts. This can create an opportunity for offenders to concentrate on the reasons why they should change their behavior and the steps they need to take, without being distracted by their impulses. These medications are already being used to some extent within the criminal justice system.[4]

20.2.4 Decreasing Racist Sentiments

Certain individuals experience a strong negative emotional reaction to members of different races. Such emotional reactions may stem from early childhood experiences, e.g. parents who taught them to fear members of a different race. Such deeply-rooted emotional reactions may help to fuel racially motivated crimes and may interfere with the racist's ability to see why racism is wrong. Some research has been undertaken into the neural basis for racial stereotyping (Hart et al. 2000; Phelps et al. 2000; Cunningham et al. 2004). Potentially this might lead to interventions that could attenuate such emotional responses. Harris has criticized this proposal on the basis that racism is likely to involve a complex network of beliefs and not merely emotional reactions (Harris 2011). Although this is almost certainly true, it does not demonstrate that ingrained emotional reactions do not contribute to the tendency to hold stubbornly onto ill-founded beliefs in the face of the evidence. Attenuating such emotional responses might help the offender to assess the issues dispassionately and realize that his racist views are ill-founded.

[4]Regarding the use of these medications in Scotland see: http://www.forensicnetwork.scot.nhs.uk/Medication%20for%20Sex%20Offenders/medication%20for%20sex%20offenders%20protocol.pdf. Regarding the use of these medications in England see: http://www.insidetime.org/resources/Publications/Use-of-Med-to-treat-SexOff_PSJ176.pdf.

20.2.5 *Delaying Gratification*

Difficulties with delaying gratification may lie behind some individuals' tendency to break the law. Cognitive enhancements could potentially help to rehabilitate criminals through enabling them to work out and implement strategies to delay gratification.[5]

20.2.6 *Increasing the Ability to Focus on Relevant Issues*

Recent studies suggest that individuals who score highly on measures for psychopathy may suffer from a kind of attention-deficit disorder which may help to explain their characteristic anti-social behavior (Newman et al. 2010). It seems that when presented with incentives for performing an action, these individuals lose sight of the reasons against performing the action. Cognitive enhancements might enable these individuals to focus on all the relevant considerations (and in particular the reasons against breaking the law).

20.3 Why Should the Criminal Justice System Attempt to Reform and Rehabilitate Offenders?

My argument is based on the assumption that reform and rehabilitation should feature among the aims of the criminal justice system. Certain theorists reject the idea that the criminal justice system should seek to achieve forward-looking goals, such as crime prevention (see e.g. Moore 1998, 84–92). For instance, according to a variety of 'pure' retributivism, the sole aim of punishment should be to ensure that the guilty suffer in proportion to their desert. On this view, the guilty should be released when (and only when) they have paid the price for their crime. This theory strikes many people as counter-intuitive. For instance, it is puzzling why the suffering of offenders should be considered intrinsically good, given that we normally view suffering as intrinsically bad. How does the addition of more suffering to the world after a crime has been committed make the world a better place? Furthermore, it seems strange to claim that the criminal justice system should not be at all concerned with preventing criminals from reoffending. Imagine a hypothetical system of punishment that actually caused most criminals to reoffend. (Imagine, for instance, that their prison experiences made them angrier at society and more motivated to engage in destructive behavior and that they learned ways of evading the authorities more successfully on release. Or imagine that the prisoners

---

[5] This issue is discussed in Kennett (2006).

became dependent on the regimented structure of prison-life and reoffended partly in order to be rearrested.) It seems natural to say that the system had failed to achieve one of its legitimate aims.

One reason why I reject pure retributivism has to do with the issue of free will and determinism. Due to space constraints, I can only touch on this issue very briefly.[6] The thesis of determinism implies that every person's actions are caused by earlier events the occurrence of which was not under the agent's control. The sense of 'cause' that is being used here does not merely mean that prior events just made the action *more likely* to occur. Rather, it means that prior events (taken together with the laws of nature) were *causally sufficient* to produce that action. Determinism does not imply that people will not modify their behavior in response to good reasons for doing so. It merely implies that whether a person recognizes and responds to one particular reason for action, rather than another on any given occasion, is determined by prior events in the manner described above.

If factors wholly beyond one's control, such as one's genes and environment are causally sufficient to produce one's conduct, then it seems that practices of attributing praise and blame for such conduct becomes like a lottery. If an agent is lucky enough to have the 'right' genes and environment, then he will develop into a virtuous agent and will receive praise for his good deeds, but if he is unlucky enough to have the 'wrong' genes and environment, then he will choose to act on immoral reasons and will be blamed for his bad actions. To argue that blaming/punishing an agent in these circumstances is justified purely on the basis that he 'deserves' it seems bizarre and as unfair as blaming someone for his height.

Some retributivists try to base their theory on libertarianism. This theory requires that human deliberations are (at least sometimes) undetermined. It also requires that they are undetermined in a way that does not merely introduce randomness into our deliberations. Most libertarians themselves concede that we lack epistemic justification for these beliefs.[7] Libertarianism therefore seems a shaky basis on which to justify imposing serious harm on law-breakers.

Determinism does not imply, however, that we should do nothing to tackle criminal behavior. As noted above, determinism is compatible with the ability to respond to reasons. Experience has shown that providing individuals with reasons for changing their behavior can be effective.

However, some theorists might question why we should try to change offenders' behavior partly through providing them with *moral* reasons to do so. According to deterrence theorists, punishment should just aim to prevent crime by providing offenders with prudential reasons to change their behavior. In simply focusing on ways of controlling offenders' problematic behavior, deterrence theorists fail to respect offenders' capacity to grasp moral reasons – a capacity which is arguably an important source of human dignity. Deterrence theory represents the criminal law

[6]My reasons for adopting this view about free will are similar to those advanced by the following writers: Smart (1961), Strawson (1986), Pereboom (2001), Honderich (2005) and Harrison (2009).

[7]This point is emphasized in Double (2002).

as a system of threats, backed by painful sanctions. This paints a worrying picture of the relationship between the citizen and the state – where the state attempts to achieve obedience solely through inducing fear in citizens.

For these reasons this chapter assumes that the criminal justice system should aim to protect society, by preventing crime, but that it should do so in a way that addresses citizens as rational beings who are capable of grasping moral reasons.

20.4 The Need to Take into Account the Interests of the Offender

My approach places considerable weight on protecting the interests of the offender. Some theorists may object that it gives the offender's interests too much weight. It might be thought that the criminal has (to a large extent) forfeited his right to our moral concern. I do not accept this forfeiture view, partly because of the considerations that I mentioned earlier about free will and determinism. However, my arguments in the rest of this chapter do not *depend* on any particular position about free will. Whatever their views on free will, many people will find it intuitive that certain basic rights are inalienable, held in virtue of being human.[8] A society that regards certain members as worthless, not only wrongs those individuals but also degrades itself by treating them as worthless. The remainder of this chapter considers three ways in which society must respect offenders' moral worth – by treating them as members of the moral community, by recognizing their status as rational agents and by refusing to subject them to needless suffering.

20.5 Membership of the Moral Community

Society's response to criminal behavior should recognize that offenders are members of the moral community, albeit members who have breached the community's norms. This principle is supported by the intuition that the state should not 'objectify' law-breakers – that offenders should be treated as persons. Objectifying a group of people can involve emphasizing that 'they' are fundamentally unlike 'us'. It can involve focusing on the idea that a deep division exists between the objectified group and the rest of society. One way of respecting offenders' membership of the moral community is to preserve connections between the offender and other moral agents. This suggests that reforming offenders should occur primarily through relationships with others. Cognitive enhancements should never be used to circumvent the need for dialogue with the offender, but rather to enable effective dialogue to occur.

[8]For a defense of this idea from the point of view of a free will sceptic (see: Vilhauer 2009).

In order to treat the offender as a member of the moral community, limits must be set on the types of biomedical intervention that are permissible. Interventions that attempt to radically re-shape the offender's basic character, goals or values are morally impermissible (Shaw 2011). The ability to employ such interventions would vastly increase the authorities' (already considerable) powers for controlling offenders' behavior. It would give the authorities a significant level of control over the individual's *inner life*. This obviously has potential for abuse, e.g. it could be used to suppress legitimate dissenters. Even if this technique were only used to prevent offenders from engaging in uncontroversially wrongful activities, it would still be morally objectionable because of the troubling *relationship* that would be created between the state and offenders. It would set the authorities on a completely different plane from offenders, greatly widening the *inequality of power* between them. It would imply that the offender is so radically morally deficient, and so unlike the rest of 'us', that the state needs to take over control of fundamental aspects of the offender's personality. Such a policy would mark a huge shift towards characterizing offenders as 'the other' and thus towards objectifying them.

In contrast, interventions that merely reduce the strength of the offender's impulses, or increase his capacity for self-control, or increase his capacity to empathize with others seem less likely to interfere with the core of the offender's personality. They do not instill in him particular values, or deprive him of the ability to decide for himself what values and beliefs he should adopt. Rather, a person who has become less impulsive and more self-controlled seems to be in a better position to think about what his values really are and to translate those beliefs into action. The capacity for empathy could also put him in a better position to appreciate the reasons that are relevant to such decisions. He can criticize the authorities on the basis of his values and beliefs. The more limited interventions that this chapter advocates do not give the state the power to guarantee that the offender will behave in one particular way. The aim of these interventions is just to put the offender in a better position to *understand* the relevant reasons for changing his behavior and to act effectively on these decisions.

It might be objected that the idea of treating the offender as a member of the moral community is unrealistic. Punishing criminals, the objection runs, necessarily involves highlighting the differences (rather than commonalities) between offenders and law-abiding citizens, by condemning the offender as a wrongdoer. Punishment, it may be argued, necessarily excludes offenders from the community. It can do this in terms of the moral stigma that attaches to a criminal conviction and sentence. It can also physically exclude the offender from the community, e.g. by putting him in prison.

It should be acknowledged that society's response to criminal behavior does involve coercion, exclusion and the highlighting of differences between offenders and law-abiding citizens. It is perhaps impossible to conceive of a practicable approach to the problem of crime that does not involve these elements to some degree. But it is submitted that society's response to criminal behavior can and should *also* involve dialogue (and not just coercion); that it should emphasize the commonalities between offenders and other moral agents (and not just the differences); and that it should preserve some connections between the offender and the rest of the community (and not exclude the offender entirely).

Another type of objection to my view states that using biomedical interventions to help change offenders' conduct inappropriately 'medicalizes' the problem of crime. According to this objection, medicalizing criminal behavior implicitly *separates* criminals from the rest of the community – sending out the message that the problem is with 'them' and not with the rest of 'us' and that offenders are the only ones who need to change. Criminal behavior is partly caused by social factors. It might be thought that giving cognitive enhancement to offenders obscures this fact, by sending out the message that the causes of crime lie solely within the individual (perhaps stemming from a biological defect) rather being the product of the offender's circumstances. This, it may be argued absolves the rest of the community from responsibility for helping to create or failing to alleviate these unfortunate circumstances.

In response, while it is important to acknowledge and seek to remedy the social causes of crime, this should not lead us to ignore the factors that can make particular individuals likely to reoffend. Pretending that these factors do not exist would distort the truth and would disadvantage the offender and wider society by putting obstacles in the way of effective rehabilitation.

20.6 Respecting the Offender's Personhood and Rationality

It is essential that any attempt to change offenders' behavior respects the offender's status as a rational human being. This chapter only endorses those biomedical interventions that do not restrict (but rather aim to positively enhance) the offender's capacities for critical reflection.

Furthermore, cognitive enhancements should only be used if the offender gives his free and informed consent. By according weight to the offender's preferences, the state treats the offender as a person who still has moral worth and whose wishes are not completely discounted. The state shows respect for the offender's rationality, by allowing him to weigh the advantages and disadvantages of enhancement versus, for instance, spending a longer time in prison, and trusting that he is able to make an appropriate decision. I will argue, below, that the state should inflict no more distress on the offender than is needed to achieve its legitimate aims. Where such aims can be achieved by different methods, it is appropriate to give the offender some choice between those methods. Provided he is given adequate information, the offender is best placed to determine which method is likely to cause him more distress.

It might be objected that even if an offender agreed to accept cognitive enhancements in preference to other methods of reform/rehabilitation, this would not amount to *genuine, free* consent, given the coercive situation in which the offender finds himself. In response, it is submitted that allowing the offender some say in the matter still shows respect for his preferences, even though the offender's options are limited. For the reasons stated in the previous section, limiting the offender's options can be justified by the need to protect society and by the value of reforming the offender and restoring him to the community. The 'consent requirement' strikes a balance between these interests and the offender's interests in not being forced to receive biomedical interventions. Provided officials do not exert *additional pressure* on offenders to receive biomedical interventions (rather than longer detention or other modes of rehabilitation), the fact that the offenders' options are limited does not seem to render their consent involuntary. After all, patients are often faced with hard choices where none of the available options are attractive and yet this does not make voluntary consent impossible. Some offenders have reported desperately wanting biomedical interventions, e.g. to help control destructive thoughts and urges and have fought hard to have access to such medication (See, e.g. Fischer 2006, 1–4). If an offender voluntarily requests a biomedical intervention, it seems more disrespectful to the offender's autonomy to refuse this request than to grant it (see Bomann-Larsen 2011, 10; Shaw 2012, 18–19).

Some theorists still find such interventions troubling, regardless of whether the offender consents. They claim that the use of biomedical interventions treats the offender as a being without rights and not as a rational agent. Eric Matthews raises this type of objection to biomedical interventions designed to suppress offenders' violent or deviant sexual impulses. He writes:

> The harm done to [the offender] would be that of treating him like a thing, not a human being.... He would have been reduced to the level of a robot... People's bad behavior can be legitimately changed only by persuasion to see that what they have been doing or proposing to do is unacceptable.... Even if [the offender] chose to undergo this treatment, that would not necessarily make it morally tolerable. To choose to be dehumanized is choosing to be in a state where one can make no more choices, where one's existence is determined not by one's own will but by the requirements of others, and that does not seem like a morally legitimate choice to make. A sex offender or someone prone to outbursts of anger can legitimately seek help in learning to control his own impulses, but not treatment designed to remove those impulses altogether: the former is compatible with his continuing humanity, the latter is not. (Matthews 2007, 181–182)

Matthews seems to begin by taking an absolutist stance against using any form of biomedical means of trying to reform criminals. He states that persuasion is the '*only*' morally acceptable technique. However, the arguments he then goes on to produce do not support such an absolutist position. According to Matthews, biomedical interventions dehumanize offenders by putting them in a state where they can "make no more choices." This is simply not true of all types of biomedical interventions and particularly not those that can be classed as forms of "cognitive enhancement."

For one thing, suppressing an offender's urges to commit violent or sexual crimes would not thereby deprive him of the ability to make choices concerning all of his other activities. If the offender were released into the community after receiving this intervention, he would certainly have greater scope for making choices about how to lead his life than if he remained in prison. Furthermore, biomedical interventions which aim to suppress offenders' destructive impulses can actually *increase* the accused's ability to make informed and meaningful choices, by reducing the impediments to effective practical reasoning. Frequent, intense surges of violent anger can cloud an individual's judgment making it extremely difficult for him to assess issues such as whether or not his anger is justified and to appreciate the reasons why he should control himself. It can also make it difficult for him to enter into the kinds of relationships that are crucial to bringing home to him the wrongfulness of criminal conduct and to assisting him to lead a law-abiding productive lifestyle in the future. He may alienate those who are trying to help him, or he may not take on board their advice because he is overwhelmed by the feeling that they are a threat to him that must be warded off. Reducing the offender's volatility through biomedical interventions can help him to focus more clearly on what he needs to do to improve his conduct.

Matthews's critique of using biomedical interventions in the criminal justice system relies on a dubious conception of what it means to be "human" as opposed to being a "thing" or a "robot." According to Matthews, an offender who received medication to suppress his urges to commit violent or sexual offences would be "dehumanized." But is a person really "reduced to the level of a robot" just because she lacks a powerful urge to commit horrific crimes of violence or sexual exploitation? Many individuals, by nature, find the idea of performing these acts utterly repulsive and disturbing. Their failure ever to experience temptations to commit such crimes does not render these individuals robot-like or less than human.

There are alternative, more plausible conceptions of what capacities are important to leading a full, human life. Distinctively human capacities include the power to accord appropriate weight to relevant reasons for action and to conform one's actions to one's considered judgments about what is the best thing to do. Biomedical interventions, including those that reduce offender's destructive urges, could potentially enhance these capacities. It shows respect for the offender's existing rational capacities to allow him to decide for himself whether to avail himself of these enhancements. Giving him this choice treats him as an agent and not as a being without rights.

The following comments by a perpetrator of violent sexual offences illustrate these points. He vividly explains how powerful urges can interfere with an individual's thoughts and reasoning powers and how medication can help to restore the offender to a freer, more rational, and more human state:

> Basically, I am plagued by repetitive thoughts, urges, and fantasies.... I cannot get those thoughts out of my mind.... The best way for the average person to try to understand this is to remember a time when a song played over and over again in your head. Even if you liked the melody, its constant repetition was quite annoying, and the harder you tried to drive it out of your head, the harder it seemed to stick. Now replace that sweet melody with noxious thoughts of degradation, rape and murder and you will begin – and only just begin – to understand what was running rampant through my mind uncontrollably....I was tired of being tormented by my own... mind. So unbelievably tired.... Having those thoughts and urges is like living with an obnoxious roommate. You cannot get away from him because he is always there. What Depo-Provera[9] did was to move that roommate down the hall to

[9]Depo-Provera is a branded drug that was originally developed as a progesterone-only female contraceptive. The active ingredient is medroxyprogesterone acetate (MPA). It is given as an

> his own apartment. The problem was still there, but it was a whole lot easier to deal with because it wasn't always in the foreground. He didn't control me anymore – I was in control of him. It was an unbelievable sense of freedom. It made me feel as if I were a human being again, instead of some sort of horrible monster.[10]

A critic of biomedical interventions might concede that a person would not be in a robot-like state just because some of his negative urges were suppressed by medication. Nevertheless, the critic might object to giving this medication to the offender because doing so would involve a *failure to recognize* the offender's humanity. On this view, the only way to treat an offender like a human being is by seeking to reform him using the power of rational argument alone, rather than trying to give him a chemical fix as if he were a broken "machine" (Freedman 2000, 136). Traditional techniques of reform and rehabilitation, such as victim-offender mediation present him with reasons to change his behavior, e.g. by trying to convince him that his behavior cannot be justified and by showing him the suffering of those affected by his crime. In contrast, biomedical interventions, it is claimed, alter the offender's thought-processes and/or behavior by directly affecting how his brain works without giving the offender any additional *reason* to think or act differently.

This criticism would have some force if biomedical enhancements were used as a *substitute* for reasoning with the offender. The idea of treating someone like a "machine" sounds so sinister mainly because it suggests that the individual is being *excluded* from rational dialogue and from relationships with others. However, this chapter advocates using cognitive enhancements to enable or facilitate rational dialogue to take place. Without the aid of cognitive enhancements the offender may never fully access certain reasons for action and may be cut off from relationships which could help him to develop as a rational human being. Furthermore, it is implausible to suggest that any method of altering behavior that does not involve rational argument is necessarily morally intolerable. Imagine that there was good evidence that putting offenders on a regular exercise regime would help to reform them, by increasing their serotonin levels, enabling them to feel more empathetic and less defensive. Imagine that this opens up a window of opportunity for offenders to enter into relationships that they had been emotionally resistant to entering, e.g. victim-offender mediation schemes. Having begun such relationships, the offenders could become better able to appreciate why their criminal actions were wrong and why they should change their conduct. The exercise regime *per se* would not give offenders a reason for changing their behavior. But this does not seem to make it morally intolerable.

Showing compassion to an individual by helping to alleviate his distress can surely be an important way of treating him as a fellow human being. As I will discuss in the next section, cognitive enhancements have the potential to spare individuals needless suffering.

---

injection every three months. When used in males, it can reduce compulsive sexual fantasies and sex drive (see Maletzky et al. 2006; Maletzky and Field 2003, who also list possible side-effects).

[10]Michael Ross, quoted in Fischer (2006, 2–3).

20.7 Suffering

In the future, cognitive enhancements may have the potential to reduce the offender's suffering significantly. As noted above, the offender himself may find the factors that impede his practical reasoning, such as repetitive thoughts and powerful, irrational urges intensely distressing. Cognitive enhancements might help to relieve this distress. Such interventions may also make the process of reform and rehabilitation itself less burdensome to the criminal. For instance, attempts to reform criminals that do not involve enhancements may require a longer time of incarceration than if enhancements were used. However, the offender's interest in not being made to suffer can count against using enhancements that might expose the individual to serious side-effects.[11]

But why should the state prefer methods of dealing with criminal behavior that involve less suffering for the law-breaker? According to retributivists, it is intrinsically *good* that the offender is made to suffer in proportion to his moral guilt. Retributivists would oppose giving enhancements to offenders if doing so would diminish the distress involved in punishment to a level that is lower than the amount of suffering that the offenders 'deserve' (see, e.g., Moore 1998). In response, it should be recalled that retributivism faces considerable difficulties for the reasons indicated in the first section of this chapter. Secondly, many of those who believe in retribution also believe that this is only *one* of the functions of the criminal justice system, alongside reform and rehabilitation. Once the offender has served the part of his sentence that is designed to inflict the suffering he deserves, there is no reason, on this view, why the part of the sentence which is aimed at reform and rehabilitation should purposely aim to impose still more suffering on the criminal.

Pure consequentialist theories place certain limitations on the amount of suffering which should be imposed on offenders. They state that the offender should only suffer to the extent that this is necessary in order to prevent crime or to promote the general welfare in some other way (Bentham 2011). This is sometimes referred to as "economical prevention." However, the protection that this principle affords to the offenders' interests does not seem to go far enough. The principle of economical prevention is compatible with inflicting levels of suffering on offenders which are intuitively far too severe. It is compatible with inflicting an extremely harsh penalty on the offender if this will prevent each of very many other people from suffering some very slight hardship. What matters for the consequentialist is the *total* level of distress to be prevented. For example, if subjecting a group of offenders to an extremely distressing form of treatment would prevent a much larger number of

[11]For examples of potential side-effects see references in footnote 1 above. Loss of bone density is one possible long-term side-effect of anti-androgens that are currently given to sex offenders (Greely 2012). As Greely notes, the possibility of this side-effect is known due to studies on the use of this medication as a form of birth control for women: "Sex offenders receive the drug at much higher doses. What is effect on their bones? No one knows, as it was never tested on men, and because the recipients are sex offenders, almost no one cares" (163).

people from each suffering a tiny inconvenience, then the total amount of distress to be prevented could, according to the utilitarian calculation, be enough to justify forcing the offenders to undergo the painful treatment.

This chapter advocates providing greater protection for the offender's interests than that implied by traditional consequentialism. It is unacceptable to impose a sentence on the offender which would cause him much greater distress than the distress which any particular individual would suffer if the sentence were not imposed. The proposed level of protection is justified by considerations of equality. It is *prima facie* wrong to create a situation where individuals suffer grossly unequal levels of distress (I will refer to this as 'the equality principle').[12] This principle could have a similar effect to the retributive principle that the severity of the penalty must be proportionate to the gravity of the crime. For instance, both principles imply that more burdensome interventions may be imposed on dangerous violent offenders than on shoplifters. However, the 'equality principle' has a distinctly different basis from the retributive doctrine. Unlike retributivism, the principle that I advocate is not based on the moral responsibility of the individuals concerned. For instance, the equality principle applies even if the law-breaker is mentally ill. It is permissible to detain dangerous psychotic individuals in an institution for relatively long periods of time, despite the fact that they are not morally responsible, if this is necessary in order to prevent them from being seriously violent to those around them. However, it would not be justifiable to take such an extreme measure to prevent a mentally ill person from committing relatively minor disturbances. This can be explained by the equality principle. It is better that each of a larger number of individuals should suffer a slight hardship than that one individual should bear an extremely heavy burden in order to prevent others from suffering this slight hardship. This principle should act as a constraint when deciding between more or less burdensome methods of reform and rehabilitation.

Some theorists may be concerned that attempting to limit the amount of suffering that the offender undergoes may prevent him from genuinely reforming. Experiencing remorse, it may be argued, which is necessarily painful, is an essential element of the process of true reform. In reply, it should be acknowledged that, in order to achieve the legitimate aim of bringing the offender to recognize that his actions were wrongful and that he needs to change, it will indeed be necessary for the offender to experience some distress. Cognitive enhancements could potentially play an important role in helping some offenders to appreciate the wrongfulness of their conduct and to experience remorse. It should also be noted that certain offenders experience greater distress than others, not because they have greater cause to feel remorseful for their crimes, but because the process of reform and rehabilitation is more prolonged and difficult for them due to factors that impede their powers of practical reasoning. Cognitive enhancements could reduce these impediments, helping to ensure that offenders go through no more distress than is necessary for genuine reform.

---

[12]This principle is defended in Honderich (1984, 78).

It might be objected that if the process of reform and rehabilitation is more difficult for some offenders (e.g. because they are prone to outbursts of anger), this is due to their own moral short-comings, and so it is fair that they suffer more distress than more even-tempered offenders. In response, this claim seems to rest on the assumption that individuals are responsible for creating their own flawed characters. For the reasons indicated in Section Three of this chapter, this is a dubious claim. Furthermore, it is important to bear in mind that it can be an extremely difficult and slow process to try to undo character traits that have been laid down early in life – a process with many relapses along the way and that, in some cases, despite the individual's considerable efforts, is never wholly successful. While the offender is learning to control his anger without medication, those around him may be at risk from or actually suffering the consequences of his outbursts. An offender who seeks to receive cognitive enhancements to facilitate the process of rehabilitation, rather than expose others to this increased risk, shows a willingness to take responsibility for his conduct, which ought to be encouraged.

A related concern is the idea that the experience of struggling with conflicting desires and resisting temptation is intrinsically valuable and that this might be lost if biomedical interventions were used in order to reduce the strength of offenders' urges. JM Olsen writes:

> So, then, what is inherently valuable in moral effort? The answer is that moral effort is required if we are to have morality at all. Morality, I maintain, requires agency, and if no moral action ever requires any effort, then we would be, in Kantian terms, mere slaves of inclination. Put another way, there is something inherently valuable about agency, but agency is empty without resistance—that is, temptation. (Olsen 2006, 289)

It is important to bear in mind, however, that it is highly unlikely for it to become technically possible to *eliminate* all of an individual's temptations to do wrong. The question is whether it is permissible to use biomedical interventions to reduce somewhat the force or number of these temptations, or to increase the ability of the offender to deal with them. As noted above, many people due to their upbringing and/or natural predispositions experience little or no temptation to commit serious crimes. This does not seem to indicate that their agency is somehow deficient in comparison with someone who feels strong temptations to break the law. Furthermore, Olsen puts forward a doubtful interpretation of the Kantian idea of being a slave to inclination. Olsen suggests that a person's good deeds are morally 'empty' unless she feels tempted to perform bad actions. Kant, in contrast, required that, to have moral worth, a good deed must not be motivated merely by an inclination to do it, but by the recognition that it is the right thing to do. If this recognition is sufficient to motivate the agent to do the good deed, then she does not seem to be enslaved to her inclinations. Feeling tempted to do bad actions, however, does not seem to be strictly necessary to enable the agent to recognize or be motivated by the reasons for doing good actions.

20.8 Conclusion

Reforming and rehabilitating offenders should be central aims of the criminal justice system. It is possible to identify several kinds of cognitive enhancement which may in the future, if sufficiently refined, play a useful role in pursuing these aims. However, certain restrictions must be placed on the means that the state may employ to achieve its goals. Society's approach to dealing with criminal behavior must treat the offender as a member of the moral community and as a rational agent and must respect the offender's interests in not being made to suffer unnecessarily.

Acknowledgements My research was funded by the Clark Foundation for Legal Education and by the Arts and Humanities Research Council. I am grateful to Elisabeth Hildt, Antony Duff and James Chalmers for helpful comments on earlier drafts of this chapter.

References

Bentham J (2011) Principles of penal law. ebooks@adelaide, Adelaide, South Australia. http://ebooks.adelaide.edu.au/b/bentham/jeremy/principles_of_penal_law/. Accessed 26 June 2012
Blair J et al (2005) The psychopath: emotion and the brain. Blackwell, Oxford
Bomann-Larsen L (2011) Voluntary rehabilitation? On neurotechnological behavioural treatment, valid consent and (in) appropriate offers. Neuroethics. doi:10.1007/s1215201191059
Crockett M et al (2010) Serotonin selectively influences moral judgment and behavior through effects on harm aversion. Psychol Cogn Sci 107:17433–17438
Cunningham W et al (2004) Separable neural components in the processing of black and white faces. Psychol Sci 15:806–813
Double R (2002) The moral hardness of libertarians. Philo 5(2):226–234
Douglas T (2008) Moral enhancement. J Appl Philos 25(3):228–245
Farah M (2004) Emerging ethical issues in neuroscience. Nat Neurosci 5(11):1123–1130
Ferari P et al (2005) Escalated aggressive behavior: dopamine, serotonin and GABA. Eur J Pharmacol 526:51–64
Fischer J (2006) My way: essays on moral responsibility. Oxford University Press, Oxford
Freedman C (2000) Aspirin for the mind? Some ethical worries about psychopharmacology. In: Parens E (ed) Enhancing human traits: ethical and social implications. Georgetown University Press, Washington, DC, pp 135–150
Greely H (2012) Direct brain interventions to 'treat' disfavored human behaviors: ethical and social issues. Clin Pharmacol Ther 91(2):163–165
Harris J (2011) Moral enhancement and freedom. Bioethics 25(2):102–111
Harrison G (2009) Hooray! We're not morally responsible! Think 8:87–95
Hart A et al (2000) Differential response in the human amygdala to racial outgroup vs. ingroup face stimuli. Neuroreport 11:2351–2355
Honderich T (1984) Punishment: the supposed justifications. Penguin, Middlesex
Honderich T (2005) On determinism and freedom. Edinb University Press, Edinburgh
Kendell H (2002) The distinction between personality disorder and mental illness. Br J Psychiatry 180:110–115
Kennett J (2006) Do psychopaths really threaten moral rationalism? Philos Explor 9(1):69–82
Maibom H (2008) The mad, the bad and the psychopath. Neuroethics 1:167–184

Maletzky B, Field G (2003) The biological treatment of dangerous sexual offenders: a review and preliminary report of the Oregon pilot depo-Provera program. Aggress Violent Behav 8: 391–412

Maletzky B, Tolan A, McFarland B (2006) The Oregon depo-provera program: a five-year follow-up. Sex Abuse 18:303–316

Matthews E (2007) Body-subjects and disordered minds. Treating the whole person in psychiatry. Oxford University Press, Oxford

Mealey L (1997) The sociobiology of sociopathy: an integrated evolutionary model. In: Baron-Cohen S (ed) The maladapted mind: classical readings in evolutionary psychopathology. Psychology Press, East Sussex, pp 133–188

Mitchell E (2003) Self-made madess. Ashgate Publishers, Aldershot

Moore M (1998) Placing blame. A theory of the criminal law. Oxford University Press, Oxford

Newman J et al (2010) Attention moderates the fearlessness of psychopathic offenders. Biol Psychiatry 67:66–70

Olsen J (2006) Depression, SSRIs, and the supposed obligation to suffer mentally. Kennedy Inst Ethics J 16(3):283–303

Penney S (2012) Impulse control and criminal responsibility: lessons from neuroscience. Int J Law Psychiatry 35:99–103

Pereboom D (2001) Living without free will. Cambridge University Press, Cambridge

Phelps E et al (2000) Performance on indirect measures of race evaluation predicts amygdala activation. J Cogn Neurosci 12:729–738

Shaw E (2011) Free will, punishment and neurotechnologies. In: Van den Berg B, Klaming L (eds) Technologies on the stand. Legal and ethical questions in neuroscience and robotics. Wolf Legal Publishers, Nijmegen, pp 177–194

Shaw E (2012) Direct brain interventions and responsibility enhancement. Crim Law Philos. doi:10.1007/s11572-012-9152-2 (online first)

Smart J (1961) Free will, praise and blame. Mind 70:291–306

Strawson G (1986) Freedom and belief. Clarendon, Oxford

Vilhauer B (2009) Free will skepticism and personhood as a desert base. Can J Philos 39(3): 489–511

Chapter 21
Enhanced Control and Criminal Responsibility

John Danaher

Abstract Should agents be held criminally responsible for the consequences of failing to make use of enhancement technologies? This chapter argues that they should, provided such technologies would have allowed them to avoid the risks associated with the state of abnormal agency. The argument comes in three parts. First, the current position with respect to abnormal agents in the criminal law is discussed. As is made clear, we are usually quite willing to hold abnormal agents criminally responsible for their acts, provided certain conditions are met, but traditional capacitarian theories of criminal responsibility fail to account for and justify this willingness. Second, a theory of criminal responsibility – the distributive theory – is presented that does account for and justifies this willingness. And third, this theory is applied specifically to the use of enhancement technologies to avoid states of abnormal agency.

Keywords Enhancement technologies • Law • Criminal justice system • Criminal responsibility • Agency

21.1 Introduction

If enhancement technologies afford us greater control over our behavior, should we be held criminally responsible for the consequences of failing to make use of them? This chapter addresses that question in three parts.

Section 21.2 reviews the current (Anglo-American) position on the issue of criminal responsibility for actions performed during a state of abnormal agency. Such a state can be said to exist whenever the capacities that are usually constitutive

J. Danaher (✉)
School of Law, Keele University, Keele, UK
e-mail: j.danaher@keele.ac.uk

E. Hildt and A.G. Franke (eds.), *Cognitive Enhancement: An Interdisciplinary Perspective*, Trends in Augmentation of Human Performance 1,
DOI 10.1007/978-94-007-6253-4_21,

of responsibility are switched off, inactive or impaired. This fact alone might be thought to negate the possibility of responsibility in such cases. But a consideration of a range of cases seems to suggest that we are quite willing to hold some abnormal agents criminally responsible. Traditional capacity-based theories of responsibility seem unable to account for these cases.

Section 21.3 proposes an alternative theoretical framework within which liability for acts committed in a state of abnormal agency can be both justified and properly assessed. Two devices, both derived from game theory, are employed in this section. The first device is that of the bargaining game, and it is used to develop the "distributive theory of criminal responsibility." It proposes that the imposition of criminal responsibility can be seen to arise from a bargaining game in which a risk burden is distributed across the members of a society. The second device is that of the precommitment game. This is a game that an agent plays with a future version of itself. The agent learns that by undertaking certain activities at one moment in time, they can precommit to avoiding other activities at a later time. It is argued that the ascription of responsibility to abnormal agents is attributable to the presence of precommitment failures, *i.e.* failures to avoid the risks attached to a future state of abnormal agency. The device of the precommitment game is then integrated into the distributive theory of responsibility by arguing that a precommitment failure can be seen as a failure to discharge one's share of the distributed risk burden.

Section 21.4 applies the framework developed in part two to the enhancement debate. Initially the distinction is made between two varieties of enhancement – restorative and supernormal. It is then argued that the distributive theory developed in Sect. 21.3 already deals with the restorative use of enhancement. The question pursued in the remainder of the third section is whether the same theory can be used to cover supernormal enhancement.

21.2 Can Abnormal Agents Be Responsible

In this section, I explore the possibility of being held criminally responsible for actions committed while in a state of abnormal agency. I begin by distinguishing between states of normal and abnormal agency. I then consider cases in which it seems legitimate to hold people responsible for actions committed in a state of abnormal agency.

21.2.1 Normal and Abnormal Agency

Typically, theories of criminal responsibility maintain that agents are responsible for their actions in virtue of the fact that they possess certain capacities that direct, control or sustain the performance of those actions (see Horder 1993).

Three classic theories of responsibility help to illustrate the idea. The first is the *character theory* of responsibility. According to this theory, an agent is responsible for an action if that action is causally linked to their character. Character is usually defined as some set of stable traits, dispositions or second-order values, and it is the capacity to establish those stable traits and dispositions and to link them to an action that makes the agent responsible (for examples, see Bayles 1982; Tadros 2005). The second theory is the *choice theory* of responsibility. According to this theory, a person is responsible for those actions that are the product of a choice mechanism (Hart 1968; Moore 1997; Alexander and Ferzan 2009). This is a mechanism, possessed by the agent, that allows him to imagine various possible future actions, sift through those actions with the use of some weighting principle and select one of those actions for performance. The third theory is the *volitional theory*, according to which an agent is responsible for an action if he consciously represents and intends that action (MacKay 1995, chapter 1; McAuley and McCutcheon 2000, chapter 2; Wilson 2002, chapter 4).[1]

For the purposes of this chapter, we need not be wedded to any particular one of these theories; what matters is the general idea of responsibility attaching to agent in virtue of certain capacities. We shall call such theories of responsibility "actual-capacity" theories.

Consideration of these actual-capacity theories allows us to distinguish between actions performed in normal and abnormal states of agency:

(i) **Normal Agency**: Actions performed while certain key capacities of agency are "active," "switched on" or "functioning normally" are actions that are performed in a ***normal state of agency***. Examples would include actions of which the agent is consciously aware of choosing or actions linked to certain character traits and so forth.
(ii) **Abnormal Agency**: Actions performed while certain key capacities of agency are "inactive," "switched off" or "impaired" are actions that are performed in an ***abnormal state of agency***. Examples would include actions performed while in a state of impaired consciousness, unconsciousness or in a state that is dissociated from one's character traits and so forth.

Our concern here is whether someone can be held criminally responsible for actions that are performed while in a state of abnormal agency. Before addressing that question directly, I need to offer a couple of comments about the meaning of some of the terms included in these definitions.

First, some may wonder whether it is correct to say that agents in abnormal states can perform "actions" at all. This would be, presumably, because they feel the term "action" is only properly applied to bodily movements (and their causal *sequelae*) that are directed by the core capacities of agency. I would suggest this definitional argument is a distraction from the core issue. Thus, I would suggest that "action" be simply stipulatively defined, for the purposes of this chapter, as "bodily movements

[1]For a more detailed philosophical take see Moore (1993).

and their *sequelae*." This definition avoids the concerns of those who wish to argue over the concept of action as opposed to the concept of responsibility.[2]

Second, while the idea of the capacities of agency being switched "on" or "off" might seem relatively unambiguous, what does it mean to say that the capacities of agency are operating "normally" or are "impaired"? The answer to this question is going to be crucial to understanding the overall relevance of my arguments to the cognitive enhancement debate. I take it that the normality of the operation of the capacities of agency can be determined in one of two ways. Either one takes some baseline set of measurements for a particular individual and measures impairment relative to that set of baseline measurements, or one surveys the population as a whole and determines some statistical average performance of those capacities and measures an individual's impairment relative to that average. Impairment is the opposite of enhancement, so it follows that just as there are two ways to determine impairment, so too are there two ways to determine enhancement. One of those ways is relative to the baseline set of measurements for the individual; the other is relative to some statistical average for the population.[3]

With those clarifications out of the way, let us proceed to consider whether it is legitimate to hold someone responsible for actions performed while in a state of abnormal agency.

21.2.2 The Normalcy Argument and Some Counterexamples

First, let us consider how proponents of an actual-capacity theory of responsibility might be inclined to argue about cases of abnormal agency. I suggest that the argument that is most consistent with their position would look something like the following:

1. X is responsible for A, iff at T1 (at or close to the time of A) X's responsibility-relevant capacities were active, switched on and functioning within normal parameters.
2. If X performed A in an abnormal state of agency, then X's responsibility-relevant capacities were either inactive, switched off, or impaired.

---

[2]Although some will insist they are linked, I argue that the range of cases discussed below cast doubt on the idea that action – at least when understood to require conscious control – is necessary for responsibility. On the link between awareness and responsibility, see Sher (2009). Relatedly, see Capes (2012), which looks at one of the principles (principle of possible refrainment) that has been used to distinguish an action from non-action and also as a necessary precondition for responsibility. Capes, in my opinion, casts sufficient doubt on the plausibility of the principle of possible refrainment.

[3]In terms of the connection to the enhancement debate, definitions like this can be accused of being arbitrary and morally unnecessary. On this point, see Bostrom and Roache (2007). The distinction between enhancement and treatment is discussed in Sect. 21.3 of this article.

3. Therefore, X would not be responsible for A if A was performed while X was in an abnormal state of agency.

Call this the "pro-normalcy" argument. The problem with this argument is that it seems vulnerable to a number of counterexamples. Take the following five scenarios; although the descriptions are my own, all of them, with the exception of the fifth, are derived from actual legal cases in which a person was ultimately held criminally responsible for their actions. The fifth scenario is an extrapolation from a recent set of experiments involving manipulation of moral reasoning through the use of transcranial magnetic stimulation. It is included merely to show how general the class of cases actually is and how technological advances might affect our perception of responsibility.

(a) Alan is a diabetic who works as a nurse. If he does not take his medication he is liable to go into a hyperglycaemic state. This can lead to severely impaired consciousness. One morning, after breaking up with his girlfriend, he forgets to take his medication. Later that morning, he enters a state of impaired consciousness and attacks a patient. He is charged with assault. At trial he argues that he cannot be held responsible for his actions because they were not performed voluntarily.[4]
(b) Martin is driving home from work one evening. It has been a long and tiring day. He feels incredibly drowsy. His eyelids get heavier and heavier ... He awakes to a scene of chaos. He has crashed the car and run down a pedestrian in the process. He is charged with causing death by dangerous driving. At trial he argues that he cannot be held liable for this offense because "driving" is a voluntary act. And since he was unconscious at the time, he could not have been "acting."[5]
(c) James is a down-on-his-luck bank manager. To solve some of his financial woes, he decides to carry out an armed robbery at the local betting office. After weeks of careful planning, the big day arrives. Initially, everything goes well. The terrified workers are handing over the money. However, when one of them makes a sudden move, James pulls the trigger on his firearm and kills the worker. At his trial, James argues that his action was reflexive, automatic and involuntary.[6]
(d) Hank is a melancholic and socially awkward Ph.D. student. He is infatuated with a fellow student. She spurns his advances. This is too much for him to bear and, traumatized, he strangles her to death. At trial he describes how, after being

[4]Relevant case law includes: R v. Hennessy (1989) 2 All ER 9; R v. Bailey (1983) 1 WLR 287; R v. Quick [1973] 3 All ER 347; Bratty v. A-G for NI [1963] AC 396 (this latter case involved epilepsy, not diabetes but did address the idea of negligently-induced automatism).

[5]Relevant case law includes Hill v. Baxter (1958) 1 QB 277; Jiminez v. The Queen (1992) 173 CLR 572.

[6]Ryan v. The Queen (1967) 121 CLR 205. And also Elliot (1968).

spurned, he seemed to become dissociated and disconnected from his actions – as though he were a mere spectator to proceedings. This, he argues, means his actions were not under his voluntary control.[7]

(e) Grace is suffering from a rare brain disease that affects her moral reasoning. To overcome this unfortunate malady, her psychiatrist advises her to purchase a new device that uses powerful magnetic waves to correct for such errors in brain function. She must start every day with a 15 min session on this machine. One day, she forgets to do so and tragedy ensues: she kills her best friend. At trial she argues that she was not in control of what she did, that her disease was the guilty party.[8]

Although this might be too obvious to state explicitly, note that in each of these scenarios, the criminally impugned conduct (the "external" element of the offense) takes place while the agent is in what was previously defined as an "abnormal" state.

Now the question arises: do we think, in line with the pro-normalcy argument, that the agents in each of the cases (a)–(e) are excused from responsibility? Although some of these cases are more difficult than others, and although I cannot presume to speak for all, I suspect that our answer to that question is "no." So for instance, I see no good reason why a man who falls asleep at the wheel of his car should be excused simply in virtue of the fact that he was asleep. Nor do I see any reason why an armed robber who meticulously plans a robbery should escape liability on the grounds that firing his gun was not under his immediate conscious direction.

Although these are simply my own intuitive reactions, I can offer some mild support for them from two directions. The first of these is from the body of case law dealing with such situations. Starting with the easiest case, that of Grace, I am unaware of any authority suggesting that responsibility can be avoided solely in virtue of the fact that one is intoxicated (other surrounding circumstances might be relevant). Similarly, case law seems to suggest that being asleep while driving does not, by itself excuse someone from responsibility. And although the case of the diabetic who neglects to take his medication is more uncertain, I think there is some support for the idea that such a person will be deemed criminally responsible under certain conditions.[9] I shall say more about such cases below.

The second line of support for my intuitions comes from the work of George Sher and his recent book *Who Knew? Responsibility without Awareness* (2009). As the title suggests, Sher's focus in this book is on the necessity of one particular capacity to our ascriptions of responsibility. The capacity in question is the capacity to be consciously aware of all the morally and prudentially relevant aspects of our

---

[7]R v. Rabey (1980) 15 CR (3d) 225.

[8]This is inspired by Young et al. (2010). The experiment involved an induced impairment to moral reasoning; my example involves an enhancement or improvement.

[9]See R v. Bailey (note 4) and Bratty v. A-G for Northern Ireland (note 4).

actions. According to some traditional theories, most noticeably Aristotle's Meyer (1994), an agent cannot be held responsible for those things of which he or she was unaware. Sher disputes this by detailing nine separate scenarios involving agents who lack awareness but who seem, nonetheless, responsible for what they do. Here is an example of one of these nine scenarios:

Joliet, who is afraid of burglars, is alone in the house. Panicked by sounds of movement in her kitchen, she grabs her husband's gun, tiptoes down the stairs, and shoots the intruder. It is her son, who has come home early for the holidays (Sher 2009, 26).

In this scenario, Joliet lacks awareness of the wrongness of her act. She thinks she is legitimately protecting herself from an intruder. But, argues Sher, she would surely be responsible for killing her son in these circumstances.

One of the considerations offered by Sher to justify the ascription of responsibility in these kinds of scenarios (*i.e.* scenarios involving lack of awareness) is to distinguish between first-person and third-person perspectives on action and responsibility. He argues that from the first-person perspective it is true that our actions must be guided by those things of which we are consciously aware. That is the essence of the first-person perspective. It follows that, if we adopt a first-person perspective on responsibility, we could only be held responsible for those things of which we are aware. But the problem is that this does not hold true when we switch to the third-person perspective on action and responsibility. When we adopt that perspective, many of our failures to be aware of something can seem blameworthy and that blameworthiness can carry over into how we perceive the actions performed in the state of impaired awareness (Sher 2009, chapter 3. See also the final condition of responsibility offered by Sher, 143). Sher further contends that the third-person perspective is the correct one when it comes to determinations of responsibility. So, his arguments support my intuitive belief that we can legitimately hold people responsible for at least some actions that are performed in a state of abnormal agency.

To sum up, the situations described in scenarios (a)–(e) are counterexamples to the pro-normalcy argument. They are counterexamples because intuitive reactions to them, backed-up by some current legal and philosophical considerations, suggest that agents can be held responsible for actions performed while in an abnormal state. Formally, this gives us the following, "contra-normalcy" argument.

4. If intuitive considerations are correct, then agents can be held responsible for actions performed in an abnormal state of agency.
5. If the active-capacity theory of responsibility is correct, agents can only be held responsible for actions performed in a normal state of agency.
6. Intuitive considerations are correct.
7. Therefore, agents can be held responsible for actions performed in an abnormal state of agency (from 4 and 6).
8. Therefore, the actual capacity theory is wrong (from 5 and 7).

This argument is far from watertight.[10] But it is, I submit, somewhat plausible. Taking this plausibility onboard, the goal for the remainder of this chapter is threefold: first, to see whether any deeper theoretical reasons can be adduced in support of the intuitive considerations just mentioned; second, if such reasons can be adduced, to determine the precise circumstances in which agents can be held responsible for actions performed in abnormal states; and, third, to see how this relates to the enhancement debate.

21.3 A Distributive Approach to Responsibility for Abnormal Agents

The purpose of this section is twofold: (i) to adduce some deeper theoretical reasons for thinking that abnormal agents can be held responsible for their acts; and (ii) to establish some criteria for holding abnormal agents responsible. In relation to the first of these tasks, a theory of responsibility, derived from the concept of an agency norm, is presented. In relation to the second of these tasks, the agency norm theory is operationalized using some concepts from game theory.

The discussion proceeds as follows. Section 21.3.1 briefly introduces the basic concept of an agency norm. Section 21.3.2 presents some concepts from game theory that can help to structure how we think about agency norms. Finally, Sect. 21.3.3 outlines the precommitment game and uses this to explain and justify our ascriptions of responsibility in cases of abnormal agency.

21.3.1 Agency Norms and Responsibility

An agency norm is some standard of agency. Unlike a traditional norm, which is usually directed at conduct and the consequences thereof, an agency norm is directed at the constitutive aspects of agency. What, you might ask, are the "constitutive aspects of agency"? Put simply, they are the agency-capacities discussed earlier. So, for example, a character norm might set down, not an ideal set of character traits (although it might do so), but an ideal standard of ownership over actions. In other words, it might demand that agents see their characters and personalities as more deeply connected to their actions. Similarly, a volitional norm might set down an ideal standard of conscious awareness over one's actions. That is to say, it might

[10]Specifically, it falls prey to the "one man's modus ponens is another man's modus tollens"-problem. In other words, a sufficiently adamant proponent of the active-capacity view could flip the argument on its head and suggest that because the actual-capacity theory is true, the intuitive considerations mentioned must be false.

demand that agents be aware of more of the morally and prudentially relevant features of their actions.

It is my contention that agency norms already play an important part in how we think about criminal responsibility. For example, consider the role of "reasonable person" tests in the criminal law. These tests direct one's attention away from how agents actually perceived and responded to their circumstances, towards more normatively ideal agent. The agents under consideration are then compared to this ideal: if they fail to match it, then they are deemed responsible.[11] So the act of ascribing responsibility can be seen to involve the prescription of agency norms.

21.3.2 A Distributive Approach to Agency Norms

The next question is: how exactly should one think about the prescription of agency norms? I propose a risk-distributive model that is based on a game theoretic conception of social and legal norms. I will briefly outline this model here.

The first step towards understanding the risk-distributive model is to appreciate the equivalence between game theoretic equilibria and social norms.[12] Game theory is a tool for studying the interactions of two or more purposive agents. A formal "game" consists of players, actions or strategies, payoffs, and information.[13] The basic goal of the game theory is to find the set of strategies in such interactions that are in equilibrium. Although there are precise mathematical formalizations of equilibria, we can say in simple terms that a set of strategies constitutes an equilibrium if no rational agent has an incentive to deviate from those strategies.

---

[11]This glosses over some of the history surrounding the operation of reasonability standards of this sort. The standard of reasonability can be assessed in subjective or objective terms. If it is assessed in objective terms, then there truly is a general normative standard of agency that is at the heart of the test. If it is assessed in subjective terms, then the normative standard is relativized to the individual in question. My personal feeling is that a highly subjectivized version of the test is absurd. The absurdity is well-covered in Norrie (2002). Norrie notes that the reasonability component from the defense of provocation has become highly relativized over the past 50–60 years. This leads to the absurd conclusion that even a substance abuser can be reasonable about his addiction. It is noteworthy that in a recent report on criminal defenses, the Irish Law Reform Commission (2009) recommended a more objective test.

[12]For more fulsome defenses of the "social norm = equilibrium"–theory, see Binmore (1993/1998), Posner (2000), Bicchieri (2006), and Gintis (2009a).

[13]The **p**layers, **a**ctions, **p**ayoffs and **i**nformation (PAPI) terminology is employed by Rasmusen in his textbook account of game theoretic modeling (Rasmussen 2004). While the acronym is useful, different terminology is employed by other writers in this area. For example, they may refer to "agents" instead of "players" or "outcomes" instead of "payoffs." For the most part, the differences are of no great significance; however, the distinction between an "action" and a "strategy" is important. An action is a possible choice at one round or move in the game; whereas, a strategy describes a set of choices for every round or move in the game. Some games have multiple rounds; some games have only one. In the latter, actions are equivalent to strategies; in the former, they are distinct.

The most famous equilibrium concept in game theory is the Nash Equilibrium, which arises whenever the rational agents are playing best responses to each other. Since a social norm is some regularized, agreed-upon, or widely-observed behavioral standard,[14] equating norms with equilibria makes a lot of sense. If one wishes to differentiate legal norms from social norms, one could use the concept of "signalling" and say that, instead of leaving it up to the agents themselves, the legal system is that which signals to different players the strategies they must play.[15]

The next step towards understanding the risk-distributive approach to agency norms is to consider the formal structure of a bargaining game. There are two parts to this structure. First, there is a group of players who can achieve some gain or surplus if they manage to cooperate and coordinate their activities. Second, there is a potential impediment to cooperation due to disputes over the appropriate division of this surplus.

A simple example would be the *divide-the-cake* game. It involves two players, A and B, who are either given a chocolate cake which they must divide between themselves or who have the option of pooling their resources in order to make a chocolate cake. The rules of the game are as follows. They must each, simultaneously, make a demand for some fraction of the cake. If the sum of the fractions demanded is greater than one, then the cake will be taken away from them (or they will simply fail to pool their resources to make it); if the sum of the fractions demanded is less than or equal to one, then they each get whatever fraction of the cake they happened to demand.[16] The basic structure behind this game is replicated in many critical areas of distributive justice such as "who is entitled to which benefits and welfare payments?"; "what is a fair system of taxation?"; "who is entitled to which legal rights?" and so on. Problems like this are at the heart of the "social contract," which is the moral core of human society.

In terms of the solution to the divide-the-cake game, we begin with a straightforward game theoretic analysis and try to find the potential Nash equilibria. Without walking through every step in that analysis, we should be able to see that this game has a large number of Nash equilibria. In fact, every pair of fractional demands that adds up to exactly one, is a potential Nash equilibrium.[17]

---

[14]Elster (2007, chapter 22) distinguishes between social, legal and moral norms. He does so on the grounds that different enforcement mechanisms are associated with each of them. I do not follow this taxonomy. For my purposes, a social norm is simply any behavioral standard governing social relationships. As such, many legal and moral norms are just specific types of social norm.

[15]One could appeal to the correlated or choreographed equilibrium concept to support this theory. See Aumann (1974, 1987), and Gintis (2009b, chapter 7) discussing correlated equilibria and their role in understanding social norms.

[16]As discussed in many sources, for more on this version, see Skyrms (1996).

[17]Why is this? Well, obviously, because in every one of those cases each player is playing a best response to the other player's best response. Imagine that A thinks B is going to demand 2/3 of the cake. In that case, A should demand 1/3: demanding less than that would fail to fully use up the surplus, and demanding more would result in the surplus being wasted. For similar reasons, if B thinks A will demand 1/3, then his best response will be to demand 2/3. As a result, the pair of

This basic analysis is somewhat disappointing. It suggests that in order for a social norm of distribution to exist, something must be done to select one of the potential equilibria from the possible set. Such equilibrium selection is, I submit, the function of the law. In deciding which distribution to pick, some moral or prudential principle will need to be applied by the legal authorities. Which principle is used is relatively unimportant for present purposes; what is important is that a principle must be adopted and that any decision about a proposed division of the surplus must be referred back to that principle. This will always involve a decision procedure in which the parties ask whether the division is one they could agree to under conditions of fair bargaining (*i.e.* away from conditions that might distract them from their chosen moral principle). The most famous example of such a decision procedure in modern times is, of course, Rawls's proposed system of bargaining from behind a veil of ignorance. The veil of ignorance helps to direct the bargainers' attention away from anything that might distract them from Rawls's preferred egalitarian principle of distribution (Rawls 1971).

This distributive theory of criminal responsibility begins by noting the double-edged sword that is involved in the distributive question outlined above. For not only is there a surplus to be divided among the participants to the bargain, there is also a correlative burden. And it is this burden that, according to my proposal, forms the basis of all agency norms.

We can see this if we look once more to the divide-the-cake game. The analysis undertaken above focused entirely on the distribution of the surplus (the cake itself). But if we assume that the two players are not simply gifted the cake and, instead, must combine their resources in order to make the cake, then we see that there is both a surplus and a burden that must be distributed. After all, both players must put in some sort of effort when making the cake – *e.g.* by purchasing or growing the ingredients or combining and baking the ingredients – otherwise the surplus will not be obtained.

Therefore, there must be norms dictating how much effort the participants are expected to expend in order to achieve that surplus. These will be agency norms, and they will ensure that people *bear the effort burden* associated with the surplus. This is the core idea of the agency-norm theory of responsibility, *i.e.* that responsibility is not, primarily, a question of whether certain capacities were exercised when an agent performed an act, but, rather, it is a question of whether it is fair to impose an effort burden on the agent.

There are two different ways in which we can give meaning to the concept of an effort burden. The first would be to think of it in a *positive* sense. From this perspective, the effort burden is something we must bear *in order to achieve* the desired surplus. This is to be contrasted with the *negative* sense, which views the

---

fractional demands (2/3, 1/3) is a Nash equilibrium. The same argument can be made for all other pairs of fractional demands that add up to one.

burden as one that must be borne *in order to avoid the risk* of losing the desired surplus. Because we who live in modern, reasonably well-functioning societies tend to treat the social surplus afforded to us by the legal system as a given (*i.e.* as part of the existing status quo), it might be more natural to think of the effort burden in the negative sense. That is to say, we think of the effort burden as something we bear in order to avoid the risk of losing that stability. We can then simply call this the "risk burden."

Now that the basic idea of the distributive theory of responsibility has been outlined, we can proceed to slot it in with our analysis of the responsibility of abnormal agents. As you recall, the contra-normalcy argument effectively gives reason to doubt the typical actual-capacity theory of responsibility. As a result, it leaves us looking for a new theory of responsibility that can support our intuitive reactions to certain scenarios. It is my contention that the distributive theory of responsibility can support those intuitive considerations. This can be shown by presenting a simple formal argument that works from the premise that the distributive theory is correct to the conclusion that at least some abnormal agents, such as those in scenarios (a)–(e) above, are responsible for their actions. As follows:

9. An agent is responsible for an action *A*, iff the agent bears the burden of avoiding the risk of *A*.
10. An agent bears the burden of avoiding the risk of *A*, iff the avoidance of *A* is something we would expect of the agent under conditions of fair bargaining.
11. Therefore, an agent is responsible for *A* iff the avoidance of *A* is something we would expect of the agent under conditions of fair bargaining (from 9 and 10).
12. In scenarios (a)–(e) outlined above in Sect. 21.2 involving abnormal agents, the respective agents would have been expected to avoid their states of abnormality (which created the risk of certain criminal acts taking place) under conditions of fair bargaining.
13. So the abnormal agents in scenarios (a)–(e) are responsible for their actions (from 11 and 12).

It is important to be clear about what this argument does and does not do. The argument *does* provide us with a general theory for assessing when it is appropriate to hold people responsible for their actions. The theory is distinct from the actual-capacity theory in that it switches focus away from the status of the agent at the time of the offense and to the burdens placed on the agent under conditions of fair bargaining. The argument *does not*, however, fully answer all questions about the legitimacy of holding abnormal agents responsible for their actions. The problem is that premise (12), in particular, seems like nothing more than a bare assertion. We are forced to ask the deeper question: why exactly is it that we would expect agents to avoid the risk of those actions that result from the states of abnormality present in scenarios (a)–(e)?

21.3.3 The Precommitment Game

We begin to answer that question by developing some relatively uncontroversial propositions about the temporal nature of agency. These propositions will then be used to construct an argument that allows us to reach a conclusion about one of the things we might expect of agents under conditions of fair bargaining.

21.3.3.1 Three Propositions About the Temporal Nature of Agency

The first proposition about the temporal nature of agency is obvious but still needs to be stated clearly[18]:

(i) Human agents perdure.

That is to say, when we see an individual human being acting at one moment in time (t1) and then see him acting at a later moment in time (t2), we see him as the same human being. Although there are different theories concerning the mechanics of this perdurance, and although the matter is not uncontroversial, I take it as a given for the remainder of this chapter.[19]

The second proposition about the temporal nature of agency complicates the picture somewhat by saying:

(ii) Although agents perdure, their interests (preferences) can change[20] in a manner that is dependent upon their constitution, their experiences and the particular context in which they find themselves.

For the purposes of this discussion, we will take a simple example of the problems that can arise from the change of interests over time. The example involves addiction, *i.e.* a short-term interest in gratification that is both (a) antithetical to long-term interest and (b) dominant over that long-term interest in the short-term. Most people will be able to think of examples from their own lives but this one involves the decision to start smoking. The decision to start smoking involves the trade-off of a long-term interest in one's future health and well-being against a short-term desire to smoke. When making the decision whether to smoke in the short-term, one must

[18]There is a longstanding debate between perdurantists and endurantists concerning the proper account of persistence through time. I adopt the language of perdurantism here because it is the position I find myself, conditionally, endorsing. Despite this, I do not think anything I say here hinges greatly on the distinctions between perdurance and endurance. One could substitute "perdurance" and "endurance" throughout my discussion. On the debate, see Hawley (2010).

[19]Since I am rejecting the actual-capacity theory of responsibility, I need not assume any continuity of personal identity or memory over time. All I need assume is the continuity of the same physical entity. It is the perdurance of personal identity, as opposed to the perdurance of the same physical entity that tends to be philosophically contentious.

[20]"Change" is here taken to include: substitution of one interest for another, elimination or subtraction of an interest, and addition of an interest.

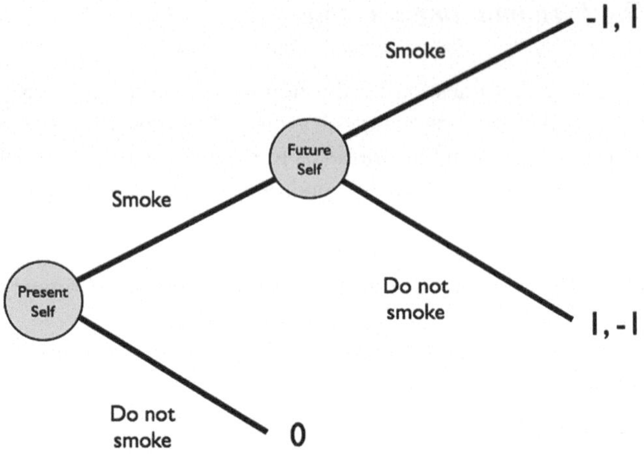

Fig. 21.1 The smoking game

take into consideration the fact that one's future self may be addicted to nicotine and find itself unable to quit.

We can illustrate this interaction with one's future self by means of a game tree. We assume that there are two players (present self, future self) and two actions (smoke, do not smoke). Your present self gets a payoff of 0 for not smoking in the short-term, a payoff of 1 for smoking in the short-term but quitting in the long-term, and a payoff of −1 for continuing to smoke in the future. Your future self, who likes his/her nicotine, will get a payoff of 1 for smoking and a payoff of −1 for not smoking.[21] That gives us the following tree (Fig. 21.1):

Solving this by means of backward induction, it is clear that your present self cannot count on your future self to support your long-term interest in health and well-being. Indeed, your future self, given the opportunity, is always going to choose to smoke. Thinking about the decision to start smoking in this way should help you to realize that you are better off if you never start smoking.

This brings us to the third of our propositions about the temporal nature of agency, which is:

(iii) Treating one's future self as a (potentially competitive) agent in a strategic interaction can help one to identify the ways and means for avoiding the possibility that one's future self will make decisions or perform actions that are contrary to one's long-term interests.

This is the most significant proposition, and it is worth spending some time exploring its significance.

[21]This example is taken from Dixit et al. (2009, 52–57).

Strategists have identified three basic tools that one can use to counteract the effects of a future self with counterproductive interests: (i) threats; (ii) promises and (iii) precommitments. A threat is something that punishes a future self for engaging in counterproductive actions (or for failing to engage in productive actions); a promise is something that rewards a future self for avoiding counterproductive actions; and a precommitment is something that a present self can do to eliminate (or reduce) the possibility of a future self who has the choice or chance to engage in counterproductive activity.[22] The argument here is that abnormal agents can be held responsible if they fail to avail themselves of viable precommitment strategies that would have helped them to avoid the risks associated with a state of abnormality.

The classic example of a precommitment strategy in action comes from the story of Ulysses (or Odysseus) and the Sirens from Homer's *Odyssey.* As you no doubt recall, Odysseus knew that if he heard the Sirens' song, he would fall under their spell. In other words, he would enter a state of abnormal agency that had a certain level of risk associated with it. To enable him to hear their song, without realizing the associated level of risk, he forced his shipmates to put wax in their ears and to bind him to the mast of the ship. This way, he was unable to do any harm to the ship.

21.3.3.2 An Argument

We can now use the three propositions about the temporal nature of agency as the basis for an argument defending the legitimacy of holding people responsible for actions committed in a state of abnormal agency. The argument runs as follows:

14. Agents (sometimes) have the means at their disposal to reduce the probability of their future selves engaging in counterproductive actions through the use of viable precommitment strategies.
15. An agent bears the burden of avoiding the risk associated with A, iff the avoidance of A is something we would expect of the agent under conditions of fair bargaining (previous argument).
16. If an agent has the means at his disposal to reduce the risk of A during a future state of abnormality, then the avoidance of A is something we might expect of the agent under conditions of fair bargaining.
17. Therefore, an agent might be responsible for A iff he could have reduced the risk of A through the use of a viable precommitment strategy.

The argument can be applied directly to the scenarios (a)–(e). To take but one example, Martin the drowsy driver could be said to have failed to avail himself of an obvious precommitment strategy that would have avoided the risks associated

---

[22] More correctly, promises and threats can be either compellent or deterrent. A compellent threat or promise is designed to encourage (compel) an agent to do something; a deterrent threat is designed to prevent (deter) an agent from doing something. The classic discussion of this distinction can be found in Schelling (2008, chapter 2 "The Art of Commitment.")

with driving while asleep. The obvious precommitment strategy would have been to stop driving when feeling drowsy. To stop driving when feeling drowsy is surely something we would expect of agents under conditions of fair bargaining, and, so, this can legitimize our belief that Martin is responsible for causing death by dangerous driving.

That much seems uncontroversial. There are, however, some worries about this argument. A particular worry arises from the new principle introduced in premise (16). This principle, when combined with (15), suggests that agents can be held responsible for the downstream consequences of their failure to avail themselves of *viable precommitment strategies*. This raises the question of what counts as a viable precommitment strategy. Surely, we need to be sensitive to the potential side effects associated with precommitment strategies when deciding upon whom the risk burden should fall. Further refinements are thus needed to make premise (16) fully defensible.

This brings us to something we might call the "generic problem of precommitment"; namely, the problem of hyperbolic discounting.[23] This problem arises from the way in which people seem to value future rewards. It obviously makes some sense to discount or downplay the value of rewards that will accrue in the future when compared with rewards that will accrue in the present. But the formula through which this discounting is performed can lead to significant differences in behavior. In particular, a hyperbolic discounting formula leads to an inconsistent treatment of future rewards. Evidence suggests that people do indeed engage in hyperbolic discounting and so prefer "smaller sooner" rewards when compared to "larger later" rewards.[24]

Because of this inconsistent treatment of the future, we experience all sorts of problems in regulating and controlling our future behavior. This needs to be incorporated into the analysis of the precommitment problem and responsibility. The ideally rational agent has consistent preferences over time; the boundedly rational agent does not. If we are trying to construct a normative system that encourages people to precommit, we should take into account their biases and foibles when doing so. In other words, we should be careful not to impose excessive burdens on people in the effort to avoid risk and achieve social stability. The benefits of encouraging people to avail themselves of precommitment strategies, *i.e.* the volume or magnitude of the risk being avoided, need to be weighed carefully against the costs of doing the same. This leads me to propose the following refinement to premise (16) from the preceding argument:

[23]See Elster (2007, chapter 6) for more on this. See also Ainslie (1992, 2001).

[24]See Ainslie (2001, chapter 3) for a review of the evidence. See also Loewenstein et al. (2003) and McClure et al. (2004). McClure's neurobiological work only supports the idea of quasi-hyperbolic discounting where preferences are definitely present-biased and follow hyperbolic curves, but some features of exponential discounting remain.

16a. Under conditions of fair bargaining, an agent would be expected to avoid the risk associated with A at T2, if at T1 the following conditions apply:

(i) The agent has the means at his disposal to avoid performing A at T2;
(ii) These means have a reasonably high probability of successfully avoiding the risk associated with A;
(iii) A has a high probability of risk associated with it or a low probability of high magnitude risk;
(iv) A failure to encourage precommitment would give rise to moral hazard (or risk spreading);
(v) Any side effects associated with the available precommitment technique are outweighed by the risk associated with A.

A word or two must be said about each of these conditions. Condition (i) is obvious: if it is right and proper to hold agents responsible for the downstream consequences of their failures to precommit, this can only be because the agents have the means at their disposal to actually precommit.

Condition (ii) is a sop to the fact that there are no certainties in this world. No precommitment strategy will have a 100 % guarantee of success, so we must live with whatever probability assessments we can muster up. Note that this implies that where objective (frequentist) data is unavailable, subjective (Bayesian) assessments may have to be utilized.

Condition (iii) introduces a sensitivity to the level of risk that is associated with the future state of abnormality. If there is a very high probability of some damaging state of affairs being realized in the future (e.g. as there might be if one drives while asleep), then it would seem right to expect people to avoid it, provided the other conditions hold. This is true even if the state of affairs is not too damaging. On some occasions, however, even a low probability of a highly damaging state of affairs might be enough to warrant the imposition of a risk burden. For example, even if the probability of a terrorist attack is low, the magnitude of the damage associated with such an attack may warrant extra caution.

Condition (iv) introduces a sensitivity to how people react to incentives. This, it should be noted, is one reason why legal systems seem to have accepted the fact that agents in states of abnormality can be held responsible for what they do. The worry in some jurisdictions is that failing to hold such people responsible out of loyalty to an active-capacity theory of responsibility may encourage others to fail to avoid (risky) states of abnormality. This, it is felt, is not something that the courts should encourage.

Condition (v) forces us to consider the side effects that might be associated with available precommitment strategies. This would seem to be particularly pertinent when the strategies involve the use of medication or new technologies with unknown risks and side effects.

It is important, when deciding whether a particular agent should be held responsible for actions performed while in an abnormal state of agency, to consider all five of these conditions. To see what this might look like, I offer the following analysis of the case of Alan the diabetic nurse who forgot to take his insulin (scenario a). As you recall, Alan subsequently attacked a patient while in a state of impaired consciousness. Conditions (i) and (ii) obviously hold true for Alan's case: he had means to avoid the state of impaired consciousness and those means had a high probability of success. Whether condition (iii) holds is less obvious, but I suspect it does, given the facts about Alan's profession and the conditions of the people he is entrusted with looking after. Condition (iv) seems to hold because the population of people who may face similar situations is not insignificant. Finally, condition (v) would also seem to hold true: although there are side effects associated with the use of insulin, they would not appear to be such as to outweigh the risks associated with the impaired state of consciousness. Thus, based on the principle embodied in premise (16a), Alan is responsible for the actions he performed while in the abnormal state of agency.

I submit that when (16a) is incorporated back into the previous argument, we have an argument that explains (i) why it is sometimes legitimate to hold agents responsible for actions committed in abnormal states and (ii) specifies when exactly this is legitimate.

21.4 From Normal Agency to Supernormal Agency

In this section I explore the implications of the preceding theory of responsibility for the cognitive enhancement debate. To set the scene, we must begin by defining enhancement and distinguishing between its forms. For present purposes, I follow the earlier definition and describe cognitive enhancement as an improvement in the functioning of a cognitive system relative to some baseline or statistical average; or, more exotically, the creation of a new system.[25] There is then a distinction to be drawn between two varieties of enhancement:

(i) **Restorative Enhancement (or "Treatment")**: when the enhancement technology or technique is used to restore the functioning of an agent's cognitive capacities to a normal state (usually measured against some population-wide statistical average).
(ii) **Supernormal Enhancement (or "Enhancement Proper")**: when the enhancement technology or technique is used to increase the functioning of an agent's cognitive capacities beyond the normal state (measured against some statistical average).

The discussion over the course of Sects. 21.2 and 21.3 has effectively dealt with restorative enhancement. The end result of that discussion was to develop a formal test for deciding whether people can be held responsible for failing to avail themselves of some enhancement technology that could have helped them avoid a

[25]Similar to the definition offered by Bostrom (2008).

future state of abnormality. The question for the remainder of this section is whether the same test applies to cases of supernormal enhancement.

The concern is that somehow there might be some important, normative or principled reasons for treating supernormal cases in a distinct or separate manner from restorative cases. I want to consider two arguments that respond to this concern.[26] The first is what I shall call the "no difference"-argument; the second is what I shall call the "inverse prisoners' dilemma"-argument. The first of these arguments relies on the use of analogies to point out that there is no principled reason for treating supernormal cases any differently than restorative cases. The second argument points out that encouraging or demanding supernormal enhancement might lead to a prisoners' dilemma-type situation in which an excessive, socially-disruptive burden is placed on everybody.

21.4.1 The "No Difference"-Argument

The "No Difference"-argument is analogical in form. Since this is a widely studied form of argument, and since the goal of this discussion is to evaluate an argument of this form, it will be worthwhile to briefly consider analogical arguments in the abstract. The appeal here is to the work of the argumentation theorist Douglas Walton. As Walton points out, analogical arguments have the following form:

(i) Case 1 is similar to Case 2.
(ii) Proposition *P* is true in Case 1.
(iii) Therefore, *P* is (probably) true in Case 2.

Walton goes on to point out that there are three key critical questions to be asked of all such arguments. First, are Case 1 and Case 2 really that similar (i.e. are there any *disanalogies*)? Second, is *P* really true in Case 1? And third, is there a third case (Case 3) that is similar to Case 1 but in which *P* is not true (*i.e.* are there any *counter-analogies*)? We will make use of these critical questions in due course.

John Harris's (2007) book *Enhancing Evolution* contains, perhaps, the most emphatic use of analogical arguments in an effort to defend the moral rectitude of enhancement. Indeed, the book opens with the following analogy (note: it does not appear in this argumentative form in the text) (Harris 2007, 1–2)[27]:

18. We do not have any moral objection to schools trying to enhance the cognitive capacities of their students through different educational programs.

---

[26]There are, of course, numerous arguments focusing on the importance (if any) of this distinction. I limit my focus here in the interests of brevity and relevancy. See the contributions from Lisa Forsberg and Kirsten Brukamp in this volume for more.

[27]Harris uses this example as an introduction to the themes of his book. However, I think the argumentative reconstruction is fair since he employs such analogies throughout his book in an effort to defend the moral legitimacy (and indeed obligatoriness) of enhancement.

19. The use of educational programs to enhance cognitive capacities is similar to the use of, say, genetic reengineering or drugs to enhance cognitive capacities.
20. Therefore, we should have no moral objection to the use of such enhancement technologies.

This is a good example of how these kinds of analogical arguments work in the enhancement context. However, it is not a no-difference argument. It simply compares two cases of enhancement. A no-difference-argument, as I define it, needs to point to a case of restorative enhancement and a case of supernormal enhancement, argue that there is no principled difference between our moral response to the two cases and, then, suggest that this applies to all restorative/supernormal comparisons.

Harris does employ a no-difference argument later in his book when he discusses the use of spectacles and binoculars (again this is a reconstruction, not something appearing in the original text) (Harris 2007, 19–20):

21. Those who use spectacles are using a restorative enhancement technology; those who use binoculars (or telescopes or microscopes) are using a supernormal enhancement technology.
22. No one thinks that they cross an important moral divide when they switch from using spectacles to using binoculars.
23. All cases involving technologies that can be used for restorative and supernormal enhancements are like the spectacle-binocular case.
24. Therefore, there is no important moral difference between restorative and supernormal uses of enhancement.

No doubt some people would be inclined to dispute this argument. In particular, they might be inclined to argue that there are significant differences between cases involving chemical or genetic enhancement and the spectacle/binocular case.[28] I will not pursue these kinds of objections here because my goal is simply to use this as inspiration for constructing a no-difference argument that is directly relevant to the present concern with tests for responsibility.

A no-difference argument that is relevant to the responsibility-case will first point to two enhancement cases (one restorative and one supernormal), in which the same basic principle or theory of responsibility is used, and then claim that all analogous cases will use the same basic theory of responsibility. Here is the example we will use, which builds upon the binocular/spectacle motif that has been established:

25. We would assess the responsibility of a person who needs to wear glasses while driving in accordance with the distributive theory of responsibility set down above.
26. We would assess the responsibility of a doctor who needs to use a microscope to make a diagnosis in accordance with the distributive theory of responsibility set down above.

[28]Harris discusses some of these objections and offers what are, to my mind, fairly persuasive refutations of them.

27. All responsibility-relevant cases involving restorative/supernormal enhancements are similar to these two cases.
28. Therefore, we would use the same theory to assess responsibility in supernormal cases as we would in restorative cases.

Since this is the version of the no-difference argument that is directly relevant to the responsibility case, we cannot avoid evaluating it. In doing so, we will make use of Walton's critical questions. First, we will ask whether the claims made about the spectacle and microscope cases are actually true. Second, we will ask whether there are significant disanalogies between the use of spectacle/microscope enhancement technologies and other varieties of enhancement technology.

In regard to the first of these questions, I take it to be a relatively uncontroversial fact that the driver and doctor would be held responsible for the downstream consequences of their failure to make use of enhancement technologies. I presume the reader agrees that it would be absurd for the doctor to excuse himself on the grounds that the microscope is supernormal, not merely restorative in effect. This might seem to suggest that no further analysis of these cases is needed. But note that the question is not whether it is true that the individuals in these cases are responsible; the question is whether the same theory of responsibility is used in both cases. The answer to this subtly different question is, I submit, "yes." The distributive theory that was set down in Sect. 21.3 is flexible enough to cover both types of case because it assesses whether it might be reasonable to expect an agent to use whatever means are at his disposal to enhance control over certain future states of affairs; it does not provide any reason to treat restorative enhancement differently from supernormal enhancement.

This is borne out when we refer back to some of the original cases that inspired the development of the distributive theory. For example, it was previously argued that Martin, the drowsy driver, could be held responsible for the consequences of failing to stop driving. This was because we felt it was reasonable to expect Martin to exert enhanced control over his future state of being. Although the net effect of this was to get Martin to avoid a state of abnormality, it would not seem accurate to say that this involved a restorative use of enhancement. After all, Martin is, so far as we are aware, an otherwise cognitively and physically normal human being (*i.e.* lying close to the center of the statistical distribution of sleepiness). At the same time, demanding that Martin stop driving when he feels sleepy is not quite the same thing as asking him to develop some supernormal faculty of conscientiousness. Still, it is pushing him in that direction, and the fact that we use the same theory of responsibility to do this, suggests that premises (26) and (27) are defensible.

The second critical question is whether there is any disanalogy between the restorative and supernormal cases. Consistent with my answer to the first critical question, I do not think there can be a categorical reason for not employing the same theory of responsibility in the restorative and supernormal cases. But I am open to the possibility that there are some consequentialist or prudentialist reasons for treating the cases separately, or for treating the supernormal case with more caution than the restorative case. The inverse prisoners' dilemma argument, which I address next, tries to make the case for such differential treatment.

21.4.2 The Inverse Prisoners' Dilemma-Argument

The argument presented here is inspired by a recent article by Saskia Nagel (2010). Nagel's article presents a set of considerations that urge caution when it comes to the use and regulation of enhancement technologies. Among these considerations is an interesting discussion of the (potentially negative) impact of enhancement on perceptions of responsibility and self-determination. But, despite addressing the issue of responsibility, Nagel is not – nor would I expect her to be – directly concerned with tests for criminal responsibility. So to fit with the subject matter of this chapter, I will simply take parts of her discussion and refashion them into a seemingly relevant argument.

Before presenting the argument, I need to briefly summarize what Nagel has to say. She points out that the increased presence of enhancement technologies presents agents with more options (choices) over the future direction of their lives. Although, many may think that more choice is a good thing – perhaps on the grounds that it gives individuals more opportunity to choose their own path to the good life – Nagel notes that there is a dark side to increased choice.

The dark side has several elements to it. Using the basic concepts of decision-theory (*i.e.* viewing decision problems as attempts to navigate a branching tree of possible choices in an effort to maximize expected utility), Nagel makes two initial observations: (i) psychological evidence suggests people do not have the computational resources to perform the maximization calculations required when there are many choices (Nagel 2010, 111–112)[29] and (ii) more choice leads to more opportunities for regret. Both of these facts indicate that more choice can actually reduce an agent's feeling of well-being.

Turning to the issue of responsibility, Nagel notes that not only does more choice lead to more regret, it also leads to a greater feeling or perception of responsibility, particularly for one's failures or shortcomings. Why so? Well, if one has the opportunity to choose a beneficial path, but one fails to do so, one is left with the lingering feeling that one "could have done otherwise." This feeling becomes more pronounced when, as is the case with enhancement technologies, we are given the option to not only change our actions, but to change, say, the particular balance of neurochemicals in our brains (Nagel 2010, 114). In such cases, we separate ourselves from the neural structures by which we are constituted, and view those structures as something over which we can exercise control and for which we have responsibility. In the end, as enhancement technologies become more pervasive and precise, there will be no part of our lives for which we do not feel we have responsibility.[30]

[29]Discussing evidence suggesting human are satisficers, not maximizers; and that some companies (e.g. Apple) improved sales by reducing the number of choices available to customers.

[30]Problems arising from the total manipulability of life are explored in Owens (2007).

With the increased personal feeling of responsibility may come increased social pressure (in the form of blame or punishment) to enhance. This would be because people are no longer thought to have excuses for failure (they had the choice to enhance after all). This will create a greater burden that people will feel the need to discharge, but which they may not, in fact, be able to discharge (Nagel 2010, 115–116).

Pulling these thoughts together – *i.e.* more choice leads to regret, greater feelings of responsibility, and greater social pressure to enhance – we can construct the "inverse Prisoners' Dilemma"-argument. Although the argument is straightforward, the name might be slightly confusing and so merits some explanation.

As most people reading this will be aware, a Prisoners' Dilemma (PD) is a particular type of game (strategic interaction) wherein the incentives are structured in such a way that rational actors are persuaded to act in a manner that is counterproductive or mutually destructive. In the classic example, two prisoners are encouraged to confess to their crimes, despite the fact that remaining silent would be better for both of them. Similarly, agents making use of a common resource might have an incentive to pollute or plunder the resource in the interests of short-term gain, despite the fact that this choice could lead to long-term destruction of the resource.

What is the relevance of the PD to this discussion? One of the major motivations behind the distributive theory of responsibility is the belief that without the imposition of a risk burden on the members of a society, society can end up in a type of PD.[31] This happens because, without the norms of responsibility and the associated legal institutions, people are not incentivized to avoid certain readily avoidable types of risk. So, as a result, one of the functions of the system of responsibility norms is to get society out of a PD by imposing a risk burden. This leads to what might be called the "normal Prisoners' Dilemma"-argument in favor of the distributive theory of responsibility.

The inverse PD arises when there is a runaway incentive to avoid all types of risk. This could arise for two reasons. First, as Nagel pointed out, it could be the case that the widespread availability of, and knowledge of the existence of enhancement technologies that can be used to avoid states of risk lead to social pressure to enhance. Second, it could be the case that the individual perceives there being some advantage or gain to be accrued from enhancement.

An analogy to competitive sports lends support for these reasons. Consider, for example, professional cycling. At least some analyses of blood-doping in that sport have argued that its prevalence is attributable to the prisoners' dilemma-style incentives faced by professional cyclers: they are tempted by the short-term gains of doping, and they are pressured by the potential losses of not keeping up with their competitors. The analogy between competitive sports and human society as a whole

[31]Perhaps the classic example of this, although not discussed in terms of the risk burden, comes from Thomas Hobbes in *Leviathan* (1651).

is not perfect, but there are, I submit, a sufficient number of competitive elements to human society to make us worry about the possibility of runaway enhancement.

The analogy between competitive sports and human society might really break down when we consider exactly why runaway enhancement is a bad thing. The powers-that-be within every sport are free to set arbitrary conditions and restrictions on those who choose to participate. So, for example, they do not allow cyclists to participate on motorcycles because the goal is not simply to get to the finish line in the fastest possible time; it is to get to the finish line in a particular way. The problem is that the freedom to impose arbitrary restrictions does not apply to society as a whole. Any restrictions that are imposed must be morally justifiable. And someone could easily argue that if enhancement can reduce risk, and if a reduction of risk is generally agreed to be a good thing, then runaway enhancement might be a good thing.

This is where the inverse-PD-argument begins to show its teeth. The argument is intended to point out that runaway enhancement is not only a serious possibility – it is a serious possibility with counterproductive consequences. These would arise from the fact that the avoidance of all risk is not possible (*i.e.* the riskless society is a myth); one can only hope to minimize risk within reasonable boundaries and then get on with the business of living. To live in a society whose incentives were skewed in such a way that avoidance of all risk became the primary goal would ultimately be contrary to our long-term interests.

Bringing this argument to bear directly on the subject matter of this chapter, we could say that encouraging the use of supernormal enhancements, which would be used by all members of society not just those with impairments, might create a society in which avoidance of risk becomes the primary goal. This would lead to the counterproductive scenario envisaged by the inverse-PD-argument. And, since we have reason to avoid this scenario, so too do we have reason to treat the restorative and supernormal cases differently.

What are we to make of this argument? I suggest the following: it certainly gives us some reason to be cautious about encouraging the use of supernormal enhancements. So, to the extent that ascriptions of responsibility under the distributive theory might encourage the use of such enhancements, there is reason to be cautious about the use of that theory. But in saying that, I do not believe the argument gives us any reason to adopt a different theory of responsibility and that, after all, was the real question being pursued in this section. Why might I believe this?

My contention is that the distributive theory, with the conditions and qualifications described in Sect. 21.3, has the resources to accommodate the concerns at the heart of the inverse-PD-argument. After all, the distributive theory does not actually demand that all possible states of risk be avoided. It says that only those states of risk for which it is *reasonable* to expect people to avoid – taking into consideration the feasibility, costs and benefits of doing so – need be avoided. By taking into consideration such factors, the distributive theory allows the community to reflect on the overall desirability of imposing a risk burden and to avoid being caught in the trap of runaway enhancement.

In conclusion, then, although there may be reason to be cautious about supernormal enhancement, there would not appear to be any reason to employ a different theory of responsibility to deal with it.

Acknowledgement The author would like to thank the Irish Research Council for the Humanities and Social Sciences for supporting his research over the period 2007–2010. He would also like to thank Dr. Mary Donnelly for her comments on earlier drafts of this chapter. Finally, he would like to thank the organizers and participants at the Cognitive Enhancement Conference for many stimulating discussions on the subject of cognitive enhancement.

References

Ainslie G (1992) Picoeconomics. Cambridge University Press, Cambridge
Ainslie G (2001) Breakdown of will. Cambridge University Press, Cambridge
Alexander L, Ferzan K (2009) Crime and culpability. Cambridge University Press, Cambridge
Aumann R (1974) Subjectivity and correlation in randomizing strategies. J Math Econ 1:67
Aumann R (1987) Correlated equilibrium and an expression of Bayesian rationality. Econometrica 1:55
Bayles M (1982) Character, purpose and criminal responsibility. Law Philos 1:5
Bicchieri C (2006) The grammar of society: the nature and dynamics of social norms. Cambridge University Press, Cambridge
Binmore K (1993 and 1998) Game theory and the social contract, vols 1 and 2. MIT Press, Cambridge, MA
Bostrom N (2008) Dignity and enhancement. In: The President's Council on Bioethics Human dignity and bioethics. Government Printing Office, Washington, DC, p 179
Bostrom N, Roache R (2007) Ethical issues in human enhancement. In: Ryberg J (ed) New waves in applied ethics. Palgrave Macmillan, London
Capes J (2012) Action, responsibility and the ability to do otherwise. Philos Stud 158:1–15. doi:10.1007/s11098-010-9662-5
Dixit AK, Skeath S, Reiley D (2009) Games of strategy, 3rd edn. Norton, New York
Elliot ID (1968) Responsibility for involuntary acts: Ryan v. the Queen. Aust Law Rev 41:497
Elster J (2007) Explaining social behavior. Cambridge University Press, Cambridge
Gintis H (2009a) Social norms as choreography. Philos Polit Econ 9:251
Gintis H (2009b) The bounds of reason. Princeton University Press, Princeton
Harris J (2007) Enhancing evolution. Princeton University Press, Princeton
Hart HLA (1968) Punishment and responsibility. Oxford University Press, Oxford
Hawley K (2010) Temporal parts. Stanford encyclopedia of philosophy. Available at: http://plato.stanford.edu/entries/temporal-parts
Hobbes T (1651) Leviathan. Andrew Crooke, London
Horder J (1993) Criminal culpability: the possibility of a general theory. Law Philos 12:193
Irish Law Reform Commission (2009) Report on defences in the criminal law (LRC 95, 2009). Available at: www.lawreform.ie
Loewenstein G, Read D, Baumeister RF (eds) (2003) Time and decision. Russell Sage, New York
MacKay R (1995) Mental condition defences in the criminal law. Oxford University Press, Oxford
McAuley F, McCutcheon P (2000) Criminal liability: a grammar. Round Hall, Dublin
McClure SM et al (2004) Separate neural systems value immediate and delayed monetary rewards. Science 306:503–507
Meyer S (1994) Aristotle on moral responsibility. Blackwell, Oxford
Moore M (1993) Act and crime: the theory of action and its implications for criminal law. Oxford University Press, New York

Moore MS (1997) Placing blame. Oxford University Press, Oxford
Nagel S (2010) Too much of a good thing? Enhancement and the burden of self-determination. Neuroethics 3:109–119
Norrie A (2002) From criminal law to legal theory: the mysterious case of the reasonable glue sniffer. Mod Law Rev 65:538
Owens D (2007) Disenchantment. In: Antony L (ed) Philosophers without gods. Oxford University Press, Oxford
Posner E (2000) Law and social norms. Harvard University Press, Cambridge, MA
Rasmussen E (2004) Games and information, 4th edn. Blackwell, Oxford
Rawls J (1971) A theory of justice. Harvard University Press, Cambridge, MA
Schelling T (2008) Arms and influence, new edn. Yale University Press, New Haven (originally published in 1966)
Sher G (2009) Who knew? Responsibility without awareness. Oxford University Press, Oxford
Skyrms B (1996) The evolution of the social contract. Cambridge University Press, Cambridge
Tadros V (2005) Criminal responsibility. Oxford University Press, Oxford
Wilson W (2002) Central issues in criminal theory. Hart Publishing, Oxford
Young L et al (2010) Disruption of the right temporoparietal junction with transcranial magnetic stimulation reduces the role of beliefs in moral judgments. Proc Natl Acad Sci 107:6753–6758

Index

E. Hildt and A.G. Franke (eds.), *Cognitive Enhancement: An Interdisciplinary Perspective*, Trends in Augmentation of Human Performance 1,
DOI 10.1007/978-94-007-6253-4,

The manufacturer's authorised representative in the EU is Springer Nature Customer Service Centre GmbH, Europaplatz 3, 69115 Heidelberg, Germany. If you have any concerns regarding our products, please contact ProductSafety@springernature.com

Printed and bound by CPI Group (UK) Ltd, Croydon, CR0 4YY
25/09/2026
02222898-0012